ALLOSTASIS, HOMEOSTASIS, AND THE COSTS OF PHYSIOLOGICAL ADAPTATION

The concept of homeostasis, the maintenance of the internal physiological environment of an organism within tolerable limits, is well established in medicine and physiology. In contrast, allostasis is a relatively new idea. Allostasis explains how regulatory events maintain organismic viability, or not, in diverse contexts with varying setpoints of bodily needs and competing motivations. Allostasis accounts for wide variation in function, adaptation, and cephalic involvement in systemic physiological regulation. It provides a conceptual framework for both the protective and the damaging effects that occur in overall regulation of physiological and behavioral systems. This book, the first edited volume to focus on allostasis, orients the reader by addressing basic physiological regulatory systems and examining bodily regulation under duress. It integrates the basic concepts of physiological homeostasis with disorders such as depression, stress, anxiety, and addiction and will therefore appeal to graduate students, medical students, and researchers working in related areas.

Jay Schulkin is a research professor in the Department of Physiology and Biophysics at Georgetown University, the Director of Research of the American College of Obstetricians and Gynecologists, and a research associate at the Clinical Neuroendocrinology Branch of the National Institute of Mental Health. He is the author of several books for Cambridge University Press, including *Calcium Hunger* (2001), *The Neuroendocrine Regulation of Behavior* (1999), and *Sodium Hunger* (1991).

Allostasis, Homeostasis, and the Costs of Physiological Adaptation

Edited by

JAY SCHULKIN

Georgetown University

CAMBRIDGE
UNIVERSITY PRESS

PUBLISHED BY THE PRESS SYNDICATE OF THE UNIVERSITY OF CAMBRIDGE
The Pitt Building, Trumpington Street, Cambridge, United Kingdom

CAMBRIDGE UNIVERSITY PRESS
The Edinburgh Building, Cambridge CB2 2RU, UK
40 West 20th Street, New York, NY 10011-4211, USA
477 Williamstown Road, Port Melbourne, VIC 3207, Australia
Ruiz de Alarcón 13, 28014 Madrid, Spain
Dock House, The Waterfront, Cape Town 8001, South Africa

http://www.cambridge.org

First published 2004

Printed in the United States of America

Typefaces ITC Stone Serif 9/12.5 pt. and ITC Symbol   *System* $\LaTeX\,2_\varepsilon$ [TB]

A catalog record for this book is available from the British Library.

Library of Congress Cataloging in Publication Data
Allostasis, homeostasis and the costs of physiological adaptation / edited by Jay Schulkin.
    p. cm.
  Includes bibliographical references and index.
  ISBN 0-521-81141-4
  1. Allostasis. 2. Homeostasis. 3. Adaptation (Physiology) I. Schulkin, Jay.
QP82.A36A457 2004
612′.022 – dc22   2003062975

ISBN 0 521 81141 4 hardback

Dedicated to Mary Dallman, Ralph Norgren, and Larry Swanson

Contents

Preface   *page* ix

Contributors   xi

Introduction   1
Jay Schulkin

1   **Principles of Allostasis: Optimal Design, Predictive
    Regulation, Pathophysiology, and Rational Therapeutics**   17
    Peter Sterling

2   **Protective and Damaging Effects of the Mediators of Stress
    and Adaptation: Allostasis and Allostatic Load**   65
    Bruce S. McEwen

3   **Merging of the Homeostat Theory with the Concept of
    Allostatic Load**   99
    David S. Goldstein

4   **Operationalizing Allostatic Load**   113
    Burton Singer, Carol D. Ryff, and Teresa Seeman

5   **Drug Addiction and Allostasis**   150
    George F. Koob and Michel Le Moal

6   **Adaptive Fear, Allostasis, and the Pathology of Anxiety
    and Depression**   164
    Jeffrey B. Rosen and Jay Schulkin

7 A Chronobiological Perspective on Allostasis and Its
   Application to Shift Work   228
   Ziad Boulos and Alan M. Rosenwasser

8 Allostatic Load and Life Cycles: Implications for
   Neuroendocrine Control Mechanisms   302
   John C. Wingfield

Commentary: Viability as Opposed to Stability: An
Evolutionary Perspective on Physiological Regulation   343
Michael L. Power

Index   365

Preface

The fun part of science includes the discoveries that we make, the people who we meet and befriend, and our exploration of the larger world. The practice of science ought to cut across narrow boundaries of self-enclosure. I have enjoyed working with both old friends and colleagues and new ones in the context of putting together this book.

Two concepts essential for research in which I have been involved are homeostatic and allostatic regulation. The first is well known, the second is not. There are many books on homeostasis. This is the first edited book on allostasis, which is the volume's primary focus. It became clear that something more than traditional homeostasis would be needed to account for the diverse forms of adaptation to changing circumstances that animals exhibit. Many investigators have noted this fact. Allostasis does not have a univocal meaning for the authors in this book. Two defining features of allostasis are its emphasis on (1) adaptive changes and diverse range of physiological and behavioral options that emerged with central nervous system involvement in peripheral physiological regulation and (2) the breakdown of regulatory systems when pushed beyond adaptation.

The authors in this volume, in one way or another, have been thinking for some time about behavioral and physiological regulatory systems. The topics are diverse but not exhaustive of the literature on regulatory physiology and systems neuroscience. It is hoped that these essays will invite others to revisit the topic toward the goal of understanding the mechanisms that underlie physiological and behavioral adaptation in the regulation of the internal milieu.

I apologize in advance to those who may not have been mentioned but who have contributed to the field. This book is but a small-scale searchlight on the field of regulatory physiology and behavioral neuroscience.

I was first introduced to the concept of allostasis because Peter Sterling and I were in the same department at the University of Pennsylvania. I was giving a departmental talk as a new professor, and in those days my rate of speech far exceeded that necessary for the content I needed to explain. I did not know Peter Sterling then, but it got back me to me that Peter was dismayed by my lecture. I went up to him to talk, and eventually we became friends. This has been an important relationship for both of us.

I took his allostasis paper with me to Italy in 1987 and spent much time critiquing it. It was only after I came to Washington and the National Institute of Mental Health in 1992 that I started to integrate the concept into my scientific research. This continued with my long-term collaboration with Bruce McEwen.

I want to thank my family and friends.

Contributors

Ziad Boulos
New York State Psychiatric Institute and Department of Psychiatry
Columbia University
New York, New York

David S. Goldstein
Clinical Neurocardiology Section
National Institute of Neurological Disorders and Stroke
Bethesda, Maryland

George F. Koob
Department of Neuropharmacology
The Scripps Research Institute
La Jolla, California

Bruce S. McEwen
Harold and Margaret Milliken Hatch Laboratory of Neuroendocrinology
The Rockefeller University
New York, New York

Michel Le Moal
Psychobiologie des Comportements Adaptatifs
Institut National de la Sante et de la Recherche Medicale
Bordeaux, France

Michael L. Power
Smithsonian National Zoological Park
Washington, D.C.

Jeffrey B. Rosen
Department of Psychology
University of Delaware
Newark, Delaware

Alan M. Rosenwasser
Department of Psychology
University of Maine
Orono, Maine

Carol D. Ryff
Department of Psychology
University of Wisconsin
Madison, Wisconsin

Jay Schulkin
Department of Physiology and Biophysics
Georgetown University
Washington, D.C., and
Clinical Neuroendocrinology Branch
National Institute of Mental Health
Bethesda, Maryland

Teresa Seeman
Department of Genetics
University of California, Los Angeles
Los Angeles, California

Burton Singer
Office of Population Research
Woodrow Wilson School of International Affairs
Princeton University
Princeton, New Jersey

Peter Sterling
Department of Neuroscience
University of Pennsylvania
Philadelphia, Pennsylvania

John C. Wingfield
Department of Zoology
University of Washington
Seattle, Washington

Introduction

Jay Schulkin

The purpose of the book is to introduce the concept of allostasis to the reader and to place it within the context of traditional conceptions of homeostasis. Both these regulatory conceptions – homeostasis and allostasis – are broadly conceived within biological adaptations in which behavior and physiology figure prominently. It is within this context of biological adaptation that regulation of the internal milieu is understood and in which both homeostasis and allostasis have scientific legitimacy.

Allostasis reflects longer-term regulatory conceptions and organismic viability in diverse contexts with varying set points of bodily needs and competing motivations. Importantly, allostatic regulation reflects neural involvement in systemic physiological and behavioral adaptation. Allostatic regulation through cephalic involvement reflects the greater flexibility of biological adaptations to maintain internal viability (Sterling and Eyer, 1988; Schulkin, 2003).

In other words, the concept of allostasis is tied to the fact that one role of the central nervous system is to coordinate regulatory responses. The brain is intimately involved in regulatory events and cephalic anticipatory responses in the regulation of the internal milieu (Pavlov, 1902; Powley, 1977, 2000; Smith, 2000). Homeostatic concepts also emphasize the role of the central nervous system in the regulation of the internal milieu, but allostasis offers a new dimension of understanding by emphasizing the *extensive* nature of central nervous system involvement in behavioral and physiological regulation.

HOMEOSTASIS

Homeostasis is a common term within the biological sciences. A variety of well-known examples of behavioral and biological regulation for maintaining homeostasis have been characterized (Table I.1). Bernard (1859, [1865] 1957), amid a biological revolution taking place in the 19th century, offered

1

Table I.1: Paradigmatic examples
of homeostasis

Temperature
pH
Glucose
Protein
Oxygen
Sodium
Calcium

two clear ideas in his studies on the regulation of the internal milieu. One is whole-body physiological regulation; the other is breakdown of tissue under duress and bodily defense. These two themes would later resonate for many investigators who came after Bernard (e.g., Goldstein, 1995; Chrousos, 1998; see Table I.2).

Bernard ([1865] 1957) and Cannon ([1915] 1929a, 1929b, 1932) understood that stability is the key, both in the short and long term, with regard to both low and high taxation of the body's resources. For Cannon homeostatic regulation reflected "a condition, which may vary but which is relatively constant" (1932, p. 24). Bernard and Cannon emphasized that the body's main defense is through physiological mechanisms.

Table I.2: Behavioral and physical adaptations following an acute challenge (Chrousos, 1998)

Behavioral adaptation	Physical adaptation
Increased arousal and alertness	Oxygen and nutrients directed to the CNS and stressed body site(s)
	Altered cardiovascular tone, increased blood pressure and heart rate
Increased cognition, vigilance, and focused attention	Increased respiratory rate; increased gluconeogenesis and lipolysis
Euphoria or dysphoria	Detoxification from endogenous or exogenous toxic products
Suppression of appetite and feeding behavior	Inhibition of growth and reproductive systems
Suppression of reproductive behavior	Inhibition of digestion, stimulation of colonic motility
Containment of the stress response	Containment of the inflammatory/ immune response

Table I.3: Short and long term consequences of glucoconticoid hormones (Adapted from Wingfield and Romero, 2001)

Short-term adaptation	Long-term disruption
Inhibition of sexual motivation	Inhibition of reproduction
Regulate immune system	Suppress immune system
Increase glucogenesis	Promote protein loss
Increase foraging behavior	Suppress growth
Increased activation of brain	Neuronal loss

Within a biological context one distinguishes long-term adaptation (successful reproduction) from short-term goals (eating, not eating to avoid being eaten by a predator). Homeostasis is traditionally couched within short-term goals and physiological adaptations. Thus distinguishing short- from longer-term adaptive responses is critical to the regulation of the internal milieu (Mrosovsky, 1990; Wingfield and Romero, 2001; Schulkin, 2003). Biological flexibility in the context of environmental demands is a fundamental adaptation in regulatory physiological systems. Steroids such as cortisol have both short- and long-term consequences. Short-term effects, following food deprivation, reflect facilitation of neuropeptide and neurotransmitter gene expression that is essential for behavioral adaptation (e.g., increased foraging behavior). Long-term effects of elevated cortisol include the breakdown of metabolic homeostasis (e.g., Dallman et al., 2000; Table I.3).

STRETCHING THE CONCEPT OF HOMEOSTASIS UNDER DURESS AND DAILY, SEASONAL, AND ECOLOGICAL CONTEXTS

Hans Selye (1956, 1974) integrated his theory of biological adaptation to duress with the findings of Bernard and Cannon and homeostasis. The body's reactions are in the service of maintaining equilibrium. Viability is endangered when, as Selye understood it, one "subtracts" stabilization from the reaction response to the specific event. This can be seen, for example, in "the increased production of adrenocortical hormones, the involution of the lymphatic organs or the loss of weight" (p. 17).

But Selye (1956, 1975) was also searching (as many have; see Moore-Ede, 1986; Mrosovsky, 1990; Bauman and Currie, 1980) for a concept beyond homeostasis, and he called it "heterostasis" (p. 85). He stated, "When faced with unusually heavy demands, however, ordinary homeostasis is not enough" (p. 85), because the "homeostat" has been raised to a level beyond its capacity, perhaps to a higher level (e.g., increased production) of function (see also Goldstein, 1995, 2000). "Resetting" of homeostatic systems is an essential function for long-term survival (Mrosovsky, 1990; Goldstein, 1995, 2000; Bauman, 2000; Wingfield and Romero, 2001).

Thus Moore-Ede formulated the term "predictive homeostasis" (Moore-Ede et al., 1982; Moore-Ede, 1986; see also Mrosovsky, 1990). Predictive homeostasis is an anticipatory adaptation and is distinguished from "reactive homeostasis." One adaptation anticipates and modulates in the context of future needs and resource allocation, whereas the other is a reaction to immediate physiological demands. This distinction arose in the context of considerations of circadian timing systems in the brain and their role in behavioral and physiological regulation in anticipation of future needs when they appear (e.g., Mistlberger, 1994).

"Rheostasis," like the term "predictive homeostasis," was coined, to account for the variation in physiological systemic systems, depending on season, time of day, and context (Mrosovsky, 1990). It was clear to those studying the role of clocks on behavior that there was wide variation in systemic physiological systems (Bauman, 2000; Wingfield and Ramekofsky, 1999).

Even within traditional laboratory contexts, it was clear that a concept beyond traditional homeostasis was needed to account for a number of behavioral responses (e.g., ingestive behaviors: Fitzsimons and Le Magnen, 1969; Fitzsimons, 1979; Stricker, 1990; Toates, 1979, 1986). Within behavioral neuroscience and whole-body physiology, a conceptual expansion was required beyond a simple set-point regulatory system. Moreover, anticipatory systems are endemic to cephalic innervations of peripheral physiological regulation of bodily needs (e.g., Woods, 1991).

BEHAVIORAL REGULATION OF THE INTERNAL MILIEU

Behavioral regulation of internal physiology was championed by Curt Richter (1942–3). For example, various animals make nests for safety, reproduction, and warmth. Richter (1942–3, 1956) demonstrated in the laboratory what had been noted in the field – the behavioral regulation of thermal physiology. Rats will build nests under cold conditions (Fig. I.1). In the laboratory, goldfish will bar press to gain access to certain ambient temperature conditions (Rozin and Mayer, 1961). Rats, in response to excess heat, will perform operants to cool their hypothalamus directly (Satinoff, 1964; Stellar and Corbit, 1973). Such behavioral regulation of the internal milieu is impressive, as Richter demonstrated in diverse laboratory contexts. Two classical examples are the ingestion of sodium by adrenalectomized rats and calcium by parathyroidectomized rats. The chronic loss of sodium or calcium in these animals results in a life-threatening situation (e.g., Richter, 1942–3, 1956). The behavior of sodium or calcium ingestion

Figure I.1: Behavioral regulation of temperature – building a nest (Richter, 1942/1943, 1956).

serves to maintain viable amounts of the element to remain alive and to function effectively.

To some extent, the motivational mechanisms to approach and avoid objects are tied to biological needs. Physiological regulation is linked to sensory pleasure and displeasure, which serves as a fundamental motivator of behavior (Stellar, 1954, 1960; Cabanac, 1971; Berridge et al., 1984). The central state of the brain orchestrates the behavioral responses in adapting to environmental events, adjusting for conflicting motivations. One major factor in the evolution of the nervous system is the organization of physiology and behavior to orchestrate anticipatory responses to events that precede the necessity of having to react to them (Pavlov, [1927] 1960; Powley, 1977; Schulkin, 1999). Cephalic innervations of peripheral physiology increased flexible and variable physiological and behavioral responses.

NEUROSCIENTIFIC PERSPECTIVE
The anatomic revolution of the 1970s and 1980s (Swanson, 1999, 2003) demonstrated the outstanding ways in which peripheral organs are in direct contact via neural pathways; for example, the gastrointestinal tract and the cardiovascular system are in direct neural contact with both the brainstem and the forebrain (e.g., Norgren, 1995; Saper, 1982). Again cephalic innervations of peripheral physiology perhaps reflect greater regulatory variation depending on the ecological context. In other words, neural innervations of peripheral physiology add another dimension in the regulation of the

NAPi-2 Immunofluorescence in the Amygdaloid Area
of the Brain
Low Pi diet + 3V Vehicle   Low Pi Diet + 3V Pi

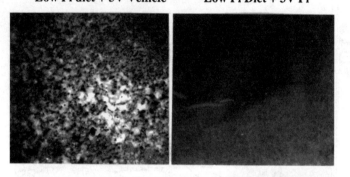

NAPi-2 Immunofluorescence in Cortical Renal Tubules
Low Pi diet + 3V Vehicle   Low Pi Diet + 3V Pi

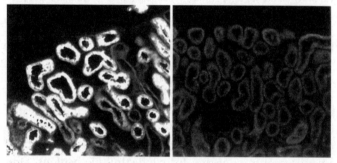

Figure I.2: Renal and central nucleus (top) phosphate transporter changes to central infusions of phosphate in phosphate-deficient rats, Pi = phosphate (Mulroney et al., 2004).

internal milieu; both behavioral and physiological mechanisms are serving the same end – namely, to ensure the presence of viable systems that function adaptively (Nicolaidis, 1977; Kuenzel et al., 1999).

Consider phosphate regulation and cephalic influence on physiological regulation. The kidney increases phosphate absorption when an animal is deprived of phosphate. A behavior, increased phosphate ingestion, emerges to complement the actions of the kidney in the conservation and maintenance of phosphate levels (Sweeney et al., 1998; Mulroney et al., 2004). The brain, in addition to affecting behavior, influences the kidney's phosphate regulatory function. In phosphate-deprived rats, infusions of phosphate into the third ventricle reduces the expression of phosphate transporter at the levels of the kidney and brain, despite the fact that plasma phosphate remains low (Mulroney et al., 2004). The brain, in other words, has a profound effect on systemic (kidney) physiological regulation (e.g.,

Matsui et al., 1995). Again, the brain sends and receives neural projections directly to and from most peripheral organs (Powley, 2000) and is therefore involved in diverse systemic regulatory functions (Fig. I.2).

THE CONCEPT OF ALLOSTASIS

For Sterling and Eyer (1988), "allostasis...involves whole brain and body rather than simply local feedback," and this is "a far more complex form of regulation than homeostasis" (p. 637). Of course, what they mean by homeostasis in this regard is a narrow notion of the concept. Nonetheless, the notion of allostasis (Sterling and Eyer, 1988) was introduced to account for some of the limitations in the concept of homeostasis. The term was introduced to take account of regulatory systems in which (1) the set point is variable; (2) there are individual differences in expression (McEwen and Stellar, 1993); (3) the behavioral and physiological responses can be anticipatory, although they need not be (Sterling and Eyer; 1988; Schulkin et al., 1994); and (4) there is a vulnerability to physiological overload and the breakdown of regulatory capacities.

There are at least three distinguishable meanings associated with the concept of allostasis (see Sterling and Eyer, 1988; McEwen, 1998; Koob and LeMoal, 2001; Schulkin, 2003):

1 **Allostasis:** The process by which an organism achieves internal viability through a bodily change of state (especially central motive state)
2 **Allostatic state:** Chronic overactivation of regulatory systems and the alterations of body set points
3 **Allostatic overload:** The expression of pathophysiology by the chronic overactivation of regulating systems

An important theme that led to the concept of allostasis is the chronic activation of regulatory systems (Sterling and Eyer, 1988; McEwen and Stellar, 1993; Schulkin et al., 1994; Koob and LeMoal, 2001). These are finite systems, and although adaptation is a key ingredient of cephalic involvement in systemic physiological regulation, these physiological systems have limits. Therefore, the degradation of these systems is a consequence of the chronic overactivation of regulatory systems.

ALLOSTASIS, FEEDFOWARD SYSTEMS, AND THE NEURAL ENDOCRINE REGULATION OF BEHAVIOR

The concept of allostasis is importantly linked to the cephalic regulation of both behavior and systemic physiological systems. The behavioral expression depends on the physiological context, the ecological niche, and the competition with other drives. The central states are sustained, in part, by steroid induction of neuropeptides and neurotransmitter gene expression in functional circuits in the brain. Behavior serves physiology in many contexts to sustain both short- and long-term viability.

Steroids are known to facilitate a number of neuropeptides and neurotransmitters in the central nervous system (McEwen, 1995). The activation of the genes that results in the expression of diverse neurochemical signaling systems in the brain also underlies diverse behavioral regulatory systems essential for bodily viability (thirst, hunger, sex drive, sodium appetite (Herbert, 1993; Fitzsimons, 1979, 1999; Pfaff, 1980, 1999; Schulkin, 1999).

Consider one example: extracellular depletion of body fluid results in the activation of the adrenal gland. This results in the synthesis and elevation of both mineralocorticoids and glucocorticoids and their actions in diverse end organ systems (Richter, 1942–3; Wolf, 1964). Both adrenal steroids increase the expression of angiotensin in the brain, which in turn facilitates water and sodium ingestion (e.g., Fitzsimons, 1979, 1999; Epstein, 1991, Fluharty, 2002). The same hormones, in other words, that conserve sodium act in the brain to generate the central states that work to ensure body fluid stability. Moreover, this is a feed-forward mechanism: the adrenal steroids facilitate the expression of angiotensin expression. One feature that Sterling and Eyer (1988) had in mind when they formulated the concept of allostasis was the dramatic and profound contribution of the nervous system in regulatory systems; one mechanism are these diverse feed-forward systems in the brain (Schulkin, 2003).

The long-term inability to turn off the hormones of sodium homeostasis, coupled with sodium ingestion, can result in cardiovascular pathology (Denton, 1982; Denton et al., 1996). In other words, the chronic regulatory response to conserve fluid balance when the hormones of body sodium homeostasis are elevated, coupled with the central signal to ingest, can result in expanded extracellular fluid volume, exaggerated levels of sodium, and cardiovascular vulnerability.

This positive facilitation of angiotensin by the adrenal steroid hormones is not an aberration, however, but a common theme in centrally generated motivated behaviors. A number of examples of increased synthesis (or sustained synthesis) for diverse gene products is reflected in other behavioral systems: (1) estrogen induction of oxytocin or prolactin gene expression and reproductive behaviors (Pfaff, 1980, 1999), (2) testosterone effects on vasopressin and scent marking (Albers et al., 1988), (3) glucocorticoid facilitation of neuropeptide Y and food intake (Leibowitz, 1995), and (4) glucocorticoid induction of angiotensin gene expression and water intake (Sumners et al., 1991). In each case, there is a positive-feedback neuroendocrine mechanism that underlies the behavioral response.

ANXIOUS DEPRESSION: AN EXAMPLE OF ALLOSTATIC OVERLOAD

Coping with anxious depression results in the predominance of expectations of adversity; little relief is experienced, and chronic angst is a daily occurrence (Gold et al., 1988; Nemeroff et al., 1984). Sustained fear is also metabolically costly, and cortisol is well known to restrain the hypothalamic-pituitary-adrenal (HPA) axis. Negative feedback is a fundamental means by which the HPA axis is restrained from activation (Munck et al., 1984). The same hormone that mobilizes resources and regulates a variety of end organ systems restrains its own activation, but there is a limit on adaptation and the ability to cope that can be achieved.

On the endocrine side, cortisol has both suppressive and permissive actions, in addition to stimulating and preparative actions (Sapolsky et al., 2000). Homeostatic, or negative restraint, focuses on the suppressive and permissive effects of cortisol, whereas allostatic regulation focuses on the stimulating and preparative actions of cortisol.

Anxious depression is a condition in which there can be both high systemic cortisol and elevated corticotropin-releasing hormone (CRH) in the cerebrospinal fluid (e.g., Nemeroff et al., 1984; Michelson et al., 1996; Drevets et al., 2002). Anxiously depressed patients also tend to have high levels of glucose metabolic rates in the amygdala (Drevets et al., 2002). The cortisol that regulates CRH gene expression in the amygdala may underlie the fear and anxiety of the anxiously depressed person (Schulkin et al., 1998; Cook, 2002). Thus, one result of the chronic inability to turn off cortisol's effects on amygdala function is the biasing of the brain toward chronically seeing events as fearful.

Most regulatory systems work well when the systems are turned on only when needed and turned off when no longer in use. In other words, when these systems remain chronically active, they can cause wear and tear on tissues and accelerate pathophysiology, a phenomenon called "allostatic overload" (Table I.4; McEwen and Stellar, 1993; McEwen, 1998; Schulkin, 1998).

Table I.4: Indicators of allostatic overload
(Adapted from McEwen, 1998)

Cardiovascular pathology
Metabolic deterioration
Neural degeneration
Immune dysfunction and vulnerability to viral impact
Bone demineralization
Elevation of neural sensitization processes

Table I.5: Bone mineral density in 24 depressed and 24 nondepressed women (Michelson et al., 1996)

Bone measurement	Depressed women	Normal women	Mean difference (95% CI)	P value
Lumbar spine (anteroposterior)				
Density (g/cm)	1.00 ± 0.15	1.07 ± 0.09	0.08 (0.02 − 0.14)	.02
SD from expected peak	−0.42 ± 1.28	0.26 ± 0.82	0.68 (0.13 − 1.23)	
Lumbar spine (lateral)				
Density (g/cm)	0.74 ± 0.09	0.79 ± 0.07	0.05 (0.00 − 0.09)	.03
SD from expected peak	−0.88 ± 1.07	−0.36 ± 0.80	0.50 (0.04 − 1.30)	
Femoral neck				
Density (g/cm)	0.76 ± 0.11	0.88 ± 0.11	0.11 (0.06 − 0.17)	<.001
SD from expected peak	−1.30 ± 1.07	−0.22 ± 0.99	1.08 (0.55 − 1.61)	
Ward's triangle				
Density (g/cm)	0.70 ± 0.14	0.81 ± 0.13	0.11 (0.06 − 0.17)	<.001
SD from expected peak	−0.93 ± 1.24	0.18 ± 1.22	1.11 (0.60 − 1.62)	
Trochanter				
Density (g/cm)	0.66 ± 0.11	0.74 ± 0.08	0.08 (0.04 − 0.13)	<.001
SD from expected peak	−0.70 ± 1.22	0.26 ± 0.91	0.97 (0.46 − 1.47)	
Radius				
Density (g/cm)	0.68 ± 0.04	0.70 ± 0.04	0.01 (−0.01 − 0.04)	.25
SD from expected peak	−0.19 ± 0.67	0.03 ± 0.67	0.21 (−0.21 − 0.64)	

CI = confidence interval.

Consider one example: a consequence of elevated levels of chronic cortisol in anxious depression is the increased likelihood of bone loss and perhaps increased levels of bone fractures (Table I.5; Michelson et al., 1996; Cizza et al., 2001). In one study, for example, people with major depression, compared with nondepressed control subjects had bone mineral density decreases at a number of sites, including the hip, spine, and neck. Cortisol was elevated in the depressed group. Interestingly, vitamin D and parathyroid levels were equivalent in the two groups. Cortisol hypersecretion is known to have consequences on bone metabolism and to facilitate bone loss over time, in addition to creating vulnerability to other central and systemic disorders (e.g., cardiovascular pathology; see also Cizza et al., 2001) that increase the likelihood of premature cardiovascular disease. These are examples of the manifestation of physiological overload and the breakdown of adaptation. Whether or not one calls this allostatic overload, the important point is the recognition of multiple systems trying to adapt and the limits of that adaptation over time.

CONCLUSION

This is a book about current concepts regarding systemic regulation of whole-body tissue; specifically, the book addresses the biological and behavioral mechanisms that maintain internal viability amid changing conditions of the external world. In depicting physiological adaptation and homeostasis, Cannon (1934) focused on the mechanisms for sustaining "stable states." Allostasis emerged in the recognition of cephalic innervations of peripheral physiological regulation and in a broader range of flexibility of physiological systems. Other concepts emerged that either attempted to extend homeostatic forms of explanation or generate other related concepts (Nicolaidis, 1977; Bauman and Currie, 1980; Mrosovsky, 1990) to account for the wide variation in physiological adaptation in changing circumstances. The homeostatic conceptual framework emerged in a medical and laboratory context; the ecological orientation noted wide variation in physiological adaptation, depending on season, time of day, ecological obstacles, reproductive status, and behavioral options (Bauman and Currie, 1980; Mrosovsky, 1990). Seasonal variation of physiology is a common theme across a wide variety of species and adaptations to diverse and changing ecological conditions.

Whether one calls them homeostatic or allostatic, there are limits to these finite physiological systems, and over time normal physiological functions are compromised. Allostasis accounts for wide variation in function, adaptation, and cephalic involvement in systemic physiological regulation.

Explanations using this concept provide a way to characterize flexible adaptive systems and central regulation of physiological control to diverse ecological and physiological demands, providing a conceptual framework for both the protective and damaging effects that occur in the overall regulation of physiological systems. The concept of allostasis does not represent a theory, however, but an orientation toward whole body physiological systems.

There are no knock-down arguments for the concept of allostasis. The concept is offered to help generate inquiry, serve as a conceptual tool to account for diverse data, and help generate interesting experiments. Likewise, there is no mythological correspondence of the concept of allostasis with a necessary set of structures. We should not get lost in terminological arguments that bear no fruit; on the other hand, concepts such as allostasis bear fruit in so far as they provide conceptual and experimental contexts that advance biological inquiry.

For a number of reasons, any new concept, and this concept in particular, generates some resistance, perhaps with good reason. There is no univocal meaning for how it has been used; one can continue to use the concept of homeostasis to account for physiological adaptation. So too are there good reasons to be cautious about the use of the concept in scientific inquiry (Dallman, 2003). Therefore, investigators may rightfully be reluctant to use a new term. There is also the 60-second intellectual reflex that looks to dismiss something as not needed. In my view, there is no reason to abandon the concept of homeostasis. It is firmly entrenched, as it should be, in our scientific lexicon, and one can continue to expand its meaning. It does, however, begin to strain our conceptual framework for understanding bodily adaptation.

This book begins with an exposition on the concept of allostasis from the investigator that invented it, followed by the thoughts of diverse researchers who have used the concept, extending its use and at times revealing inconsistencies. In the end, what matters is the generation of interesting research amid explanations that prove sound, illuminating, and long-lasting. It is hoped that this book will provoke others to consider whole-body physiological regulation and the mechanisms that govern both normal and pathological conditions.

REFERENCES

Albers, H. E., Louis, S. Y., Ferris, C. F. (1988). Testosterone alters the behavioral response of the medial preoptic-anterior hypothalamus to micro injection of arginine vasopressin in the hamster. *Brain Res* 456:383–6.

Bauman, D. E. (2000). Regulation of nutrient partitioning during lactation: home-ostasis and homeorhesis revisited. In: *Ruminant Physiology: Digestion, Metabolism and Growth and Reproduction* (ed. P. J. Cronje). New York: CAB.

Bauman, D. E., Currie, W. B. (1980). Partitioning of nutrients during pregnancy and lactation: a review of mechanisms involving homeostasis and homeorhesis. *J Dairy Sci* 63:1514–29.

Bernard, C. (1859). *Lecons sur les proprietes physiologiques et les alterations pathologiques de l'organisme*. Paris: Balliers.

Bernard, K. C. ([1865] 1957). *An Introduction to the Study of Experimental Medicine*. New York: Dover Press.

Berridge, K. C. (1996). Food reward: Brain substrates of wanting and liking. *Neurosci Biobehav Rev* 20:1–25.

Berridge, K. C., Flynn, F. W., Schulkin, J., Grill, H. J. (1984). Sodium depletion en-hances salt palatability in rats. *Behav Neurosci* 98:652–60.

Cabanac, M. (1971). Physiological role of pleasure. *Science* 173:1103–7.

Cabanac, M., Dib, B. (1983). Behavioral responses to hypothalamic cooling and heat-ing in the rat. *Brain Res* 264:79–87.

Cannon, W. B. ([1915] 1929a). *Bodily Changes in Pain, Hunger, Fear, and Rage*. New York: Appleton-Century-Crofts.

Cannon, W. B. (1929b). Organization for physiological homeostasis. *Physiol Rev* 9:399–431.

Cannon, W. B. (1932). *The Wisdom of the Body*. New York: Norton.

Cannon, W. B. (1934). Stresses and strains of homeostasis. *Am J Med Sci* 189:1–14.

Chrousos, G. P. (1998). Stress and neuroendocrine integration of adaptive responses. *Annal of the New York Academy of Sciences* 851:124–48.

Cizza, G., Ravn, P., Chrousos, G. P., Gold, P. W. (2001). Depression: A major, unrec-ognized risk factor for osteoporosis? *Trends Endocrinol Metab* 12:198– 203.

Cook, C. J. (2002). Glucocorticoid feedback increases the sensitivity of the limbic system to stress. *Physiol Behav* 75:455–64.

Corbit, J. D. (1973). Voluntary control of hypothalamic cooling. *J Comp Physiol Psychol* 83:394–411.

Dallman, M. F. (2003). Stress by an other name? *Horm Behav* 43:18–20.

Dallman, M. F., Akana, S. F., Bhantnagar, S., Bell, M. E., Strack, A. M. (2000). Bottomed out: Metabolic significance of the circadian trough in glucocorticoid concentra-tions. *Int J Obestity* 24:540–6.

Denton, D. (1982). *The Hunger for Salt*. Berlin: Springer-Verlag.

Denton, D. A., McKinley, M. J., Wessinger, R. S. (1996). Hypothalamic integration of body fluid regulation. *Proc Natl Acad Sci U S A* 93:7397–404.

Drevets, W. C., Price, J. L., Bardgett, M. E., Reich, T., Todd, R. D., Raichle, M. E. (2002). Glucose metabolism in the amygdala in depression: Relationship to diagnostic subtype and plasma cortisol. *Pharm Biochem and Behav* 71:431–47.

Epstein, A. N. (1991). Neurohormonal control of salt intake in the rat. *Brain Research Bulletin* 27:315–320.

Fitzsimons, J. T. (1979). *The Physiology of Thirst and Sodium Appetite*. Cambridge: Cambridge University Press.

Fitzsimons, J. T. (1999). Angiotensin, thirst and sodium appetite. *Physiolo Rev* 76:583–686.

Fitzsimons, J. T., Le Magnen, J. (1969). Eating as a regulatory control of drinking. *J Comp Physiol Psychol* 67:273–83.

Fluharty, S. J. (2002). Neuroendocrinology of body fluid homeostasis. In: *Hormones, Brain and Behavior* (ed. D. Pfaff). New York: Academic Press.

Gold, P.W., Goodwin, F.K., Chrousos, G.P. (1988). Clinical and biochemical manifestation of depression: Relation to the neurobiology of stress. *N Eng J Med* 319:348–353.

Goldstein, D. S. (1995). *Stress, Catecholamines, and Cardiovascular Disease*. New York: Oxford University Press.

Goldstein, D. S. (2000). *The Autonomic Nervous System in Health and Disease*. New York: Marcel Dekker.

Herbert, J. (1993). Peptides in the limbic system: Neurochemical codes for co-ordinated adaptive responses to behavioral and physiological demand. *Prog Neurobiol* 41:723–91.

Herbert, J. (1996). Sexuality, stress and the chemical architecture of the brain. *Ann Rev Sex Res* 7:1–43.

Herbert, J., Schulkin, J. (2002). Neurochemical coding of adaptive responses in the limbic system. In: *Hormones, Brain and Behavior* (ed. D. W. Pfaff). New York: Elsevier.

Kuenzel, W. J., Beck, M. B., Teruyana, R. (1999). Neural sites and pathways regulating food intake in birds: A comparative analysis to mammalian systems. *J Exp Zool* 283:346–64.

Koob, G. F., LeMoal M. (2001). Drug addiction, dysregulation of reward, and allostasis. *Neuropsychopharmocology* 24:94–129.

Leibowitz, S. F. (1995). Brain peptides and obesity: Pharmacological treatment. *Obesity Res* 3:573–9.

McEwen, B. S. (1995). Steroid actions on neuronal signaling. *Ernst Schering Res Foundation Lecture Ser* 27:1–45.

McEwen, B. S. (1998). Protective and damaging effects of stress mediators. *New Engl J Med* 338:171–9.

McEwen, B. S., Stellar, E. (1993). Stress and the individual: Mechanisms leading to disease. *Arch Int Med* 153:2093–3101.

Matsui, H., Aou, S., Ma, J., Hori, T. (1995). Central actions of parathyroid hormone on blood calcium and hypothalamic neuronal activity in the rat. *Am J Physiol* 37:R21–7.

Michelson, D., Stratkakis, C., Hill, L., Reynolds, J., Galliven, E., Chrousos, G., Gold, P. W. (1996). Bone mineral density in women with depression. *N Engl J Med* 335:1176–81.

Mistlberger, R. E. (1994). Circadian food-anticipatory activity: Formal models and physiological mechanisms. *Neurosci Biobehav Rev* 18:171–95.

Moore-Ede, M. C. (1986). Physiology of the circadian timing system: Predictive versus reactive homeostasis. *Am J Physiol* 250:R737–52.

Moore-Ede, M. C., Sulzman, F. M., Fuller, C. A. (1982). *The Clocks That Time Us*. Cambridge: Harvard University Press.

Mrosovsky, N. (1990). *Rheostasis: The Physiology of Change*. Oxford: Oxford University Press.

Mulroney, S. E., Woda, C. B., Halaihel, N., Louie, B., McDonnell, K., Schulkin, J., Levi, M. (2000). Regulation of type-II sodium-phosphate (NaPi-2) transporters in the brain: central control of renal NAPI-2 transporters. *FASEB*.

Mulroney, S. E., Woda, C. B., McDonnell, K., Schulkin, J., Haramati, A., Levi, M. (2004). Regulation of type II sodium-phosphate transporters in the brain, *American Journal of Physiology*, in press.

Munck, A., Guyre, P. M., Holbrook, N. J. (1984). Physiological functions of glucocorticoids in stress and their relations to pharmacological actions. *Endocr Rev* 5:25–44.

Nemeroff, C. B., Widerlov, E., Bissette, G., Wallens, H., Karlsson, I., Ekluud, K., Kilts, C. D., Loosen, P. T., Vale, W. (1984). Elevated concentrations of CSF corticotropin releasing factor-like immunoreactivity in depressed outpatients. *Science* 26: 1342–4.

Nicolaidis, S. (1977). Sensory-neuroendocrine reflexes and their anticipatory and optimizing role in metabolism. In: *The Chemical Senses and Nutrition* (eds. M. R. Kare, O. Maller). New York: Academic Press.

Norgren, R. (1995). Gustatory system. In: *The Rat Nervous System* (ed. G. Paxinos). New York: Academic Press.

Pavlov, I. P. ([1927] 1960). *Conditional Reflexes. An Investigation of the Physiological Activity of the Cerebral Cortex*. London: Oxford University Press.

Pavlov, I. P. ([1897] 1902). *The work of the digestive glands* (trans. W. H. Thompson). London: Griffin.

Pfaff, D. W. (1980). *Estrogens and Brain Function*. New York: Springer-Verlag.

Pfaff, D. W. (1999). *Drive*. Cambridge: MIT Press.

Powley, T. L. (1977). The ventromedial hypothalamic syndrome, satiety and cephalic phase. *Psychol Rev* 84:89–126.

Powley, T. L. (2000). Vagal circuitry mediating cephalic-phase responses to food. *Appetite* 34:184–8.

Richter, C. P. (1956). Salt appetite of mammals: Its dependence on instinct and metabolism. In: *L'Instince dans le Comportement des Animaux et de l'Homme*. Paris.

Richter, C. P. (1957). Phenomena of sudden death in animals and man. *Psychosom Med* 19:191–8.

Richter, C. P. (1942–3). Total self-regulatory functions in animals and human beings. *Harvey Lectures* 38:63–103.

Richter, C. P. (1922). A behavioristic study of the activity of the rat. *Comp Psychol Mongr* 1:1–55.

Rozin, P., Mayer, J. (1961). Theramal reinforcement and thermoregulatory behavior in the goldfish, carassius auratus. *Science* 134:942–43.

Saper, C. (1982). Convergence of autonomic and limbic connections in the insular cortex of the rat. *J Comp Neurol* 210:163–73.

Sapolsky, R. J., Romero, M., Munck, A. U. (2000). How do glucocorticoids influence stress responses? Integrating permissive, suppressive, stimulatory, and preparative actions. *Endoc Rev* 21:55–89.

Satinoff, E. (1964). Behavioral thermoregulation in response to local cooling of the rat brain. *Am J Physiol* 206:1389–94.

Schulkin, J. (1991). *Sodium Hunger*. Cambridge: Cambridge University Press.

Schulkin, J. (1999). *The Neruroendocrine Regulation of Behavior*. Cambrige: Cambridge University Press.

Schulkin, J. (2003) *Rethinking Homeostasis: Allostatic Regulation in Physiology and Pathophysiology*. Cambridge: MIT Press.

Schulkin, J., McEwen, B. S., Gold, P. W. (1994). Allostasis, amygdala, and anticipatory angst. *Neurosci Biobehav Rev* 18:385–96.

Schulkin, J., Gold, P. W., McEwen, B. S. (1998). Induction of corticotropin-releasing hormone gene expression by glucocorticoids. *Psychoneuronedocrinology* 23:219–43.

Selye, H. (1956). *The Stress of Life*. New York: McGraw-Hill.

Selye, H. (1974). *Stress without Distress*. New York: New American Library.

Selye, H. (1982). Stress: Eustress, distress, and human perspectives. In: *Life Stress. Vol III. Companion to the Life Sciences* (ed. S. B. Day). New York: Van Nostrand Reinhold.

Smith, G. P. (2000). Pavlov and integrative physiology. *Am J Physiol* 279:R743–55.

Stellar, E. (1960). Drive and motivation. In: *Handbook of Physiology* (Series eds. J. Field & V. E. Hall). Section 1, *Neurophysiology*, Vol. 3 (section ed. H. W. Magovn). Washington, DC: American Physiological Society.

Stellar, E. (1954). The physiology of motivation. *Psychol Rev* 61:5–22.

Stellar, E., Corbit, J. D. (1973). Neural control of motivated behavior: A report based on an NRP work session. *Neurosci Res Prog Bull* 11:295–410.

Sterling, P., Eyer, J. (1988). Allostasis: A new paradigm to explain arousal pathology. In: *Handbook of Life Stress, Cognition, and Health* (eds. S. Fisher, J. Reason). New York: John Wiley & Sons.

Stricker, E. M. (1990). Homeostatic origins of ingestive behavior. In: *Neurobiology of Food and Fluid Intake*. New York: Plenum Press.

Sumners, C., Gault, T. R., Fregly, M. J. (1991). Potentiation of angiotensin II-induced-drinking by glucocorticoids is a specific glucocorticoid type II receptor (GR) mediated event. *Brain Res* 552:283–90.

Swanson, L. W. (1999). The neuroanatomy revolution of the 1970s and the hypothalamus. *Brain Res Bull* 50:397–8.

Swanson, L. W. (2003). *Brain Architecture*. Oxford: Oxford University Press.

Sweeny, J. M., Seibert, H. E., Woda, C., Schulkin, J., Haramati, A., Mulroney, S. E. (1998). Evidence for induction of a phosphate appetite in juvenile rats. *Am J Physiol* 275:R1358–65.

Toates, F. M. (1979). Homeostasis and drinking. *Behav Brain Sci* 2:95–136.

Toates, F. (1986). *Motivational Systems*. Cambridge: Cambridge University Press.

Wingfield, J. C., Ramekofsky, M. (1999). Hormones and the behavioral ecology of stress. In: *Stress Physiology in Animals* (ed. P. H. M. Balm). Sheffield, UK: Sheffield Acedmic Press.

Wingfield, J. C., Romero, L. M. (2001). Adreno-cortical responses to stress and their modulation in free-living vertebrates. In: *Coping with the Environment: Neural and Endocrine Mechanisms* (ed. B. S. McEwen). Oxford: Oxford University Press.

Wolf, G. (1964). Sodium appetite elicited by aldosterone. *Psychonom Sci* 1:211–12.

Woods, S. C. (1991). The eating paradox: How we tolerate food. *Psychol Rev.* 98:588–505.

1 Principles of Allostasis: Optimal Design, Predictive Regulation, Pathophysiology, and Rational Therapeutics[1,2]

Peter Sterling

INTRODUCTION

This chapter compares two alternative models of physiological regulation. The first model, *homeostasis* ("stability through constancy"), has dominated physiology and medicine since Claude Bernard declared, "All the vital mechanisms... have only one object – to preserve constant the conditions of... the internal environment." His dictum has been interpreted literally to mean that the purpose of physiological regulation is to clamp each internal parameter at a "setpoint" by sensing errors and correcting them with negative feedback (Fig. 1.1; Cannon, 1935). Based on this model, physicians reason that when a parameter deviates from its setpoint value, some internal mechanism must be broken. Consequently, they design therapies to restore the "inappropriate" value to "normal."

The homeostasis model has contributed immeasurably to the theory and practice of scientific medicine, so to criticize it might almost seem absurd. Yet all scientific models eventually encounter new facts that do not fit, and this is now the case for homeostasis. In physiology, evidence accumulates that parameters are *not* constant. Their variations, rather than signifying error, are apparently designed to *reduce* error. In medicine, major diseases now rise in prevalence, such as essential hypertension and type 2 diabetes, whose

I thank Joseph Eyer for many wonderful years of collaboration, Charles Kahn for suggesting the Greek roots of allostasis, Jonathan Demb for help with Figures 1.1, 1.5, and 1.13, and Jay Schulkin for his encouragement and patience. I also thank Jay Schulkin, Gerd Blobel, Mark Friedman, Paul Glimcher, Bettina Hoerlin, Neil Krieger, Simon Laughlin, Nicole Neff, Paul Rozin, Gino Segre, Ingrid Waldron, Martin Wilson, and Sally Zigmond for stimulating discussions and for valuable comments on the manuscript. I am greatly indebted to Sharron Fina for preparing both the manuscript and most of the figures.

[1] This essay is dedicated to the memory of Howard A. Schneiderman, who recruited me to experimental biology and bailed me out of a Mississippi jail.

[2] Collected essays on this and related topics are available at http://retina.anatomy.upenn.edu/allostasis/allostasis.html

17

Homeostasis   **Allostasis**

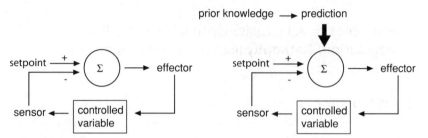

Figure 1.1: Alternative models of regulation. Homeostasis describes mechanisms that hold constant a controlled variable by sensing its deviation from a "setpoint" and feeding back to correct the error. Allostasis describes mechanisms that *change* the controlled variable by predicting what level will be needed and then overriding local feedback to meet anticipated demand.

causes the homeostasis model cannot explain. For in contrast to the hypertension caused by a constricted renal artery and the diabetes caused by immune destruction of insulin-secreting cells, these newer disorders present no obviously defective mechanism. Treating them with drugs to fix low-level mechanisms that are not broken turns out not to work particularly well. The chapter expands on each of these points.

The second model, *allostasis* ("stability through change"), takes virtually the opposite view. It suggests that the goal of regulation is *not* constancy, but rather fitness under natural selection. Fitness constrains regulation to be efficient, which implies preventing errors and minimizing costs. Both needs are best accomplished by using prior information to predict demand and then adjusting all parameters to meet it (Fig. 1.1). Thus allostasis considers an unusual parameter value not as a failure to defend a setpoint, but as a response to some prediction. The model attributes diseases such as essential hypertension and type 2 diabetes to sustained neural signals that arise from unsatisfactory social interactions. Consequently, the allostasis model would redirect therapy away from manipulating low-level mechanisms and toward improving higher levels to restore predictive fluctuation, which under this model is the hallmark of health.

This essay comprises six main sections. The first provides a capsule history of the allostasis model, which by now extends back over 40 years. The second section offers a brief critique of the homeostasis model, focusing on blood pressure because of its broad medical significance. The third section presents key principles of allostasis. Introduced are recent concepts of optimal matching and adaptive regulation, which are then used to reconsider problems of human physiology, such as blood pressure. The fourth section describes how allostasis depends on higher neural mechanisms, and the fifth section suggests how these mechanisms interact with certain aspects of modern social organization to generate some of the major modern diseases. The last section treats the question of where to intervene.

ORIGINS OF ALLOSTASIS

For several decades, I combined research and teaching in neuroscience with social activism. In the mid-1960s, canvassing door-to-door in the African American ghettos such as Central and Hough of Cleveland Ohio, I noticed that many people who answered my knock were partially paralyzed – faces sagging on one side, walking with a limp and a crutch. The cause was "stroke," a rare affliction in my own community, and one that I never encountered later when canvassing in white, upper-class Brookline, Massachusetts. What caused so many strokes, I wondered, and how might they be connected to Cleveland's racial segregation? Arriving around 1970 at the University of Pennsylvania, I found that Joseph Eyer, another biologist-activist, had assembled clear epidemiological evidence that stroke and heart disease, and their precursor hypertension, all accompany various forms of social disruption, including migration, industrialization, urbanization, segregation, unemployment, and divorce (Eyer 1975, 1977; Eyer and Sterling, 1977).

While publishing the epidemiological data, we began to investigate the possible biological mediators. The fury in Hough – which during the summer of 1966 exploded in riots and occupation by National Guard troops – would tend to activate Cannon's well-known, "fight-or-flight" system (sympathetic nerves and adrenal medulla) and Selye's "stress" system (hypothalamo-pituitary-adrenal cortex). But we were astonished by new evidence from fluorescence microscopy that all blood vessels are richly innervated by catecholamine nerve fibers and new evidence from electron microscopy that most endocrine cells are also innervated. For example, sympathetic nerves contact the kidney cells that secrete renin, and parasympathetic nerves contact the pancreas cells that secrete insulin. Recent work has shown that nerves even contact cells that form bone and scavenger cells (macrophages) that serve inflammation and immune surveillance (Flier, 2000; Bernik et al., 2002; Blalock, 2002; Takeda et al., 2002; Tracy, 2002). This suggested that the brain has close access to essentially every somatic cell.

Furthermore, John Mason measured multiple hormones in awake, behaving monkeys and found concerted shifts that made functional sense. A mild demand for focused attention raised hormones associated with catabolism and suppressed those associated with anabolism (Mason, 1968, 1971, 1972). Furthermore, prolonging these demands caused sustained elevations of blood pressure (Harris et al., 1973). Mason concluded that the broad metabolic patterns over short and long time scales, and under mild as well as emergency conditions, are controlled by the brain. Subsequently, myriad studies of neuroendocrine control have supported this conclusion (Schulkin, 1999).

Back then, standard medicine attributed essential hypertension and atherosclerosis to excessive consumption of salt and fat – as though what people choose to eat were unrelated to their internal physiological and mental states. So it was compelling to learn that the peripheral hormones that raise blood pressure, such as angiotensin, aldosterone, and cortisol, also modulate brain regions that stimulate hunger for sodium (reviewed by Schulkin, 1999; Fluharty, 2002). Similarly, peripheral hormones that increase catabolism, such as cortisol, also modulate brain regions that stimulate hunger for energy-rich substrates – fat and carbohydrates (reviewed by Schulkin, 1999; Schwartz et al., 2000; Saper et al., 2002). Of course, such findings would not have surprised Pavlov, who had demonstrated early on the brain's anticipatory control over many phases of digestion, nor Richter, who had connected specific hungers to physiological regulation (Schulkin, 2003a, 2003b).

But to a social activist this seemed immensely relevant: if the brain regulates both physiology and its supporting behavior, then treatments directed only at the peripheral physiology would tend to be countered by the behavior. So, rather than clamp blood pressure at some "normal" value by diuretics, vasodilators, and beta-adrenergic antagonists (the main antihypertensive drugs of the 1970s and '80s), wouldn't it be better to reduce social and psychological disruption? That is, wouldn't it be better to address the higher-level signals that stimulate both the physiology and the behavior? We found a perfect example at the Philadelphia Child Guidance Clinic.

Diabetic children who experience chronic bouts of ketoacidosis had been widely treated with beta-adrenergic antagonists. This often proved ineffective, and it was hypothesized that the metabolic disturbance is induced by parental conflict expressed through the child ("who is right, Daddy or Mommy?"). This was directly observed in "stress interviews." The parents' fatty acid levels would rise but soon return toward baseline, whereas the child's would remain elevated for hours (Fig. 1.2A). Clearly, potent psychological demands were driving multiple physiological mechanisms to override the beta-adrenergic mechanism. Salvador Minuchin, the clinic director, described this as a poignant demonstration that "behavioral events among family members can be measured in the bloodstream of other family members" (Minuchin, 1974).

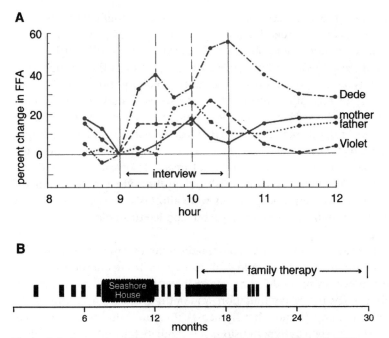

Figure 1.2: Parental conflict modulates a child's blood chemistry. **A.** While parents expressed conflict during an interview, free fatty acid levels rose in all family members. Initially the children, both diabetic, watched through a one-way mirror. At 10 o'clock, they entered the room, whereupon each parent tried to enlist Dede to take his or her side, while Violet remained aloof. Violet's free fatty acid levels followed the parents', but Dede's were greatly elevated. Reprinted from Minuchin, 1974. **B.** Child had been hospitalized (▌) for emergency treatment of ketoacidosis 23 times over 2 years, and beta-blocker treatment of her "superlabile" diabetes was unsuccessful. Family therapy that encouraged the parents to express their disagreements directly (rather than through the child) prevented further relapse. Modified from Baker et al., 1974.

Such children stabilized easily in the hospital but, upon reentering the family, soon relapsed. When the parents were helped to resolve their marital conflicts directly, the children stabilized at home without the beta-blocker (Fig. 1.2B; Baker et al., 1974). This example of successful intervention *between people*, rather than between nerve and liver, seemed of broad sociomedical significance (Sterling and Eyer, 1981). Nevertheless, the idea on which it rests, that the brain controls human physiology, remains largely outside the realm of standard teaching in biology and medicine.

Later, while summarizing this material for another essay collection, it hit us that when you *name* an idea, it has a better chance. So, we coined a new word, "allostasis," to emphasize two key points about regulation: *parameters vary,* and *variation anticipates demand* (Sterling and Eyer, 1988). The idea did spread to some degree, largely through the prolific writings of experts on stress and neuroendocrine regulation, such as McEwen, Schulkin, Sapolsky, Koob, and their colleagues (Sapolsky, 1998; Koob and Le Moal, 2001; McEwen, 2002; Schulkin 2003a, 2003b). Yet even these proponents of allostasis have been somewhat reluctant to abandon homeostasis as the core theory of regulation and have tended to view allostasis as a modulator of homeostatic mechanisms. Some simply equated it with "stress" or "fight-flight" response and suggested that it is an anachronism. For example, "Allostasis has evolved as the response for running away from a predator, escaping acute danger, or fighting off a threat.... However, a defense system that has its roots in an archaic fish can be absurd in a modern human" (Elbert and Rochstroh, 2003). If this were allostasis, it would be entirely justified to discount it as just a fancy word applied to an old idea (Dallman, 2003).

But the allostasis model has a more radical intent – to *replace* homeostasis as the core model of physiological regulation. There are solid scientific reasons: the allostasis model connects easily with modern concepts in sensory physiology, neural computation, and optimal design. Also, this model can begin to comprehend what homeostasis cannot: the main diseases of modern society, such as hypertension, obesity-diabetes, and drug addiction. There are also practical, socially relevant reasons: the allostasis model suggests a different goal for therapeutics and thus a different direction for medical education and treatment. Consequently, this essay begins by assuming that the original conjecture is proved – that physiology is indeed sensitive to social relations. The evidence for this is now vast and thoroughly summarized by McEwen (2002), Sapolsky (1998), and Berkman and Kawachi, (2000). Thus I first describe some difficulties with the homeostasis model and then set out some core principles of the allostasis model that might justify the fancy name.

PROBLEMS WITH HOMEOSTASIS AS THE PRIMARY MODEL FOR REGULATION

Constancy Is *Not* a Fundamental Condition for Life

It seems past time to acknowledge that when Bernard declared constancy to be the sole object of all vital mechanisms, he went too far. Most biologists now agree that the true object of all the vital mechanisms is not "constancy" but survival to reproduce. So what all the vital mechanisms actually serve is reproductive success under natural selection. Moreover, there is nothing magical about "constancy." We now know that the conditions extend to amazing extremes: thermophilic bacteria can thrive at 100°C, and the limit for their successful culturing extends to 113°C! (Hochachka and Somero, 2002). Cell temperatures in the desert can fluctuate by nearly 100°C, and even in complex metazoans the pH of blood and cytosol varies systematically with temperature (Hochachka and Somero, 2002).

Of course, some parameters are regulated quite closely. For example, the mammalian brain tolerates only small fluctuations in oxygen, glucose, temperature, and osmotic pressure. An acute insult that drives any one of these parameters beyond its design limit can trigger cascades of positive feedback that are quickly lethal. Such catastrophic departures from stability certainly require emergency treatment directed at low-level processes (Buchman, 2002). But the purpose of such tight regulation may not be to defend "constancy" in the abstract. Rather, it may simply reflect specific design choices that optimize overall mammalian performance for successful competition.

For example mammalian brain tissue, such as the intact retina or a slice of cerebral cortex, functions for hours in a simple medium at *room temperature*. A neuron's sensitivity is lower than for the optimal 37°C by twofold for each 10 degrees (Dhingra et al., 2003), similar to the temperature sensitivity of most biochemical reactions. So the mammalian brain's normal operating temperature apparently reflects an early design decision: to move fast, we must think fast. This decision had myriad consequences; for example, to move fast, we must also *see* fast. This requires the photoreceptors to be small, which in turn sets the design of retinal circuits (Sterling, 2004). In short, close regulation of human cerebral temperature does not exemplify *the* condition for preserving *all* life – it is just one condition set by a particular design.

A Mean Value Need Not Imply a Setpoint but Rather the Most Frequent Demand

It also seems past time to reevaluate the core hypothesis of the homeostasis model: that the average level of each parameter represents a "setpoint" that is "defended" against deviations (errors) by local feedback (Fig. 1.1). This model captured much of the experimental truth in a simple "preparation" – such as an isolated organ or an animal whose brain has been silenced by anesthesia or decerebration – which were the primary experimental models for more than 100 years. But regulation under natural conditions presents a response pattern that the homeostasis model cannot easily explain.

Consider the record of arterial blood pressure measured continuously over 24 hours in a normal adult (Fig. 1.3). Far from holding steady, pressure fluctuates markedly around 110/70 mm Hg for 2 hours. Then in correlation

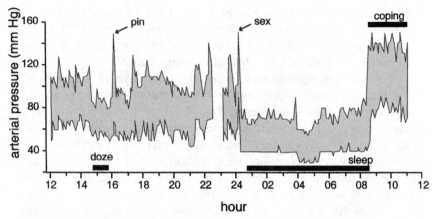

Figure 1.3: Arterial pressure fluctuates to meet predicted demand. Pressure was plotted in a normal adult at 5-minute intervals over 24 hours. Note that pressure spends about equal time above and below the steady daytime level. This pattern suggests not defense of a setpoint, but rather responsiveness to rising and falling demand. Upper trace, systolic; lower trace, diastolic. Redrawn from Bevan et al., 1969.

with identified external stimuli and mental states, it varies more extremely. As the subject dozes in lecture, pressure falls to 80/50. When he is jabbed with a pin, pressure spikes to 150/70; then, having recognized the prank, he again relaxes, and the pressure sinks to 80/50. During sexual intercourse, pressure spikes to 170/90 and then falls profoundly during sleep to ~70/40 with 1 hour as low as 55/30. In the morning, pressure steps up nearly to its level during sex and remains high for hours.

This record contains no hint that blood pressure is defended at particular setpoint. Quite the contrary, it fluctuates markedly and does so on multiple time scales – minutes, seconds, and hours. There are elevations, both brief and sustained, above the most frequent level. There are also similar depressions below the most frequent level. If this level truly represented a setpoint, we might expect it to fluctuate only mildly except when a particularly arousing signal would drive it higher (fight or flight). But the pressure spends about as much time *far below* the most frequent level as above it, and this is not predicted by a model of setpoint + arousal-evoked elevation. If fluctuations were caused by poor control, for example, by excessive or insufficient loop gain (Fig. 1.1), the deviations would show characteristic temporal patterns, such as "ringing" or lag. But the varied temporal patterns and their exquisite matching to particular behavioral and neural states imply that fluctuations arise not from poor control but from *precise* control.

Most parsimoniously, the record suggests that pressure is regulated to match anticipated demand, rising to certain signals and falling to others.

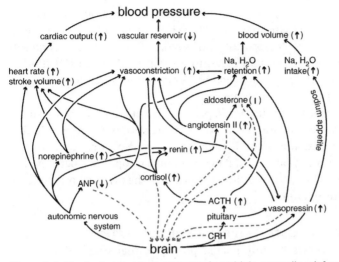

Figure 1.4: The brain sets blood pressure via multiple, mutually reinforcing mechanisms. Negative feedback mechanisms are acutely overridden. When demand persists, all mechanisms are reset to operate at the new level. Most hormones illustrated here are also sensed by the brain (dashed arrows) in specific regions that control behaviors supporting increased pressure. Thus, aldosterone and angiotensin II are sensed by brain regions that enhance salt appetite and drive salt-seeking behavior. CRH = corticotrophic releasing hormone; ACTH = corticotropin; ANP = atrionatiuretic peptide. Modified from Sterling and Eyer, 1988.

This implies that the most frequent value, 110/70, occurs not because pressure is clamped there, but because that value satisfies the most frequent level of demand (see Fig. 1.5). Indeed, were pressure actually clamped at an average value, it would match some specific need only by sheer accident. This is true for all states and all parameters: average values are useless. The essential need is to occupy distinctly different states and to move flexibly between them. But how could this occur, given local negative feedback mechanisms that do tend to resist fluctuations?

Once the brain predicts the most likely demand for oxygen, it resets the blood pressure to achieve the needed flow rate. Pressure here plays the same role as in a shower: for a given resistance, set by the caliber of all the channels, pressure sets the flow. To adjust the pressure, the brain directly modulates all three primary effectors: nerves signal the heart to pump faster, some blood vessels to constrict and others to dilate, and kidneys to retain salt and water. These direct neural messages are reinforced by additional signals acting in parallel (Fig. 1.4). For example, the neural system that excites the primary effectors also releases multiple hormones that send the same message. Hormones signaling the opposite message are suppressed. This pattern: multiple, mutually reinforcing signals acting on multiple, mutually reinforcing effectors, overrides the various feedbacks that oppose change.[3] Recognizing such fluctuation, some authors have proposed the idea of shifting setpoints, termed "rheostasis" (Mrosovsky, 1990). Shifting setpoints might seem to describe certain cases, for example, sustained elevation of body temperature in fever, but even here temperature is responding to specific signals that fluctuate adaptively.

The same is true for essentially *all* parameters: temperature, blood distribution, hormone levels, and so on. All fluctuate with different amplitudes and time constants, and these fluctuations all share a single goal. Yet the goal is not constancy, but coordinated variation to optimize performance at the least cost. This is the core idea of allostasis, the essential principles of which are addressed next.

PRINCIPLES OF ALLOSTASIS (PREDICTIVE REGULATION)

This section discusses six interrelated principles that underlie allostasis: (1) organisms are designed for efficiency, (2) efficiency requires reciprocal trade-offs, (3) efficiency requires predicting what will be needed, (4) prediction requires each sensor to adapt its sensitivity to the expected range of input, (5) prediction requires each effector to adapt its output to the expected range of demand, and (6) predictive regulation depends on behavior whose neural mechanisms also adapt.

Organisms Are Designed for Efficiency

Organisms must operate efficiently. Beyond escaping predators and resisting parasites, they must compete effectively with conspecifics. If you encounter a bear while hiking with a friend, you need not outrun the bear – just your friend. So natural selection sculpts every physiological system to meet the loads that it will most likely encounter in a particular niche plus a modest safety factor for the unusual load. No system can be "overdesigned" because robustness to very improbable loads will slow the organism and raise fuel costs. Nor can a system be "underdesigned" because, if it fails catastrophically to a commonly encountered load, well, that's it. In effect the organism

---

[3] It is probably no accident that the error-correction model that Bernard adopted for physiology mimicked the simple device that inaugurated the 19th century's industrial technology (the speed governor on Fulton's steam engine). But machines have evolved, and the 21st century automobile now preempts driver errors. The myriad sensors in a BMW (~100) relay data to a central mechanism (computer chip) that calculates the power and braking needed by each wheel to optimize stability and skid resistance. Data from other sensors are centrally integrated to control fuel, oxygen, and spark timing for each cylinder to optimize fuel consumption at each power level. This resembles biology, where changing gait maximizes efficiency at different speeds (Alexander, 1996; Weibel, 2000). In short, for a car with a "brain," predictive regulation produces better stability and greater efficiency.

resembles an elevator cable, which must be just sufficiently robust to prevent the cancellation of the manufacturer's insurance (Diamond, 1993).

It follows that all internal systems should mutually match their capacities. Thus our intestinal absorptive capacity supplies sufficient fuel for our most likely energy need – with modest excess to meet unusual demands (Hammond and Diamond, 1997). Our lung and circulatory capacities supply sufficient oxygen to burn the available fuel; and our muscles contain sufficient mitochondrial capacity to provide an adequate furnace (Weibel, 2000). Clearly it would be inefficient for an organ to provide more capacity than could be used downstream, or for an organ downstream to provide more capacity than can be supplied from upstream. This aspect of organismal design, where physiological capacities optimally match, is termed "symmorphosis" (Taylor and Weibel, 1981). It holds for digestive, respiratory, and muscular systems, and also for neural systems (Diamond, 1993; Weibel, 2000; Sterling, 2004).

Efficiency Requires Reciprocal Trade-Offs

Efficiency requires that resources be shared. Otherwise, each organ could meet an unusual demand only by maintaining its own reserve capacity. To support this extra capacity would require more fuel and more blood – and thus more digestive capacity, a larger heart, and so on, thereby creating an expensive infrastructure to be used only rarely. Consequently, organs can trade resources – that is, make short-term loans. Regulation based on reciprocal sharing between organs is efficient, but for several reasons it requires a centralized mechanism: (1) to continuously monitor all the organs, (2) to compute and update the list of priorities, and (3) to enforce the priorities by overriding all the local mechanisms (Fig. 1.4).[4]

For example, skeletal muscle at rest uses about 1.2 liters of blood per minute (~20% of resting cardiac output), but during peak effort it uses about 22 L/min (~90% of peak cardiac output), an 18-fold increase. Much of the extra blood comes from increased cardiac output, but that is insufficient. And although tissues may store fuel, such as glycogen and fatty acids, they cannot store much oxygen. Nor would it be useful to maintain a large reservoir of deoxygenated blood because peak demand completely occupies the pulmonary system's capacity to reoxygenate. So a reservoir of deoxygenated blood would require a reservoir of lung, heart, and so on.

---

[4] Again, industrial analogies seem pertinent. Consider the efficiencies achieved by sharing electricity in a power grid and by rapidly redistributing inventory in a factory system. This type of efficiency requires continual, rapid updating of information about current demand, plus prior knowledge of how demand will probably change with factors such as temperature, time of day, season, world market, and so on.

In turn, these would require increased capacities for digestion, absorption, excretion, and cooling. Consequently, for an unstorable resource, subject to variable demand, it is most efficient to share. So, at peak demand about 10% of the total flow to muscle is *borrowed* (Weibel, 2000).

The loan cannot come from the brain, which requires a constant supply, that, if interrupted for mere seconds, causes loss of consciousness. So muscle borrows from the renal and splanchnic circulations, whose individual shares of cardiac output drop from about 20% to 1%, and whose absolute supplies fall by four- to fivefold (Weibel, 2000, figure 8.6). The skin circulation also contributes. Kidney, gut, liver, and skin can generally afford to lend for the short term – depending on circumstances. For example, skin can postpone reoxygenation, but during exercise in a warm environment it requires more blood for cooling. The gut can also postpone reoxygenation, but following a meal it requires more blood to transport digests into the portal circulation.

Reciprocity Requires Central Control
The brain, although it represents 2% by weight in a 70-kilogram man, requires 20% of the resting blood flow. This proportion is so great that when a given brain region increases activity, the extra blood is requisitioned not from somatic tissues, but from other brain regions (Lennie, 2003). Thus, within the brain itself, resources are reciprocally shared.

Because the needs of muscle, gut, and skin can be irreconcilable, appropriate trade-offs between them (and all the organs) must be calculated. This requires a central mechanism, the brain, which must also enforce a specific hierarchy of priorities and shift them as needs change. When muscular effort is urgent, but you have just eaten and the environment is warm, the brain triggers a vomiting reflex; when cooling is more urgent than effort, the brain may reduce the priority for an erect body posture and trigger the vasovagal reflex ("fainting"): the heart slows, vessels dilate, blood pressure falls, and muscle tone collapses. In short, the brain must decide the conditions for each loan and set the schedule for repayment. Furthermore, because such conflicts potentially threaten overall stability (survival), these solutions are accompanied by unpleasant sensations, such as nausea and dizziness, which the brain also provides. These sensations are vividly remembered to reduce the likelihood of repetition.

Efficiency Requires Predicting What Will Be Needed
We have already seen that blood pressure fluctuates to match the ever-shifting prediction of what might be needed (Fig. 1.3). This is true for essentially all physiological mechanisms. Consider an additional example, control of blood glucose by insulin.

This is usually presented as a core example of homeostasis: ingested glucose raises the blood level, stimulating pancreatic beta cells to release insulin, which stimulates muscle and fat cells to take up the glucose and restore blood levels to the standard ~90 mg/dL. Indeed a pancreas placed in vitro and exposed to glucose will release insulin. But when an intact person sits down to a meal, the sight, smell, and taste of food predict that blood glucose will soon rise, and this triggers insulin release via neural mechanisms *well before* freshly ingested glucose reaches the blood (Schwartz et al., 2000). This anticipatory pulse of insulin signals muscle and fat cells to take up glucose and signals the liver to cease releasing it. Thus this prediction can prevent a large rise in blood glucose.

A different prediction can do the opposite, that is, it can elevate blood glucose above the most frequent level. For example, Cannon reported that members of the Harvard football team, anticipating a game, would elevate their blood glucose to levels that spilled into the urine (Cannon, 1920). In other words, predicting an intense need for metabolic energy can raise blood glucose to diabetic levels. Insulin and the myriad other hormones that regulate the fuel supply are modulated rigorously from the brain, which bases its predictions on a continuous data stream regarding metabolic state that arrives via nerves from the liver and sensors in the cerebrovascular organs, such as the area postrema and the hypothalamus (Friedman et al., 1998; Saper et al., 2002). The importance and challenge of predictive regulation is best appreciated by the type 1 diabetic who, to minimize surges of blood glucose, injects insulin *before* a meal, and who, to allow his muscles to admit glucose, injects insulin *before* exercise.

Sensors Must Match the Expected Range of Input

Sensors are designed to transduce a range of inputs into a range of outputs (Fig. 1.5, upper panel). Typically the input-output curve is sigmoid and set so that its midpoint corresponds to the statistically most probable input (Fig. 1.5, lower panel). The curve's steep, linear region brackets a range of inputs that are somewhat likely, and its flatter regions correspond to inputs (very weak or very strong) that are relatively unlikely (Laughlin, 1981). This design has a clear advantage: the most likely events are treated with greatest sensitivity and precision (Laughlin, 1981; Koshland et al., 1982). When input events are small relative to noise, they may be amplified nonlinearly to remove the noise by thresholding (Field and Rieke, 2002), but most sensors amplify linearly as shown here (Rieke et al., 1999). Note that the design of each sensor embodies "prior knowledge," derived via natural selection, regarding the range of the most likely inputs (Sterling, 2004).

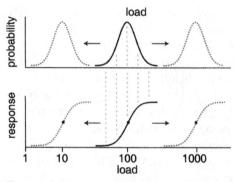

Figure 1.5: Regulatory mechanisms adapt to keep the input-output curves centered on the most probable loads. **Upper panel.** Every system confronts some distribution of probable loads (bold curve). As conditions shift, so does the distribution (dashed). **Lower panel.** The input-output curve (bold) is typically sigmoid with its most sensitive region (steep part) matched to the most probable loads. When the distribution of probable loads shifts, the input-output curve shifts correspondingly (dashed). See Laughlin, 1981.

This simple design is effective when the statistical distribution of inputs is steady. But environmental signals fluctuate enormously; for example, light intensity changes between day and night by 10-billion-fold. The linear range of a visual sensor spans only 10-fold, so over the course of a day, the sensor would frequently confront a range of inputs far too large or too small for its response curve (Fig. 1.5, lower panel). For much stronger inputs, the sensor would be too sensitive, and its output would saturate; for much weaker inputs, it would be too insensitive and would miss them.

There is a remedy: sense the altered input statististics → calculate a new probability distribution → shift the response curve to rematch its steep region to the most likely loads (Fig. 1.5, lower panel). This strategy for continually rematching outputs to expected inputs has been observed at all levels of biological organization, from bacteria and somatic (nonneural) cells to neurons (Sakmann and Creutzfeldt, 1969; Koshland, 1987). At lower levels, the process has been termed "adaptation," and recently "Bayesian" has been added to emphasize Bayes's insight that the best estimate of what is happening in the world combines data from our sensors with our prior knowledge about what is probably out there (Rieke et al., 1999). This principle operates at many levels. Thus, we rely on a single experience of the unpleasant sensations associated with regulatory conflict (dizziness, nausea) to permanently enlarge our store of "prior knowledge." And in the perceptual realm, we identify an ambiguous object by sight or touch by combining sensory inputs with our prior knowledge of what the context suggests is most likely (Geisler and Diehl, 2002, 2003). In case of conflict between vision and touch, we rely on prior knowledge of which sense is probably more accurate (Ernst and Banks, 2002).

Prediction Requires Each Sensor to Adapt Its Sensitivity to the Expected Range of Inputs

Prediction must be based on sensors that are both accurate and fast with respect to the processes that they help to regulate. How sensors maintain their accuracy and speed over a large dynamic range is now well understood for various neural systems, especially for vision. The basic principles seem likely to be generalizable to all sensors. Indeed, the conclusions from analyzing vision in the fly and vertebrate retinas are similar to those reached by analyzing chemotaxis in bacteria (cf., Koshland et al., 1982; Koshland, 1987; Laughlin, 1994; and Rieke et al., 1999). Therefore, this section summarizes the current understanding of how and why sensors adapt to their inputs.

Rate of Adaptation Matches the Rate of Changing Input

Input statistics can fluctuate rapidly – for example, light intensity on the eye of a flying insect varies over milliseconds. For a visual neuron to match its responses to such shifts, it must adapt in milliseconds; otherwise it will always be optimized for a past condition and never for the input that it will most likely encounter next. Such a neuron measures the input very briefly – just long enough to provide reliable statistics – and then shifts sensitivity accordingly. Because photons arrive stochastically, their intrinsic noise is set by the square root of the number counted; so a fly neuron that adapts over 10 milliseconds can sense a change of 100 photons captured over 5 milliseconds and then predict the most likely intensity of the next instant to within 10% – leaving 5 milliseconds to shift the response curve (Fairhall et al., 2001). In short, natural selection ensures prediction down to the limit set by physical laws (Laughlin, 1994; Rieke et al., 1999; Sterling, 2004).

The time course of predictive adaptation differs for every system and depends partly on the length of time spent under a particular load. For example, after carrying groceries from your car to the kitchen, your sense of effort is reduced briefly – a coffee cup feels "lighter" than normal. But after wearing ski boots or a backpack for hours, the sense of weightlessness lasts longer. Over hours mechanoreceptors in muscle, tendon, and ligament have reduced their sensitivities to match the persistently increased load. Then over tens of minutes, sadly perhaps, we regain our usual sense of effort as these mechanisms readapt to predict the next round of most likely loads. Even astronauts, initially exhilarated by zero gravity, gradually cease to notice either their weightlessness or even their unusual orientation within the cabin (for us, "upside down") because all their sensory mechanisms readjust to predict the most likely conditions.

There are two levels of prediction: (1) most likely state in the next moment, generally best captured by the current state and its rate of change; (2) most likely time course of the new state, generally best captured by length of time in the current state. This second factor, persistence, improves efficiency because each change requires a response, and each response has a cost. Many predictors reduce costs by anticipating regular shifts in demand. For example, circadian prediction proves so advantageous that every cell in the body uses it to regulate the expression of vast numbers of genes according to predicted demand (Roenneberg and Merrow, 2003). On a longer time scale, seasonal variation in day length predicts average environmental temperature and food availability, performing much more reliably than local temperature. Furthermore, for migratory species day length predicts the most likely temperature thousands of miles away. Consequently, predictions based on day length have been built into the brains of many species as "prior knowledge" that profoundly regulates their physiology (Mrosovsky, 1990).

Prediction Requires Each Effector to Adapt Its Output to the Expected Range of Demand

Effectors also shift their output curves to match a change in the expected range of demand (Fig. 1.5). Of course, effectors change more slowly than sensors because their adaptations are more expensive.[5] The example most familiar, because we observe it directly, is skeletal muscle. One bout of intense effort, although fatiguing, little affects the response curve. But prolonged effort over days, weeks, and months gradually evokes a panoply of gene modulations: increased synthesis of proteins for muscle, bone, and connective tissue, plus corresponding shifts of metabolic and respiratory enzymes. It would be a costly design that mobilized all these mechanisms just to deliver your groceries, or that completely *de*mobilized them after just a day's layoff from training. Even so, world-class athletes known for their superior fitness, such as Roger Clemens and Lance Armstrong, never reduce their physical demands more than momentarily, lest their effectors readapt even slightly to lower demand. Similarly, at zero gravity only a few weeks are sufficient to reduce a perfectly fit astronaut to jelly.

Internal effectors also adapt gradually. For example, although the brain's sensor of circadian time (suprachiasmatic nucleus) resets to a shift in day

---

[5] If demand rises beyond current capacity for *long* periods, new power plants get built, but, just as in the body, effector capacity lags.

length within one cycle, the liver, which synthesizes many gene products under circadian control, resets over 6 days (Roenneberg and Merrow, 2003). In fact, all cells, via diverse molecular sensors on their surfaces (receptor proteins), regulate to meet predicted demand.[6] Furthermore, these receptors themselves regulate in number and sensitivity to match predicted demand over a range of time scales. Typically, prolonged exposure to high levels of its natural ligand (signaling molecule) reduces receptor number and sensitivity.

Note that downregulation of a receptor triggered by prolonged exposure to its ligand occurs by negative feedback. But this need not be caused by an "error"; rather, the downregulation is simply a response to the anticipation of a higher level of the ligand. Thus, when blood glucose is persistently elevated and triggers persistent secretion of insulin, insulin receptors eventually anticipate high insulin and downregulate. The system learns that blood glucose is *supposed* to be high. Similarly, sustained demand for elevated blood pressure teaches all effectors to expect it and gradually adapt: arterial smooth muscle cells hypertrophy to contract against higher pressure; the carotid sinus wall thickens to reduce baroreceptor sensitivity; secretory cells hypertrophy to support the pressure rise with more renin, norepinephrine, cortisol, and so on. In short, it seems inevitable that the sustained elevation of blood glucose would gradually reduce insulin sensitivity, that is, cause "insulin resistance" and thus type 2 diabetes, and that sustained elevation of blood pressure would gradually cause essential hypertension. Such changes are the appropriate adaptations to predicted demand (Fig. 1.5).

Predictive Regulation Relies on Complex Behavior Whose Neural Mechanisms Also Adapt

Few of the raw materials needed for regulation are stored in any quantity. Most of the body's sodium is in the blood and extracellular space, and sodium is lost daily along with water in tears, sweat, and urine.[7] Calcium is stored within various intracellular compartments, but there it is needed for signaling and must not be depleted. Only bone can loan calcium for the short term, but obviously the loan must be repaid. Fuel is generally stored in modest quantity as glycogen and fat, whose rapidly mobilizable components within muscle cells are just sufficient to carry a trained runner through a marathon (Weibel, 2000). Prolonged exertion, as in the Tour de France,

---

[6] This represents an important difference from machines. The BMW's central integrator can adapt, but its effector parts do not adjust; they only wear out.

[7] The moose provides an instructive exception. Its winter diet of tree browse contains hardly any sodium, but the moose can borrow sodium from its capacious rumen (stomach for fermentation) and then repay the debt, after the ice thaws, by feeding on water plants rich in sodium (Denton, 1993).

soon depletes stored fat and is ultimately limited by the gut's maximum absorptive capacity, which can sustain energy consumption over basal levels by only about fourfold (Hammond and Diamond, 1997). In short, physiological regulation is inexorably tied to replenishing.

The most efficient way to update a grocery list is *immediately,* as an item is used.[8] There are two reasons. First, to be depleted is unpleasant – and can quickly become lethal. Second, supplies, such as salt, water, and fuel, are not always available. Consequently the brain's every command to consume a particular substance is always accompanied by parallel commands to reduce its loss and to seek opportunities to replenish. This need to replenish generally involves a rich set of cognitive and emotional experiences. Consider this concrete example from an actual hike in the Arizona desert, a 7,000-foot descent into the Grand Canyon.

This hot, arid environment demands evaporative cooling through sweat, which expends water and sodium. In anticipation, the brain triggers release of antidiuretic hormone and aldosterone to stringently suppress salt and water loss through urine. Nevertheless, we soon feel thirsty and pause to replenish.

Drinking from the water bottle, which we had anticipated would be essential, our thirst is satisfied. But watching the level drop, as our companions also drink, we feel anxious that the supply might be insufficient and that others might drink our share. Responding to this concern, one group member suggests a rule to govern further consumption, another consults the map for the next spring, and a third redistributes the weight among the packs, taking more for himself, to better match the strengths of the individual hikers. On reaching the spring, the anxiety dissipates; there is pleasure in the drinking and rejoicing in the sense of solidarity that accompanied our successful cooperation. But before long, gazing up the sheer walls to the Canyon's distant rim, we begin to worry and wonder how to carry enough water to climb out.

Such an experience illustrates that human physiological regulation depends powerfully on a host of high-level neural mechanisms: retrieval of prior knowledge, multiple emotions, perception, planning, cooperation, and altruism (Fehr & Fischbacher, 2003). Such an experience can thrust us back, before Gortex and Polartec, to the root of human evolutionary success and cause us to reflect on its neural basis. Somewhere in the brain all the critical factors must be weighed, a plan devised and forcefully executed. A critical site turns out to be the prefrontal cortex.

---

[8] Supermarkets have now adopted this practice: each item sold at the register by its bar code is automatically subtracted from the inventory and reordered.

HOW ALLOSTASIS DEPENDS ON HIGHER LEVELS

Prefrontal Cortex: Where Thought and Feeling, Past and Present, Meet

Each sensory system projects to its own primary area in the neocortex where elaborate computations begin to identify key environmental features and group them based on prior knowledge (Fig. 1.6A; Geisler and Diehl, 2002, 2003). These areas relay to various higher order areas (30 or more areas for vision) that compute still higher order features. Eventually, data from the separate senses converge at particular cortical sites, such as the intraparietal lobule, to be compared and weighed against each other (Ernst and Banks, 2002). Ultimately the higher order and multimodal regions converge in cascades on the prefrontal cortex (Fig. 1.6A), and from this cascaded pattern emerges the best estimate about the present.

To judge the significance of the present state requires knowledge of the past. Although it remains uncertain exactly where and how the brain stores specific memories, we do know that retrieval involves elaborate connections within "limbic" structures and that these also project in cascades to prefrontal cortex (Fig. 1.6B). Thus, in our example, the desert hike, the prefrontal cortex connects the visually perceived level in the water bottle with the recalled experience of running out. This cortical region, which encompasses about a dozen specific areas, can even connect the present situation to a grandfather's tale about running out of water, or to the story of Hagar and Ishmael from Genesis.

But to simply compare present versus past and calculate a plan is insufficient. To *execute* a plan, thoughts need to be driven – focused – by emotion. This is another task of the limbic system, to generate some combination of feelings, such as anxiety, rage, and love, that sustainedly connect our thoughts, which otherwise tend to flicker about. We know that prefrontal cortex participates in the focusing of thought by emotion because for about 30 years (1940–70) prefrontal lobotomy was widely performed as a treatment for mental illness. Following lobotomy, a person can conceive of a plan but cannot hold it in focus for long enough to complete it, a deficit that has been succinctly termed "instability of intent" (Nauta, 1971; Valenstein, 1973, 1986; Sterling, 1978).

Just as thought must be focused by emotion to support physiological regulation, so must the experience and expression of emotion be governed by perception and thought. Thus, the perception of a diminishing water supply triggers anxious feelings, and our persistent expression of these feelings to our companions may curtail their rate of consumption. If that fails,

A. Neocortical cascades to prefrontal cortex   B. Limbic cascades to prefrontal cortex

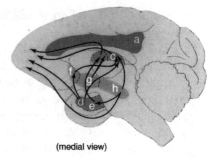

(lateral view)   (medial view)

C. Prefrontal cascades to neocortex   D. Prefrontal cascades to limbic system

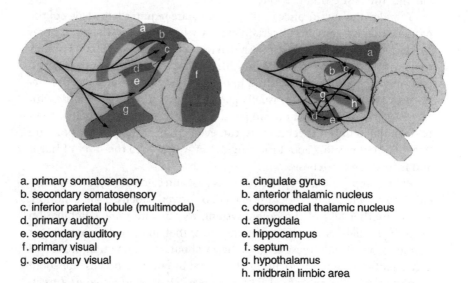

a. primary somatosensory
b. secondary somatosensory
c. inferior parietal lobule (multimodal)
d. primary auditory
e. secondary auditory
f. primary visual
g. secondary visual

a. cingulate gyrus
b. anterior thalamic nucleus
c. dorsomedial thalamic nucleus
d. amygdala
e. hippocampus
f. septum
g. hypothalamus
h. midbrain limbic area

Figure 1.6: Prefrontal cortex integrates cascaded inputs from neocortical and limbic systems, and feeds back to both. This arrangement serves two functions: to imbue intellectual calculations with urgency and focus and to modulate emotional expression by perceptual and cognitive context (Nauta, 1971). Diagram shows the brain of macaque monkey. Redrawn from Sterling, 1978.

anxiety may turn to fear – whose expression as anger will attract the attention of even the least sensitive companion. These mechanisms are served by cascaded outputs from prefrontal cortex back to the neocortical and limbic structures (Fig. 1.6C, 1.6D). Following lobotomy, a person can still express emotion, but because it is no longer modulated by shifting perceptions, affect is "flattened," and when affect does shift, it is frequently inappropriate to the circumstance.

These deficits illuminate how profoundly human physiology depends on continual modulation of emotional expression. Inappropriate affect will fail to achieve the immediate goal – conservation and sharing of a limited resource. Affect, if too mild, will fail to persuade our companions; if too strong, it will anger them and threaten the group's stability. Flattened affect also degrades even the simplest exchange, whether of substance (water) or an idea (schedule for consumption), because every exchange requires an emotional acknowledgment. A plain "thank you," unaccompanied by a wink, nod, smile, or touch, may be perceived as insincere and thus fail to provide what is essential to every exchange between humans: some reciprocal emotional recognition. Such recognition both relieves anxiety and provides pleasure. This is so fundamental that an individual with flattened affect is rapidly identified as deviant and isolated from the social interactions essential for survival.

"Stick and Carrot" for Anticipatory Regulation: Neural Mechanisms for Anxiety and Satisfaction

The feelings of anxiety, fear, and anger depend on neural activity in the amygdala, a large complex of nuclei in the basal forebrain that connects reciprocally with the prefrontal cortex (Fig. 1.6B, 1.6D). The amygdala also serves our highly developed ability to predict these feelings in others based on their facial expression and body language. We know this partly because the amygdala was ablated during the 1960s–70s as a therapy to attenuate anxiety and rage (cf. Mark and Ervin, 1970; Valenstein, 1973; Sterling, 1978) and partly because of recent magnetic resonance imaging studies (Adolphs et al., 1994; LaBar et al., 1998). The amygdala collects and integrates myriad lower-level signals concerned with physiological regulation: (1) steroid hormones and peptides that regulate blood pressure, and mineral and energy balance; (2) neural signals from the brainstem visceral areas, such as nucleus of the solitary tract and the hypothalamus; (3) signals from brainstem raphe neurons that modulate levels of arousal and mood via the neural transmitter, serotonin (Schulkin et al., 1994; Schulkin, 1999).

Serotonin serves many different functions in the brain, because the raphe neurons project very widely, for example, down to sensory and motor columns of spinal cord and up to the cerebellum and neocortex. But serotonin released in the amygdala appears to suppress transmission of anxiety signals to prefrontal cortex. Consistent with this theory, drugs that increase release of serotonin or its persistence in the synaptic cleft reduce anxiety, elevate mood, and enhance social pleasure[9] (Wise, 2003). Presumably, the powerful social signals transmitted via neocortical cascades to the prefrontal cortex (Fig. 1.6A) also feed down to the amygdala and other limbic structures to regulate serotonin release in the cortex (Fig. 1.6A, 1.6D). These mechanisms, which contribute to physiological regulation by releasing serotonin to reduce the anxiety that drives vigilance, might be considered, to simplify greatly, as behavior encouraged by the "stick."

Other behaviors that serve physiological regulation are driven less by the promise of reducing anxiety than by the expectation of "reward" – some outcome that leads to a feeling of satisfaction. Satisfaction depends on activity of neurons in the midbrain's ventral tegmental area (VTA). This limbic region connects reciprocally with the basal forebrain (nucleus accumbens) and the prefrontal cortex (Fig. 1.6B, 1.6D). The VTA, like the amygdala, integrates signals related to myriad appetites – for specific nutrients, water, fuels, sex, and so on – and uses prior experience to establish specific predictions of how each behavior should be rewarded. A VTA neuron is quiet until an outcome exceeds the expectation, but then it increases firing linearly with magnitude of the perceived reward (Schultz, 2002; Fiorillo et al., 2003). Each action potential releases a pulse of the neural transmitter, dopamine, whose binding by receptors in the nucleus accumbens and prefrontal cortex signals "satisfaction."

This mechanism provides a common pathway to sustain behaviors that serve *all* appetites (Fig. 1.7) (Montague and Berns, 2002). It works because what the accumbens–prefrontal cortex apparently "wants" is neither sodium, sugar, fat, nor sex per se, but simply a pulse of dopamine. Because this mechanism is general and plastic, it can lock onto virtually any experience or behavior that might release dopamine: music, visual art, food, or social recognition. For one mechanism to serve so many core needs, the satisfaction that it brings must be brief; that is, the mechanism must adapt quickly. Thus, the profound restlessness of the human spirit exemplified by Goethe's Faust, seems to arise mechanistically from the ceaseless call of the accumbens and prefrontal cortex for pulses of dopamine, to which they rapidly desensitize.

Consistent with this, many drugs that enhance the sense of well-being, including nicotine, ethanol, opioids, and cannabinoids, work by directly activating VTA neurons via specific molecular receptors to enhance dopamine

---

[9] MDMA (methylenedioxy-n-methylamphetamine, "Ecstasy") is used recreationally to enhance social interactions, especially at all-night dancing parties, termed "raves." Of course, similar ecstasy can be achieved during intense social interactions, as documented, for example, in the ethnographic film classic, *"Trance and Dance in Bali"* (Bateson and Mead, 1952).

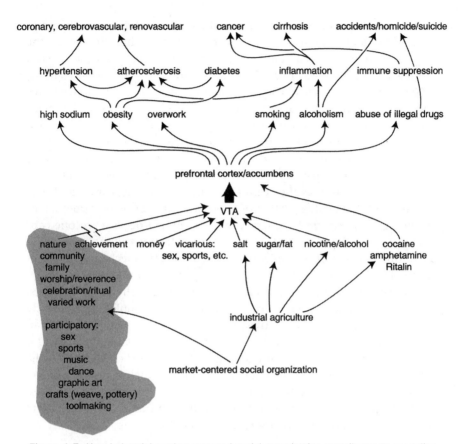

Figure 1.7: How industrial market-centered social organization contributes to mortality from hypervigilance and hyposatisfaction. Major causes of mortality stem from several co-occurring and mutually reinforcing pathogenetic processes. These arise partly from regulatory mechanisms (Fig. 1.4) designed to meet the need for hypervigilance and also from behaviors that try to meet the need for daily satisfactions. Each of many potential sources of satisfaction (shaded on left) can cause neurons in the ventral tegmental area (VTA) to deliver a pulse of dopamine to the nucleus acumbens and prefrontal cortex, thereby briefly providing a sense of well-being (Montague and Berns, 2002; Schultz, 2002). But industrial market-centered social organization narrows the sources of satisfaction;whereas satisfaction from a single source (work, food, nicotine) tends to adapt (Fig. 1.5), requiring higher levels for the same relief. A chronically elevated appetite in the context of "industrial agriculture," which provides the key substances cheaply and markets them intensively (Nestle, 2002; Schlosser, 2002), leads to a panoply of pathogenetic mechanisms that have been grouped as "metabolic syndrome" (Moller, 2001; Zimmet et al., 2001). When the reward system remains "unsatisfied" by natural inputs, drugs can "short circuit" the reward system by directly increasing cerebral dopamine (Schultz, 2002; Wise, 2003).

release in the accumbens and prefrontal cortex. Other drugs, such as amphetamine and cocaine, evoke these feelings by acting directly on the
dopamine terminals in accumbens and cortex to enhance release or block
reuptake, which allows dopamine to persist in the synaptic cleft (Schultz,
2002; Wise, 2003). When a signal from any one source is prolonged, this
system desensitizes, like all other sensors and effectors, to keep its input-
output curve centered on the most frequent demand (Fig. 1.5). Thus, just
as our enhanced sense of effortlessness fades quickly after setting down the
groceries, so does our enhanced sense of well-being fade – and for the same
computational reasons: all systems must adapt to a persistent signal.

This mechanism, as the "carrot" for anticipatory regulation and thus
key to the behavioral regulation of physiology, harbors grave potential for
pathology. The system is designed to serve myriad needs, each one contributing a small dollop of satisfaction. But satisfaction cannot be stored, so
if the number of sources shrinks, the task of driving the mechanism devolves
to the few that remain. And the more frequently one source is called on,
because of desensitization, the less satisfaction it can deliver. Yet the persistent demand of this one circuit calls for still stronger stimulation, insistently
crying out, "feed me!" Under these circumstances the "reward circuit" can
mediate addiction to essentially any source of satisfaction. But note that for
this to occur, nothing inside the body need be "broken" or "dysregulated."
The system can arrive at this state, locked on to a single source of satisfaction, simply because life circumstances have reduced all the other sources.
Now we can summarize how the "stick" of hypervigilance and the "carrot"
of satisfaction contribute to disorders of physiological regulation.

PATHOPHYSIOLOGY FROM ALLOSTASIS

Hypertension: Adaptation to Sustained Vigilance
Roughly one-quarter of U.S. adults are hypertensive (blood pressure
>140/80 mm Hg on repeated measurement). A few cases arise from identifi-
ably defective phenotypes (e.g., Wilson et al., 2001, but 95% are classified as
"essential" hypertension – cause unknown. Prevalence is nearly 40% greater
for African Americans than for whites (32% vs. 23%; Carretero and Oparil,
2000a). This difference is commonly attributed to genetics, but this seems
doubtful because the West African ancestors of New World blacks were not
hypertensive (Waldron, 1979). Furthermore, hypertension seems to be more
strongly associated with various sources of social distress, rather than race
per se. Thus, its main sequelae – death from coronary heart disease, cerebrovascular disease, and atherosclerosis – are more prevalent for divorced
versus married men and for low versus high employment grades by factors

as large or larger than for race (see Fig. 1.11). Thus, if in Cleveland I had canvassed *poor white* neighborhoods, I would also have seen men with limps and sagging faces.

The homeostasis model cannot explain essential hypertension because it attributes all pathology to a "defect" – to something "broken." But the allostasis model suggests that there is no defect. More parsimoniously, it proposes that hypertension emerges as the concerted response of multiple neural effectors to prediction of a need for vigilance (Fig. 1.4). When this prediction is sustained, all the effectors, both somatic and neural, adapt progressively to life at high pressure. The adaptations all seem entirely explicable from our general knowledge of signaling and regulation (Fig. 1.5). Although the endpoint may be tragic (Figs. 1.7, 1.11), every step along the path seems perfectly "appropriate."

Vigilance starts when a child is delivered from its mother's protection to the care of strangers. Thirty years ago this occurred when U.S. children first entered school at around age 6. Correspondingly, blood pressures were constant from the first year of life until age 6 (median systolic level ~100 mm Hg). Then commenced a steady rise, so that by age 17 half of all boys showed systolic pressures above 130 mm Hg, and about 20% showed pressures above 140 mm Hg (hypertensive). The rise for girls was similar, although slightly milder (Fig. 1.8).

But now blood pressures begin to rise in the first year of life (National Institutes of Health, 1997). This startling change might be associated with the rise of "day care" and the shift of mothers away from their infants and into the workforce. Consistent with this hypothesis, rat pups detached from their mothers show an eightfold rise of corticosterone over 24 hours, and human toddlers detached from their parents show increased cortisol (Schulkin, 1999). As shown in Figure 1.4, the neural signals that call for increased blood pressure also call for salty foods. These the fast-food industry ("industrial agriculture") provides in prodigious quantity, both in the supermarket and as part of the federally funded school lunch program (Fig. 1.7; Nestle, 2002; Schlosser, 2002). Industrial agriculture does not *cause* hypertension by excessively salting prepared foods; it merely obliges the public's appetite for sodium, which is driven quite appropriately by intact regulatory systems. Indeed, if under present conditions of life, the food industry were to restrict sodium, we might see the development of public "salt licks" like those used to attract deer.

In a younger person, if the predicted need for vigilance declines, effector adaptations can reverse promptly. But persistent demand leads to more profound and persistent effector adaptations. Over decades the constant call for vigilance adapts arterial muscle and carotid sinus to thicken and stiffen

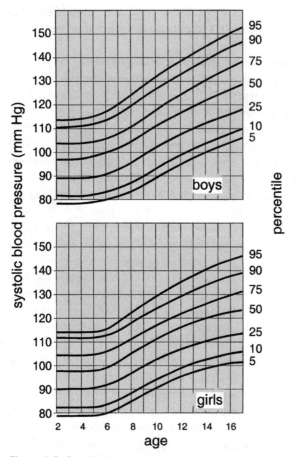

Figure 1.8: Systolic blood pressures were steady until school age and then rise continuously. Diastolic pressures also rise. Recent data show pressures rising in the first year, perhaps associated with the increase of "day care" (National Institutes of Health, 1997). Redrawn from Blumenthal et al., 1977.

so that pressure rarely returns to normal levels. Probably there are also corresponding adaptations in the brain. We know now that adult synapses continuously adjust their molecular components and that "memories" are stored at all levels, even in the spinal cord (Lücher and Frerking, 2003; Ikeda et al., 2003). So the many hormones that feed back to the brain to sustain high pressure (Fig. 1.4) probably entrain many levels to expect and support high pressure.

Thus, coordinated somatic and brain adaptations generate response patterns of "established" hypertension (Fig. 1.9). The hypertensive pattern, like the normal pattern (Fig. 1.3), does not seem to be "defended" at a

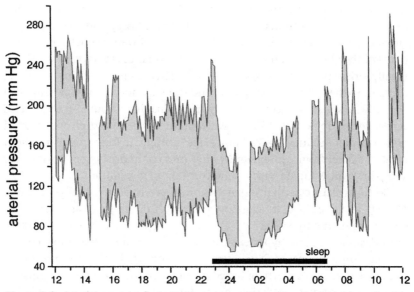
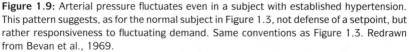

Figure 1.9: Arterial pressure fluctuates even in a subject with established hypertension. This pattern suggests, as for the normal subject in Figure 1.3, not defense of a setpoint, but rather responsiveness to fluctuating demand. Same conventions as Figure 1.3. Redrawn from Bevan et al., 1969.

particular level. Rather it is modulated up and down, apparently according to demand, with an overall range of 140 points. This pattern suggests adaptation to chronic vigilance and, consistent with this the hypertensive pattern, is absent in undisrupted preindustrial societies where children remain in contact with their parents and strangers are rare (Eyer and Sterling, 1977).

Established hypertension is most common in segments of modern society where family structure is most disrupted, where children are least protected, and where they are marked from birth for suspicion and various forms of ill treatment. One example stamped into memory is of the teenager I tutored in Hough who refused an opportunity to attend summer camp. As she explained, "I'd be in the sun a lot, and I'm already dark enough." Not to mention the universal experience of African American men who are regularly stopped by the police for the offense known as "DWB" – driving while black.

To such "prior knowledge" – the well-founded expectation of suspicion or contempt – hypervigilance is a natural response. Although the most overtly repressive forms of racism, such as lynching and legally enforced segregation of public facilities, have declined since the 1960s, de facto segregation of neighborhoods and schools has actually expanded, as has the disparity in wealth between rich and poor (Massey and Denton, 1994; Orfield, 2001). These trends intensify the sense of wariness between inhabitants of the ghetto and the outside – which enhances everyone's vigilance. Finally, since "prior knowledge" is embodied in communal attitudes to serve regulation, the communal suspicion and fear engendered by experience can not only be stored for decades within individual brains but also transferred across generations (Fig. 1.7). So we should expect that even when overt racial conflict declines, "prior knowledge" among African Americans about what to expect from white people, built into communal expectations over 400 years, will dissipate with a time constant of generations.

So to explain essential hypertension, there is no need to postulate a "defect" in any particular regulatory pathway. Certainly we can create a hypertensive mouse by knocking out one gene or another (Zhu et al., 2002), but we can also create hypertension and atherosclerosis in a whole colony of mice simply by introducing a stranger (Henry et al., 1967). Certainly we recognize that the variance of blood pressure within a community must be partially caused by genetic differences, but this cannot explain why blood pressures of essentially *all* our children rise with age. Nor why the rise is largest and most persistent in the poorest and most socially disrupted communities. Nor why African Americans are more hypertensive than genetically similar populations in West Africa. These observations certainly point to an environmental cause. Furthermore, hypertension represents only one of many similar threads from the quilt of predictive regulation.

Obesity and Metabolic Syndrome: Adaptation to Hyposatisfaction

Roughly half of U.S. adults are obese, a condition that contributes to type 2 diabetes. Obesity and type 2 diabetes jointly contribute to a constellation of pathologies, recently termed "metabolic syndrome," which includes hypertension, glucose intolerance (diabetes), hyperinsulinemia, dyslipidemia, visceral obesity, atherosclerosis, and hypercoagulability (Zimmet et al., 2001). Together these factors create a profoundly lethal cascade (Fig. 1.7), and all follow the familiar epidemiological pattern, elevated with divorce, low socioeconomic status, and in disrupted preindustrial communities (Figs. 1.10 and 1.11; Zimmet et al., 2001; Diamond, 2003). Like blood pressure, these conditions are rising in children, in whom the rate of obesity has reached 15% (Hill et al., 2003).

The homeostasis model cannot explain the prevalence of obesity. If metabolism were primarily controlled by negative feedback, then decreased energy expenditure would lead to decreased food intake. Yet presently in the United States the opposite is so: the less we exercise, the more we eat. This

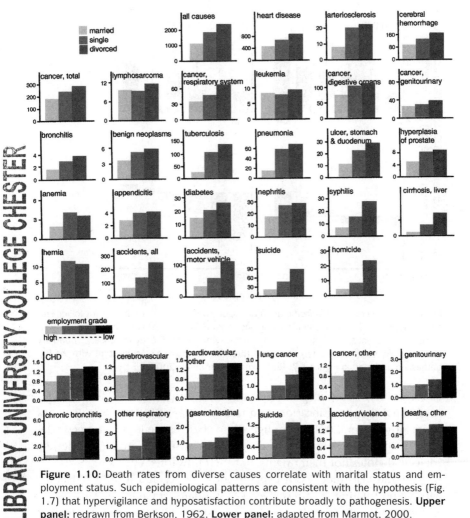

Figure 1.10: Death rates from diverse causes correlate with marital status and employment status. Such epidemiological patterns are consistent with the hypothesis (Fig. 1.7) that hypervigilance and hyposatisfaction contribute broadly to pathogenesis. **Upper panel:** redrawn from Berkson, 1962. **Lower panel:** adapted from Marmot, 2000.

conundrum could be caused by defects in the regulatory chain. For example, certain obese individuals are deficient in leptin, an important negative regulator of feeding, and when administered leptin, their weight returns toward normal (Farooqi et al., 2002). But, just like hypertension, specific defects in energy regulation are rare. They account for only a minor fraction of obesity and for none of its striking increase (Hill et al., 2003).

Nor can homeostasis explain the growing prevalence of type 2 diabetes. Its core feature, insulin "resistance," involves changes at many levels, including decreased concentrations of insulin receptors, kinase activities,

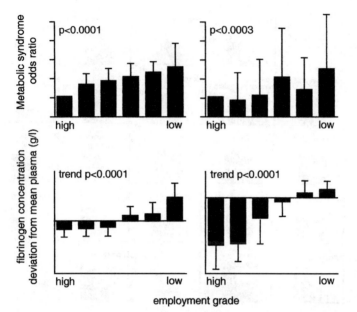

Figure 1.11: Metabolic syndrome and fibrinogen concentration are higher for lower employment grades. Metabolic syndrome includes many factors that contribute to cardiovascular and cerebrovascular disease, and elevated fibrinogen directly increases probability of myocardial or cerebral infarction. Thus these patterns are also consistent with the hypothesis diagramed in Figure 1.7. Redrawn from Brunner, 2000.

concentration and phosphorylation of IRS-1 and -2, PI(3)K activity, glucose transporter translocation, and the activities of intracellular enzymes (Saltiel and Kahn, 2001). Although these changes are loosely termed "defects" (Saltiel and Kahn, 2001), they do not arise from mutant alleles, so "defect" denotes not their origins, but their unwanted effects.

The allostasis model can explain both obesity and insulin resistance without postulating any true defect. The standard signals for vigilance (such as cortisol), which raise the appetite for sodium, also raise the appetite for carbohydrate and fat (Schulkin, 1999). This makes functional sense – if we will soon need more salt, we will also soon need more fuel. Elevated cortisol also shifts the distribution of fat deposits toward the viscera – one feature of metabolic syndrome. And when chronically high levels of carbohydrate evoke chronically high levels of insulin, its receptors and their downstream mechanisms naturally reduce their sensitivities (just as every signaling system responds to prolonged, intense stimulation) – the input-output curve inexorably shifts to the right (Fig. 1.5). This does not explain mechanistically how myriad inter- and intracellular signals cause resistance, a problem now pursued by thousands of molecular and cellular biologists. The point to recognize is that bad things can happen as the natural outcome of predictive regulation.

Cortisol and related signals are elevated not only during hypervigilance, but also during states of hyposatisfaction – when outcomes prove less than expectations. Because satisfaction cannot be stored, it must be continuously renewed. So if its potential sources become constricted, the brain must inevitably rely on those that remain: people needing a pulse of satisfaction will try to find it somehow (Fig. 1.7). For those of higher socioeconomic status, there are opportunities for satisfaction in work, achievement, and money. Singular pursuit of such opportunities tends to spiral out of control ("workaholism," "type A" behavior, etc.). This may occur especially when expectations inculcated by the family as "prior knowledge" are so high as to be intrinsically unsatisfiable.[10] Another likely factor is that a stimulus that initially releases dopamine adapts, limiting the satisfaction obtainable from its repetition.

For people of lower socioeconomic status, potential sources of satisfaction are less available, but food is abundant and cheap. So the allostasis model suggests that the brain overrides local negative feedback (metabolic satiety signals), just as it overrides the negative feedback that would counter commands to raise blood pressure, and people eat. For the reasons just cited, satisfaction is fleeting, so people eat even more (Saper et al., 2002; Schultz, 2002; Dallman et al., 2003).

Alcohol and drug addictions follow a similar pattern and apparently share many of the same mechanisms (Wise, 2003). For example, the neuropeptide NPY enhances feeding and is also abundant in brain areas mediating drug addictions. The acute effect of NPY resembles alcohol in reducing anxious behavior, and it is also associated with developing alcohol and cocaine dependence. Similarly leptin, identified primarily with feeding and energy balance, contributes to hypertension. Thus, there is considerable cross-talk between these systems along brain pathways that serve satisfaction (VTA-amygdala-accumbens-prefrontal cortex).

The rise in obesity and type 2 diabetes has recently been attributed to "thrifty genes." Noting that the most explosive increases are in populations that have suddenly changed from food scarcity to plenty, especially among Pacific Islanders, Diamond recalls Neel's hypothesis that certain human groups were selected to "eat up" in times of plenty to protect against times of famine (Diamond, 2003). This implies that body fat is not regulated to a setpoint, but varies according to some prediction – in this case, future

---

[10] My mother in her eighties likes to tell my friends that she still feels guilty about smoking during her first pregnancy because, otherwise, "Peter might have been smarter."

hunger. This theory would be entirely consistent with the allostasis model, but there may be an additional explanation.

Consider that for these groups the sudden appearance of plentiful food is accompanied by the equally abrupt dissolution of the entire culture. Consequently, obesity is only one disorder of many that accompany disruption of a preindustrial society. Among Native Americans, Australian Aborigines, Inuit, and so on, the rise in obesity and type 2 diabetes invariably accompanies rises in essential hypertension, alcoholism, drug addiction, suicide, and murder (Eyer and Sterling, 1977). Furthermore, the same correlations are found in modern societies: the highest rates of all these afflictions appear in the most disrupted populations, those with the worst life experience, the lowest expectations, and the least hope. Over the period of rising racial segregation in urban neighborhoods and schools (Denton, 1993; Orfield, 2001), the prevalence of obesity in predominantly black elementary and middle schools has tripled (Gordon-Larsen et al., 1997).

In summary, the allostasis model attributes the pathogenesis of hypertension and metabolic syndrome to prolonged adaptation to hypervigilance and hyposatisfaction. The impact is strongest among populations with the best reasons for vigilance, the narrowest range of satisfactions, and expectations that are least often met (Figs. 1.10, 1.11). Intensified genetic screening will undoubtedly identify defective or alternative alleles that render individuals more sensitive to one of these conditions than another. Such screens will help to explain *who* in a particular population gets *which* disorder. But, because gene frequencies change over centuries, whereas the prevalence of these syndromes has risen over decades, their explosive increases cannot be attributed to genetic defects. So in considering where to intervene to prevent and treat this constellation of disorders, we should ask first, what changes in preindustrial societies accompany the appearance of plentiful food.

Large-Scale Social Organization: Reasons for Hypervigilance and Loss of Satisfactions

Preindustrial communities, which upon disruption are so susceptible to modern disorders, have the following shared features. First, the communities are small, so most human encounters are between people who are completely familiar with each other and who are often closely related. Consequently each person can pretty well predict how every other person will behave, and all are constrained by clear rules to behave well. Second, every transaction between individuals is governed by the near certainty that there will be more transactions in the future. So, fairness and generosity tend to be rewarded, and cheating tends to be punished (Glimcher, 2002). Third, in a small community every significant transaction is widely known, discussed, and judged – providing another incentive to play fair. Finally, goods tend to be sparse and their ownership is well known, so theft is pointless and consequently rare. All these circumstances reduce the need to sustain vigilance.

For industrial, market-dominated societies, essentially the opposite is true. Communities are large and comprise mostly strangers who come from different cultures with different rules and often with ample historical reasons for mistrust. Furthermore, many transactions occur between strangers who will never meet again, so trust is less rewarding, and the need to sustain hypervigilance is greater. Any urban skeptic, who might consider this a baseless speculation, should recall how remarkable it seems when on a trip to the country we find that rural people don't bother to lock the door or set a car alarm – and that enough trust persists that you can pump gasoline first and pay later.

What about satisfactions? People in preindustrial communities connect strongly with nature. Daily they experience dawn and dusk, the rising of the moon, the night sky, the murmur of a stream, the season. An adult's labor is highly varied: hunting, fishing, and gathering; clearing, sowing, weeding, and reaping; making tools and pots; spinning and weaving; building shelter (Wilson, 1984). Each person develops many skills – all of the sort that in large, urban communities, we now seek as "*re*-creation." Consider that each connection with nature, the practice of each skill, and the different rhythm of each activity might provide just the right signal to release the small pulse of dopamine that we experience as a satisfaction.

These small satisfactions are lost in industrial, market-dominated societies, which tend to shrivel the richness of an individual's labor down to a few repetitive motions. Today a worker may be less likely to serve on an assembly line than to sit isolated in a cubicle and stare at a computer screen, but the variety and challenge of labor have not been recovered. This was thoroughly anticipated by early observers of the capitalist economy. Here is Adam Smith:

The man whose life is spent in performing a few simple operations has no occasion to exert his understanding, or to exercise his invention in finding out expedients for difficulties which never occur. He naturally loses, therefore, the habit of such exertion and generally becomes as stupid and ignorant as it is possible for a human creature to become. (Wealth of Nations, 1776, pp. 734–735)

And here is F. W. Taylor, the father of "scientific management," describing pridefully his contribution to this process of stupidification:

Owing to the fact that workmen . . . have been taught . . . by observation of those immediately around them, there are . . . perhaps 40, 50, or hundred ways of doing each act in the trade. Now . . . there is always one best method . . . which is quicker and better than any of the rest. And this . . . can only be discovered through a scientific study and analysis of all the methods . . . together with minute, motion and time study. This involves the gradual substitution of science for rule of thumb throughout the mechanic arts. (Principles of Scientific Management, 1911, pp. 24–25)

Again, the urban skeptic might object that all these losses are simply the cost of today's many comforts, which are surely satisfying. But here is Herman Melville:

We felt very nice and snug, the more so since it was so chilly . . . out of bed-clothes . . . seeing that there was no fire in the room. The more so, I say, because truly to enjoy bodily warmth, some small part of you must be cold, for there is no quality in the world that is not what it is merely by contrast. . . . If you flatter yourself that you are all over comfortable, and have been so a long time, then you cannot be said to be comfortable any more. But if, like Queequeg and me in the bed, the tip of your nose or the crown of your head be slightly chilled, why then, indeed . . . you feel most delightfully and unmistakably warm. For this reason, a sleeping apartment should never be furnished with a fire, which is one of the luxurious discomforts of the rich. (Moby-Dick, 1851, p. 59)

These examples fit a hypothesis that our regulatory systems were selected to seek satisfactions that are small and brief. The best satisfactions occur when an experience exceeds prior expectations, for example, when it contrasts to a prior discomfort. Because of this, when an experience is sustained, we soon adapt, and it ceases to satisfy. Furthermore, because it is expectation (or hope) that counts, when we achieve a certain level of comfort, it will cease to satisfy when we learn that others have more. Although these conclusions can all be learned from the classics of religion and literature, their neural basis is just now being elucidated by recordings from VTA neurons and measurements of dopamine release (e.g., Fiorello et al., 2003; Phillips et al., 2003). These studies may help understand why industrial, market-centered society loses the full range of satisfactions, and also why preindustrial societies abandon theirs so quickly when brief exposure to modern goods (shotgun, chainsaw, outboard motor) instantly raises expectations. This renders such groups profoundly vulnerable to hyposatisfaction – just as they are vulnerable to new germs (Diamond, 1999) – and thus prey to the standard addictions of modern life: fast food, alcohol, and drugs.

Note that "adaptation" refers simply to the resetting of response sensitivity to a signal. Although it may turn out badly over time, the outcome is not caused by any low-level error or defect. Consequently, it should not be considered as "inappropriate" or as "dysregulation." Rather, disordered human relationships can drive perfectly normal adaptations of internal physiology into mass-scale pathogenesis – just as disorder within a family can drive a child into diabetic acidosis (Fig. 1.2). The allostasis model clearly identifies a paradox: people are dying, but their internal regulatory mechanisms are intact. So where should we intervene?

RATIONAL THERAPEUTICS: WHERE TO INTERVENE?

Homeostasis Treats Low-Level Targets
Following the homeostasis model, physicians try to restore each parameter to what they consider an "appropriate" level. Therefore, hypertension is treated with drugs that target the three primary effectors of elevated pressure: (1) diuretics to reduce blood volume, (2) vasoconstrictor antagonists to dilate the vascular tree, (3) heart rate antagonists to reduce cardiac output. The pharmaceutical industry continues to test myriad molecules that regulate these three mechanisms, and fundamental research widely promises to identify new targets. Thus, WNK kinases and their associated signaling pathways "may offer new targets for the development of antihypertensive drugs" (Wilson et al., 2001); as might the β1subunit of the calcium-activated potassium channel (Brenner et al., 2000) and gene targets of the estrogen receptor β (Zhu et al., 2002).

The same is true for obesity. A recent review lists 6 neuromodulators that increase feeding and 10 that decrease feeding and then concludes, "a multidrug regimen that targets multiple sites within the weight-regulatory system may be necessary to achieve and sustain weight loss in many individuals" (Schwartz et al., 2000). Similarly, a study using RNAi in the nematode identifies 305 gene inactivations that reduce body fat and 112 gene inactivations that increase it – and concludes that many of these genes "are promising candidates for developing drugs to treat obesity and its associated diseases" (Ashrafi et al., 2001). The same strategy is proposed for type 2 diabetes and metabolic syndrome (Moller, 2001) and for drug addictions (Laakso et al., 2002).

There are three problems with targeting low-level mechanisms. First, each signal evokes multiply cascaded effects, so even the most specific molecular antagonist will cause a cascade of effects. For example, in hypertension the angiotensin converting enzyme affects all of angiotensin's myriad downstream targets (arteriolar muscle, kidney, and multiple brain sites; see Fig. 1.4), and so also does its widely prescribed inhibitor. Similarly, in type 2 diabetes one effect of hyperglycemia is to elevate the signaling

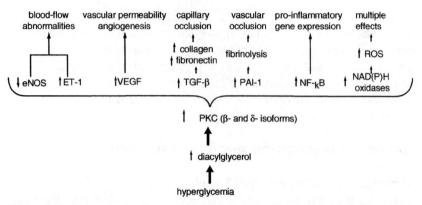

Figure 1.12: How one signaling molecule (diacylglycerol) can stimulate a cascade of pathogenesis. Hyperglycemia, part of the "metabolic syndrome," affects numerous signaling molecules, each of which activates its own cascades. As exemplified here, diacylglycerol triggers protein kinase C, and thereby a host of signals, all of which contribute to vascular pathology. eNos = endothelial nitric oxide synthase; ET-1 = endothelin-1; VEGF = vascular endothelial growth factor; TGF-b = transforming growth factor-b; PAI-1 = plasminogen activator inhibitor. Adapted from Brownlee, 2001.

molecule, diacylglycerol. This triggers a host of cascades, some of whose bad effects are shown in Figure 1.12. Although it might seem advantageous to antagonize an early step, such as the activation of protein kinase C, myriad other cascades with beneficial effects would also be affected (Fig. 1.12). It turns out that because of such cascading effects, low-level inhibitors and antagonists tend to be strongly iatrogenic (Sterling and Eyer, 1981; Buchman, 2002).

Second, the variables targeted for treatment are being driven to their particular levels by concerted signals from the brain (Fig. 1.4) in response to predicted needs (Fig. 1.13A). Consequently, if one signal is suppressed by a drug, the brain compensates by driving all the others harder. Thus, when blood pressure is treated by a diuretic to reduce volume, there are compensatory increases in heart rate and vasoconstriction. These can be treated in turn by beta-adrenergic antagonists, calcium channel antagonists, and so on

Figure 1.13: Where to intervene? **A.** Healthy system. As demand distribution shifts upward briefly, the response distribution follows to maintain variation centered on most probable demand (see Fig. 1.3). As demand distribution returns to its initial state, the response distribution follows. **B.** Unhealthy system. When high demand predominates for long periods, the system adapts to this expectation. When demand is reduced briefly, the system does not return to the initial state. **C.** Standard pharmacotherapy. While demand stays high, drugs that antagonize key effector mechanisms force the response distribution back toward its initial mean; but this reduces responsiveness and evokes iatrogenic

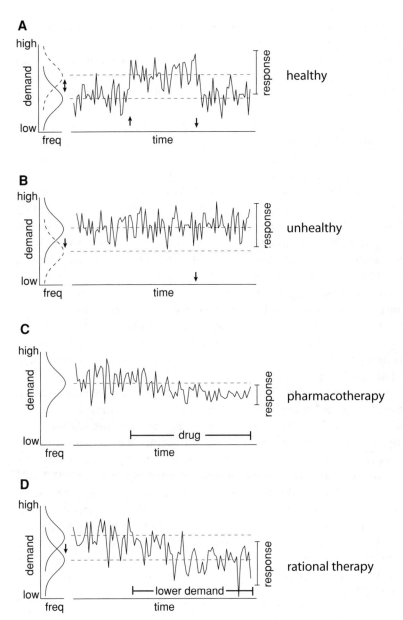

effects. This should be expected because the organism must continue to meet elevated demand but with fewer or weaker effectors. This is a common complaint of patients on antihypertensive and psychotropic medications. **D.** Rational therapy. When demand is reduced for long periods, the system readapts to the initial demand distribution. The mean response returns to its initial level while responsiveness is maintained.

(Sterling and Eyer, 1981; Carretero and Oparil, 2000b). But adding more drugs to a complex system increases the frequency of iatrogenesis. This is why proposals to treat obesity with a multidrug regimen at multiple brain sites or to screen 417 genes as drug targets for obesity seem implausible.

Third, there is a cost to performance in clamping a variable to some target level by blocking the effectors designed to modulate it. Clamping renders that variable insensitive to predicted need, which opposes the whole point of physiological regulation (Fig. 1.13A). Thus clamping blood pressure low with a beta-blocker commonly causes "exercise intolerance" – inability to increase cardiac output when it is needed (Fig. 1.13C).

For all of these reasons, less than 25% of hypertensive patients in the United States are controlled. The major problem is considered to be "the very high rate of discontinuance or change in medications: 50–70%... within the first six months" (Carretero and Oparil, 2000b). These high discontinuance rates are considered to reflect, among other factors, "a combination of adverse drug effects, cost of drugs, and poor efficacy" (Carretero and Oparil, 2000b). Consequently, despite their remarkable in-genuity, 30 years of low-level pharmacological treatments for hypertension have not worked. For the same reasons, it seems doubtful that low-level treatments for obesity and metabolic syndrome will be more successful, and already there have been serious adverse effects – for example from fen-fluramine and amphetamines.

These problems also apply to pharmacotherapy for mental disorder. Certainly drugs are better than lobotomy: they can be titrated and are re-versible over the short run. When applied for long periods, however, the "antipsychotic" drugs, which primarily antagonize various dopamine re-ceptors, cause motor disorders. These "tardive" (late appearing) dyskinesias eventually occur in most patients and persist after the drugs are withdrawn (Gelman, 1999). Beyond this devastating iatrogenic effect, drugs that work by antagonizing the major modulators of the nucleus accumbens, amyg-dala, and prefrontal cortex will, like beta-blockers for blood pressure, re-duce responsiveness (Fig. 1.13C). Such drugs would be predicted to cause in stability of intent and to flatten affect (Fig. 1.6). In fact they do, and this is a major reason why patients often refuse to take them (Sterling, 1979; Gelman, 1999).

Allostasis Emphasizes Higher-Level Interventions

The allostasis model defines health as *optimal predictive fluctuation*. A shift in the probability of demand should shift the response, and when the pre-diction reverses, so should the response (Fig. 1.13A). A system becomes unhealthy when, during long periods of high demand, effectors adapt so strongly that they cease to follow promptly when the prediction reverses (Fig. 1.13B). Drugs can force the response back to the original level, despite continued prediction of high demand, but this compresses responsiveness (Fig. 1.13C). A more rational goal of intervention would be to shift the predicted distribution of demand back toward its original level. This would allow the effectors to naturally reestablish flexible variation around the predicted lower demand, thus preserving the range of responsiveness (Fig. 1.13D).

Can this work for hypertension? Consider that the current authoritative recommendations for treatment are no longer drugs but (1) weight loss; (2) exercise; (3) moderate alcohol consumption; (4) diet reduced in sodium and fat and increased in calcium, potassium, and fiber; (5) smoking cessation (Carretero and Oparil, 2000b). Weight loss is strongly correlated with reduced blood pressure and is considered to be the most effective of all nonpharmacological treatments. Moderate exercise, such as brisk walking or bicycling three times per week, may lower systolic pressure by 4–8 mm Hg. The dietary recommendation is based on the DASH study, which found overall reductions in blood pressure of 11.4/5.5 mm Hg to a diet rich in fruits, vegetables, and low-fat dairy products, with further reductions of pressure to reduced sodium intake (Sacks et al., 2001). These reductions are said to be "comparable to or greater than those usually seen with monotherapy (i.e., 1 drug) for stage 1 hypertension" (Sacks et al., 2001).

These studies document that the distribution of responses can indeed return to lower levels (Fig. 1.13D). But as the DASH study notes, long-term health benefits "will depend on the ability of people to make long-lasting dietary changes, including the consistent choice of lower-sodium foods" and "upon (their) increased availability" (Sacks et al., 2001). This requires, in effect, a sustained victory in the prefrontal cortex of abstract knowledge about what is "good for you" over all the unsatisfied appetites that cause the problem in the first place. Hold onto your McDonald's stock.

The most successful interventions do not deny the sense of need. Rather, they find ways to satisfy it by enlarging positive social interactions and revivifying the sense of connectedness. In the case of coronary heart disease, when patients combined diet and exercise in a group context with a charismatic leader, atherosclerotic plaques regressed over a year, as established by angiography (Ornish et al., 1990). Other outstanding examples are "therapeutic communities," such as the "Twelve-step" programs for treating addictions to alcohol and various illegal drugs.

Therapeutic communities formed the basis for the first mental asylums, such as the Quaker-organized York Retreat (England in 1796) and the first American asylums of the early 19th century, Pennsylvania Hospital and

Worcester State Hospital in Massachusetts. These institutions offered "moral therapy," which included physical labor (farm work), well-lit rooms, good food, porter in moderation (alcohol), and lectures on varied topics such as astronomy and literature. Every feature of the program had but one goal: to enhance the patient's sense of well-being (Bockoven, 1956; Tuke, 1964; Grob, 1966). Extraordinarily detailed follow-up studies published in the late 19th century showed these programs to be highly effective, as were subsequent programs with similar goals, for example, that of the Boston Veteran's Administration in treating posttraumatic stress disorders after World War II (Greenblatt et al., 1955).

This is *not* to argue against treating any mental disorder with a drug. Almost certainly, some disorders will be found to arise from specific molecular defects, like the specific mutations of ion channels, gap junctions, and signaling enzymes, and so on that are being identified as causing various neurological disorders (Rosenberg et al., 1997). But just as those defects are fairly rare, and just as molecular defects account for a minor proportion of hypertension, there is likely to be a rather large residual group that will be considered "essential mental illness," arising from the same core problems of social disruption and disconnection.

This seems particularly applicable to the large group of boys now diagnosed with "attention-deficit/hyperactivity disorder" (ADHD). The prevalence of this diagnosis among boys in the United States has reached $\sim$10–30%, and it varies inversely with socioeconomic status. The standard drug treatment is methylphenidate (Ritalin) – an amphetamine analog – or dextroamphetamine. These drugs do help a rambunctious youngster to settle down in the classroom and concentrate for longer periods than he could normally manage. Should this surprise us?

These are the drugs that a street addict takes to obtain his small satisfactions – to quiet his restless prefrontal cortex (Fig. 1.7). And these are the drugs that the long-distance trucker takes to concentrate on the road. So it seems entirely consistent that a boy dosed with amphetamine can concentrate on the assigned task. But over the long term, these drugs will certainly cause brain adaptations with specific consequences that cannot be foreseen.

This example seems especially poignant because it arises from disrespecting our greatest evolutionary advantage: our intrinsic diversity of talent and temperament. A proto-scholar might sit still effortlessly in a classroom, whereas a proto-navigator or proto-comedian might not. The allostasis model would not administer the very drugs on which (outside the classroom) we have declared "war." Rather, it would investigate the possible causes of a youngster's restlessness and would intervene by finding activities – beyond sitting still with a book – that *would* absorb him.

Physiological Mechanisms of High-Level Intervention

Although healing has always relied on the power of positive human inter-action (Sterling and Eyer, 1981), to proponents of the homeostasis model this has often seemed like so much mumbo-jumbo. But modern imaging begins to reveal some specific underlying mechanisms. For example, ad-ministration of levo-dopa has been an important therapy for Parkinson's disease, apparently because it is converted to dopamine by midbrain neu-rons whose release of dopamine in the striatum relieves Parkinsonian symp-toms. Yet it turns out that placebo can relieve these symptoms about as well as levo-dopa. Apparently the expectation of therapeutic benefit evokes re-lease of endogenous dopamine, as measured by positron emission tomog-raphy (PET; de la Fuente-Fernández et al., 2001; de la Fuente-Fernández and Stoessl, 2002).

Similarly, placebos can relieve pain, just like opioids. It was suggested that the placebo evokes release of endogenous opioids in the brain. This hypothesis is now supported by PET, which shows placebo and opioid anal-gesia activating the same brain regions (Petrovic et al., 2002). Finally, despite the common wisdom that exercise improves mood and sharpens the mind, there was no clear physiological basis. Now it is found that exercise raises levels of brain-derived neurotrophic factor and other growth factors known to serve synaptic plasticity and learning (Cotman and Berchtold, 2002).

Some will object to the allostasis model by pointing out that death rates have plummeted from many causes, including cardiovascular disease, and that life expectancy has steadily risen. Certainly, diagnosis has benefited enormously from new imaging methods (computed tomography, magnetic resonance imaging, angiography, ultrasound), and biological chemistry has benefited from monoclonal antibodies, polymerase chain reaction, fast as-say by mass spectrometry, and so on. And surgery has developed less in-vasive approaches and better management of acute physiology and shock (Buchman, 2002). So, if you are shot or stabbed or injured in an automobile, your chances of survival are greatly improved, as they also are if you suffer myocardial infarction or stroke.

Yet by all accounts the financing and organizing of health care are a big problem. The roughly 15% of the gross national product spent for health goes largely for high-tech treatments, leaving a pittance for the low-tech, human contributions, like nursing and rehabilitation, that require time and emotional contact (Levit et al., 2000). Although high-tech medicine can be amazing, medical errors are frequent and lethal, causing an esti-mated 50,000 deaths annually (Kohn et al., 2000). Furthermore, high-tech care is unevenly distributed along racial and socioeconomic lines; whereas mammoth pharmaceutical and insurance companies take huge bites for relatively small contributions. Finally, much ingenuity and expense is directed at treating casualties of excessive haste (accidents), aggression, and unsatisfied need.

The allostasis model hints that the biggest improvements in health might be achieved by enhancing public life. The guiding principle would be this: do everything that promises to reduce the need for vigilance and to restore small satisfactions. Enhance contact with nature by building more parks and by providing communal opportunities to garden – that is, not just to look at but to grow flowers and vegetables. Enhance opportunities to walk and cycle by restricting automobile traffic. Prevent this restriction from becoming an annoyance by improving public transportation. Encourage broader participation in sports, especially among youth, by constructing public facilities for gymnastics, skating, skateboarding, climbing, and swimming.

Improve work by acknowledging that no human can be satisfied by performing an unvarying task for 8 hours a day, 40 hours per week, 50 weeks per year. Companies, and now even our National Institutes of Health, play a recording, "this phone call may be monitored for purposes of quality control." What this implies, of course, is that the task is so uninteresting that the operators need to be threatened with every call that their supervisor might be listening. For workers at the computer, every keystroke can be similarly monitored. Such humiliating and alienating procedures were introduced recently and could easily be eliminated. The astronomical disparities of income are also recent and could be narrowed while still preserving incentives for the more energetic and clever. Such proposals are well within our capacity to organize and implement, for they would benefit the rich as well as the poor by reducing everyone's need for vigilance and by expanding everyone's range of small satisfactions.

REFERENCES

Adolphs, R., Tranel, D., Damasio, H., Damasio, A. (1994). Impaired recognition of emotion in facial expressions following bilateral damage to the human amygdala. *Nature* 372:669–72.

Alexander, R. M. (1996). *Optima for Animals*. Princeton: Princeton University Press.

Ashrafi, K., Chang, F. Y., Watts, J. L., Graser, A. G., Kamath, R. S., Ahringer, J., Ruvkun, G. (2001). Genome-wide RNAi analysis of *Caenorhabditis elegans* fat regulatory genes. *Nature* 421:268–72.

Baker, L., Minuchin, S., Rosman, B. (1974). The use of beta-adrenergic blockade in the treatment of psychosomatic aspects of juvenile diabetes mellitus. In: *Advances in Beta-Adrenergic Blocking Therapy* (ed. J. A. Snart), pp. 67–80. Princeton: Princeton Excerpta Medica.

Bateson, G., Mead, M. (1952). *Trance and Dance in Bali*. Documentary.

Berkman, L. F., Kawachi, I. (2000). Social Epidemiology. New York NY: Oxford University Press.

Berkson, J. (1962). Mortality and marital status. Reflections on the derivation of etiology from statistics. *Am J Pub Health* 52:1318.

Bernik, T. R., Friedman, S. G., Ochani, M., DiRaimo, R., Ulloa, L., Yang, H., Sudan, S., Czura, C. J., Ivanova, S. M., Tracy, K. J. (2002). Pharmacological stimulation of the cholinergic antiinflammatory pathway. *J Exp Med* 195:781–8.

Bevan, A. T., Honour, A. J., Stott, F. H. (1969). Direct arterial pressure recording in unrestricted man. *Clin Sci* 36:329.

Blalock, J. E. (2002). Harnessing a neural-immune circuit to control inflammation and shock. *J Exp Med* 195:F25–8.

Blumenthal, S., Epps, R. P., Heavenrich, R., Lauer, R. M., Lieberman, E., Mirkin, B., Mitchell, S. C., Boyar-Naito, V., O'Hare, D., McFate-Smith, W., Tarazi, R. C., Upson, D. (1977). Report of the task force on blood pressure control in children. *Pediatrics* 59:797.

Bockoven, J. S. (1956). Moral treatment in American psychiatry. *J Nerv Ment Dis* 124:167–321.

Brenner, R., Peréz, G. J., Bonev, A. D., Eckman, D. M., Kosek, J. C., Wiler, S. W., Patterson, A. J., Nelson, M. T., Aldrich, R. W. (2000). Vasoregulation by the beta1 subunit of the calcium-activated potassium channel. *Nature* 407:870–6.

Brownlee, M. (2001). Biochemistry and molecular cell biology of diabetic complications. *Nature* 414:813–20.

Brunner, E. J. (2000). Toward a new social biology. In: *Social Epidemiology* (ed. L. Berkman, I. Kawachi), pp. 306–331. New York: Oxford University Press.

Buchman, T. G. (2002). The community of the self. *Nature* 420:246–51.

Cannon, W. B. (1920). *Bodily Changes in Pain, Hunger, Fear and Rage: An Account of Recent Researches into the Function of Emotional Excitement.* New York: Appleton.

Cannon, W. B. (1935). Stresses and strains of homeostasis. *Am J Med Sci* 189:1–10.

Carretero, O. A., Oparil, S. (2000a). Essential hypertension. Part I: Definition and etiology. *Circulation* 101:329–35.

Carretero, O. A., Oparil, S. (2000b). Essential hypertension. Part II: Treatment. *Circulation* 101:446–53.

Cotman, C. W., Berchtold, N. C. (2002). Exercise: A behavioral intervention to enhance brain health and plasticity. *Trends Neurosci* 25:295–301.

Dallman, M. F. (2003). Stress by any other name . . . ? *Horms Behav* 43:18–20.

Dallman, M. F., Pecoraro, N., Akana, S. F., la Fleur, S. E., Gomez, F., Houshyar, H., Bell M. E., Bhatnagar, S., Laugero, K. D., Manalo, S. (2003). Chronic stress and obesity: a new view of "comfort food." *Proc Natl Acad Sci U S A* 100:11696–701.

de la Fuente-Fernández, R., Ruth, T. J., Sossi, V., Schulzer, M., Calne, D. B., Stoessl, A. J. (2001). Expectation and dopamine release: Mechanism of the placebo effect in Parkinson's disease. *Science* 293:1164–6.

de la Fuente-Fernández, R., Stoessl, A. J. (2002). The placebo effect in Parkinson's disease. *Trends Neurosci* 25:302–6.

Denton, D. (1993). Control mechanisms. In: *The Logic of Life* (ed. C. A. R Boyd, D. Noble), pp. 113–46. Oxford: Oxford University Press.

Dhingra, N. K., Kao, Y.-H., Sterling, P., Smith, R. G. (2003). Contrast threshold of a brisk-transient ganglion cell in vitro. *J Neurophysiol* 89:2360–9.

Diamond, J. (1993). Evolutionary physiology. In: *The Logic of Life* (ed. C. A. R. Boyd, D. Noble), pp 89–111. New York: Oxford University Press.

Diamond, J. (1999). *Guns, Germs, and Steel: The Fates of Human Societies.* New York: W. W. Norton & Sons.

Diamond, J. (2003). The double puzzle of diabetes. *Nature* 423:599–602.

Elbert, T., Rockstroh, B. (2003). Stress factors. *Nature* 421:477–8.

Ernst, M. O., Banks, M. S. (2002). Humans integrate visual and haptic information in a statistically optimal fashion. *Nature* 415:429–33.

Eyer, J. (1975). Hypertension as a disease of modern society. *Int J Health Serv* 5:539–58.

Eyer, J. (1977). Prosperity as a cause of death. *Int J Health Serv* 7:125–50.

Eyer, J., Sterling, P. (1977). Stress-related mortality and social organization. *Rev Radical Political Econ* 9:1–44.

Fairhall, A. L., Lewen, G. D., Bialek, W., de Ruyter van Steveninck, R. (2001). Efficiency and ambiguity in an adaptive neural code. *Nature* 412:787–92.

Farooqi, I. S., Matarese, G., Lord, G. M.. Keogh, J. M., Lawrence, E., Agwu, C., Sanna, V., Jebb, S. A., Perna, F., Fontana, S., Lechler, R. I., DePaoli, A. M., O'Rahilly, S. (2002). Beneficial effects of leptin on obesity, T cell hyporesponsiveness, and neuroendocrine/metabolic dysfunction of human congenital leptin deficiency. *J Clin Invest* 110:1093–1103.

Fehr, E., Fischbacher U. (2003). The nature of human altruism. Nature 425:785–791.

Field, G. D., Rieke, F. (2002). Nonlinear signal transfer from mouse rods to bipolar cells and implications for visual sensitivity. *Neuron* 34:773–85.

Fiorello, C. D., Tobler, P. N., Schultz, W. (2003). Discrete coding of reward probability and uncertainty by dopamine neurons. *Science* 299:1898–1902.

Flier, J. S. (2002). Is brain smpathetic to bone? *Nature* 420:619–22.

Fluharty, S. J. (2002). Neuroendocrinology of body fluid homeostasis. In: *Hormones, Brain and Behavior*, pp. 525–569. New York: Elsevier Science.

Friedman, M. I., Ji, H., Graczyk-Milbrandt, G., Osbakken, M. D., Rawson, N. E. (1998). Hepatic sensing in the control of food intake: unresolved issues. In: *Liver and Nervous System* (ed. D. Häussinger, K. Jungermann), pp. 220–29. London: Kluwer Academic Press.

Geisler, W. S., Diehl, R. L. (2002). Bayesian natural selection and the evolution of perceptual systems. *Philos Trans R Soc Lond (Biol)* 357:419–48.

Geisler, W. S., Diehl, R. L. (2003). A Bayesian approach to the evolution of perceptual and cognitive systems. *Cogn Sci* 118:1–24.

Gelman, S. (1999). *Medicating Schizophrenia: A History.* Piscataway, NJ: Rutgers University Press.

Glimcher, P. W. (2002). Decisions, decisions, decisions: Choosing a biological science of choice. *Neuron* 36:323–32.

Gordon-Larsen, P., Zenel, B. S., Johnston, F. E. (1997). Secular changes in stature, weight, fatness, overweight, and obesity in urban African American adolescents from the mid-1950's to the mid-1990's. *Am J Hum Biol* 9:675–88.

Greenblatt. M., York, R. H., Brown, E. L. (1955). *From Custodial to Therapeutic Patient Care in Mental Hospitals: Explorations in Social Treatment.* New York: Russell Sage Foundation.

Grob, G. N. (1966). *The State and the Mentally Ill. A History of Worcester State Hospital in Massachusetts, 1830–1920.* Chapel Hill: University of North Carolina Press.

Hammond, K. A., Diamond, J. (1997). Maximal sustained energy budgets in humans and animals. *Nature* 386:457.

Harris, A. H., Gilliam, W. J., Findley, J. D., Brady, J. V. (1973). Instrumental conditioning of large magnitude. Daily, 12-hour blood pressure elevations in the baboon. *Science* 182:175–77.

Henry, J. P., Meehan, J. P., Stevens, P. M. (1967). The use of psychosocial stimuli to induce prolonged systolic hypertension in mice. *Psychosom Med* 29:408.

Hill, J. O., Wyatt, H. R., Reed, G. W., Peters, J. C. (2003). Obesity and the environment: where do we go from here? *Science* 299:853–5.

Hochachka. P. W., Somero, G. N. (2002). *Biochemical Adaptation: Mechanism and Process in Physiological Evolution.* New York: Oxford University Press.

Ikeda, H., Heinke, B., Ruscheweyh, R., Sandkühler, J. (2003). Synaptic plasticity in spinal lamina I projection neurons that mediate hyperalgesia. *Science* 299:1237–40.

Kohn, L. T., Corrigan, J. M., and Donaldson M. S. (eds) (2000). Institute of Medicine. To err is human: building a safer health system. Washington DC, National Academy Press.

Koob, G. F., Le Moal, M. (2001). Drug addiction, dysregulation of reward, and allostasis. *Neuropsychopharmacology* 24:172–89.

Koshland, D. E. Jr. (1987). Switches, thresholds and ultrasensitivity. *Trends Biochem Sci* 12:225–9.

Koshland, D. E. Jr., Goldbeter, A., Stock, J. B. (1982). Amplification and adaptation in regulatory and sensory systems. *Science* 217:220–5.

Laakso, A., Mohn, A. R., Gainetdinov, R. R., Caron, M. G. (2002). Experimental genetic approaches to addiction. *Neuron* 36:213–28.

LaBar, K. S., Gatenby, J. C., Gore, J. C., LeDoux, J. E., Phelps, E. A. (1998). Human amygdala activation during conditioned fear acquisition and extinction: a mixed-trial fMRI study. *Neuron* 20:937–45.

Laughlin, S. (1981). A simple coding procedure enhances a neuron's information capacity. *Z Naturforsch C* 36:910–12.

Laughlin, S. B. (1994). Matching coding, circuits, cells, and molecules to signals: General principles of retinal design in the fly's eye. *Prog Ret Eye Res* 13:165–96.

Lennie, P. (2003). The cost of cortical computation. *Curr Biol* 13:493–7.

Levit, K., Smith, C., Cowan, C., Lazenby, H., and Martin, A. (2000). Inflation spurs health spending in 2000. Baltimore MD, National Health Statistics Group, Office of the Actuary, Centers for Medicare and Medicaid Services.

Lücher, C., Frerking, M. (2003). Restless AMPA receptors: Implications for synaptic transmission and plasticity. *Trends Neurosci* 24:665–70.

Mark. V. H., Ervin, F. R. (1970). *Violence and the Brain.* New York: Harper & Row.

Marmot, M. (2000). Multilevel approaches to understanding social determinants. In: *Social Epidemiology* (ed. L. F. Berkman, I. Kawachi), pp. 349–67. New York: Oxford University Press.

Mason, J. W. (1968). Organization of endocrine mechanisms. *Psychosom Med* 30:565.

Mason, J. W. (1971). A reevaluation of the concept of "nonspecificity" in stress theory. *J Psychiat Res* 8:323.

Mason, J. W. (1972). Organization of psychoendocrine mechanisms. In: *Handbook of Psychophysiology* (ed. N. S. Greenfield, R. A. Sternbach), pp. 3–91. New York: Holt, Rhinehart & Winston.

Massey, D. S., Denton, N. A. (1994). *American Apartheid*. Cambridge: Harvard University Press.

McEwen, B. (2002). *The End of Stress As We Know It*. Joseph Henry Press/Dana Press. Washington DC.

Minuchin, S. (1974). *Families and Family Therapy*. Cambridge: Harvard University Press.

Moller, D. E. (2001). New drug targets for type 2 diabetes and the matabolic syndrome. *Nature* 414:821–7.

Montague, P. R., Berns, G. S. (2002). Neural economics and the biological substrates of valuation. *Neuron* 36:265–84.

Mrosovsky, N. (1990). *Rheostasis. The Physiology of Change*. New York: Oxford University Press.

Nauta, W. J. H. (1971). The problem of the frontal lobe: A reinterpretation. *J Psychiatr Res* 8:167–87.

Nestle, M. (2002). *Food Politics*. Berkeley and Los Angeles: University of California Press.

National Institutes of Health. (1997). *The Sixth Report of the Joint National Committee on Prevention, Detection, Evaluation, and Treatment of High Blood Pressure*. NIH Publications, Betaesda, MD.

Orfield, G. (2001). *Schools More Separate: Consequences of a Decade of Resegregation*. Cambridge: Harvard Civil Rights Project, Harvard University.

Ornish, D., Brown, S. E., Scherwitz, L. W., Billings, J. H., Armstrong, W. T., Ports, T. A., McLanahan, S. M., Kirkeeide, R. L., Brand, R. J., Gould, K. L. (1990). Can lifestyle changes reverse coronary heart disease? *Lancet* 336:129–33.

Petrovic, P., Kalso, E., Petersson, K. M., Ingvar, M. (2002). Placebo and opioid analgesia – imaging a shared neuronal network. *Science* 295:1737–40.

Phillips, P. E. M., Stuber, G. D., Helen, M. L. A .V., Wightman, R. M., Carelli, R. M. (2003). Subsecond dopamine release promotes cocaine seeking. *Nature* 422:614–18.

Rieke, F., Warland, D., de Ruyter van Steveninck, R., Bialek, W. (1999). *Spikes: Exploring the Neural Code*. Cambridge: MIT Press.

Roenneberg, T., Merrow, M. (2003). The network of time: Understanding the molecular circadian system. *Curr Biol* 13:R198–207.

Rosenberg, R. N., Prusiner, S. B., DiMauro, S., Barchi, R. L. (1997). *Molecular and Genetic Basis of Neurological Disease*. Boston: Butterworth-Heineman.

Sacks, F. M., Svetkey, L. P., Vollmer, W. M., Appel, L. J., Bray, G. A., Harsha, D., Obarzanek, E., Conlin, P. R., Miller, E. R. I., Simons-Morton, D. G., Karanja, N., Lin, P.-H. (2001). Effects of blood pressure of reduced dietary sodium and the dietary approaches to stop hypertension (DASH) diet. *New Engl J Med* 344:3–9.

Sakmann, B., Creutzfeldt, O. D. (1969). Scotopic and mesopic light adaptation in the cat's retina. *Pflügers Arch* 313:168–85.

Saltiel, A. R., Kahn, C. R. (2001). Insulin signalling and the regulation of glucose and lipid metabolism. *Nature* 414:799–806.

Saper, C. B., Chou, T. C., Elmquist, J. K. (2002). The need to feed: Homeostatic and hedonic control of eating. *Neuron* 36:199–211.

Sapolsky, R. M. (1998). *Why Zebras Don't Get Ulcers*. New York: W. H. Freeman.

Schlosser, E. (2002). *Fast Food Nation*. New York: HarperCollins.

Schulkin, J. (1999). *The Neuroendocrine Regulation of Behavior*. New York: Cambridge University Press.

Schulkin, J. (2003a). Allostasis: A neural behavioral perspective. *Horm Behav* 43:21–7.

Schulkin, J. (2003b). *Rethinking Homeostasis: Allostatic Regulation in Physiology and Pathophysiology*. Cambridge: MIT Press.

Schulkin, J., McEwen, B. S., Gold, P. W. (1994). Allostasis, amygdala, and anticipatory angst. *Neurosci Biobehav Rev* 18:385–96.

Schultz, W. (2002). Getting formal with dopamine and reward. *Neuron* 36:241–63.

Schwartz, M. W., Woods, S. C., Porte, D., Seeley, R. J., Baskin, D. G. (2000). Central nervous system control of food intake. *Nature* 404:661–72.

Smith, A. (1937). *The Wealth of Nations*. New York: Random House, pp. 734–735.

Sterling, P. (1978). Ethics and effectiveness of psychosurgery. In: *Controversy in Psychiatry* (ed. J. P. Bradie, H. K. Brodie), pp. 126–60. Philadelphia: W. B. Saunders.

Sterling, P. (1979). Psychiatry's drug addiction. The New Republic, December 8th. pp. 14–18.

Sterling, P. (2004). How retinal circuits optimize the transfer of visual information. In: *The Visual Neurosciences* (ed. L. M. Chalupa, J. S. Werner), pp. 243–68. Cambridge: MIT Press.

Sterling, P., Eyer, J. (1981). Biological basis of stress-related mortality. *Soc Sci Med* 15E:3–42.

Sterling, P., Eyer, J. (1988). Allostasis: A new paradigm to explain arousal pathology. In: *Handbook of Life Stress, Cognition and Health* (ed. S. Fisher, J. Reason), pp. 629–49. New York: John Wiley & Sons.

Takeda, S., Elefteriou, F., Levasseur, R., Liu, X., Zhao, L,, Parker, K. L., Armstrong, D., Ducy, P., Karsenty, G. (2002). Leptin regulates bone formation via the sympathetic nervous system. *Cell* 111:305–17.

Taylor, C. R., Weibel, E. R. (1981). Design of the mammalian respiratory system. I. Problem and strategy. *Respir Physiol* 44:1–10.

Taylor, F. W. (1967). *Principles of Scientific Management*. New York: Norton, pp. 24–25.

Tracy, K. J. (2002). The inflammatory reflex. *Nature* 420:853–9.

Tuke, S. (1964). *Description of the Retreat, 1813*. Reprinted with Introduction by R. Hunter and I. Macalpine. London: Dawson of Pall Mall.

Valenstein, E. S. (1973). *Brain Control*. New York: John Wiley & Sons.

Valenstein, E. S. (1986). *Great and Desperate Cures: The Rise and Decline of Psychosurgery and Other Radical Treatments for Mental Illness*. New York: Basic Books.

Waldron, I. (1979). A quantitative analysis of cross-cultural variation in blood pressure and serum cholesterol. *Psychosom Med* 41:582.

Weibel, E. R. (2000). *Symmorphosis*. Cambridge: Harvard University Press.

Wilson, E. O. (1984). *Biophilia*. Cambridge: Harvard University Press.

Wilson, F. H., Disse-Nicodème, S., Choate, K. A., Ishikawa, K., Nelson-Williams, C., Desitter, I., Gunel, M., Milford, D. V., Lipkin, G. W., Achard, J.-M., Feely, M. P., Dussol, B., Berland, Y., Unwin, R. J., Mayan, H., Simon, D. B., Farfel, Z., Jeunemaitre, X., Lifton, R. P. (2001). Human hypertension caused by mutations in WNK kinases. *Science* 293:1107–12.

Wise, R. A. (2003). Brain reward circuitry: Insight from unsensed incentives. *Neuron* 36:229–40.

Zhu, Y., Bian, Z., Lu, P., Karas, R. H., Bao, L., Cox, D., Hodgin, J., Shaul, P. W., Thorén, P., Smithies, O., Gustafsson, J., Mendelsohn, M. E. (2002). Abnormal vacular function and hypertension in mice deficient in estrogen receptor beta. *Science* 295:505–8.

Zimmet, P., Alberti, K. G. M. M., Shaw, J. (2001). Global and societal implications of the diabetes epidemic. *Nature* 414:782–7.

2 Protective and Damaging Effects of the Mediators of Stress and Adaptation: Allostasis and Allostatic Load

Bruce S. McEwen

INTRODUCTION

The discovery that there are gradients of health across the full range of income and education, known as socioeconomic status (SES; Adler et al., 1994, 1999; Brunner et al., 1997; Marmot et al., 1991), has emphasized the need to have a biological framework in which to conceptualize and measure the cumulative impact of social factors (i.e., social status, income, education, working and living environments, lifestyle, health-related behaviors, and stressful life experiences on physical and mental health). Biologists are very good at studying mechanisms and not so successful in providing concepts and methods that are usable by social scientists, including epidemiologists, economists, sociologists and health psychologists, for collecting data on groups of individuals in relation to psychosocial variables. In an attempt to make some progress in this direction within the framework of the MacArthur Foundation Research Network on Socioeconomic Status and Health, a number of us have developed a set of concepts and terminology that are intended to help organize the design and measurement of long-term effects of psychosocial factors on disease processes (McEwen, 1998, 2000; McEwen and Stellar, 1993; Schulkin et al., 1994; Seeman et al., 1997; McEwen and Seeman, 1999).

The author is indebted to many colleagues for their discussions and suggestions regarding the concepts and terminology contained in this essay. In particular, Teresa Seeman (UCLA), Burt Singer (Princeton), and Carol Ryff (University of Wisconsin) have made fundamental contributions not only to the concepts but also to the measurement and evaluation of allostatic load. I am also indebted to colleagues in the Research Network on Socioeconomic Status and Health for their constructive feedback: Nancy Adler, Chair, UCSF; Sheldon Cohen, Carnegie Mellon; Mark Cullen, Yale; Ralph Horwitz, Yale; Michael Marmot, University College, London; Karen Matthews, Pittsburgh; Kathryn Newman, Harvard; Joeseph Schwartz, SUNY Stony Brook; Teresa Seeman, UCLA: Shelley Taylor, UCLA; David Williams, Michigan; and Judith Stewart, network administrator. Also Jay Schulkin of the National Institutes of Health made valuable suggestions and contributed to the importance of anticipation in allostasis and allostatic load.

One of the main challenges has been to expand beyond the limitations and ambiguities imposed by the terms "stress" and "homeostasis." One purpose of this essay is to discuss the physiology of the response to "stress" in terms of a new formulation involving two concepts: "allostasis" and "allostatic load." We believe that these two terms allow for a more restricted and precise definition of "stress" that refers to an event or series of events that evoke a set of behavioral and physiological responses. These terms also clarify inherent ambiguities in the concept of homeostasis, and we also note the ways in which they replace and clarify aspects of the "general adaptation syndrome" as formulated by the late Hans Selye. A primary reason for clarifying terminology is to allow development of a more relevant set of measurements of the biological states that are related to resilience or to disease. Thus, this chapter also discusses the choice of mediators that can be measured in body fluids or by other relatively simple and noninvasive means on human subjects. The chapter by Singer, Ryff, and Seeman takes up the important issue of evaluation and validation of the allostatic load measures in human subjects.

Successful collaboration between biology and the social sciences and epidemiology requires a biological framework in which to conceptualize and measure the cumulative impact of social factors (see above). At the same time, within the biomedical sciences, the concepts of stress and homeostasis have been used in ambiguous ways that obfuscate a number of important aspects of the impact of experience and genes on health and disease. This chapter, offers three terms, "allostasis," "allostatic states," and "allostatic load," as organizing principles for understanding the influence of the aforementioned factors on the processes of physiological adaptation and the exacerbation of disease. Allostasis also clarifies an inherent ambiguity in the term "homeostasis" and distinguishes between the systems that are essential for life (homeostasis) and those that maintain those systems in balance (allostasis). An additional advantage of clarifying terminology is to facilitate development of a more relevant set of measurements of the biological states that are related to resilience or vulnerability to disease. The measurement of allostatic states and resulting allostatic load are discussed in relation to the hormonal and tissue mediators of adaptation, including those that can be measured in body fluids or by other relatively simple and noninvasive means on human subjects.

DEFINITION OF KEY TERMS

Allostasis: Achieving stability through change; a process that maintains homeostasis, defined as those physiological parameters essential for life.

Allostatic state: Chronic imbalance in the regulatory system, reflecting excessive production of some mediators and inadequate production of others, for example, hypertension; perturbed cortisol rhythm in major depression or after chronic sleep deprivation; chronic elevation of inflammatory cytokines and low cortisol in chronic fatigue syndrome (controversial); imbalance of cortisol, corticotropin-releasing factor (CRF), and cytokines in Lewis rat that increases risk for autoimmune and inflammatory disorders. (*Note:* The primary mediators, such as cortisol and catecholamines, are responsible for an allostatic state; in such a state, they are subject to dysregulation that causes them to be elevated, or suppressed, relative to normal.)

Allostatic load: Cumulative changes that reflect continued operation of the allostatic state or overactivation of allostatic responses, for example, neuronal atrophy or loss in hippocampus, atherosclerotic plaques, abdominal fat deposition, left ventricular hypertrophy, glycosylated hemoglobin, high cholesterol, low high-density lipoprotein (HDL), and chronic pain and fatigue associated with imbalance of immune mediators (These are all *secondary outcomes* that can be measured and are associated with increased risk for a disease.)

Homeostasis: Stability of physiological systems that maintain life; applies strictly to a limited number of systems, such as pH level, body temperature, and oxygen tension, that are truly essential for life and are therefore maintained over a narrow range, as a result of their critical role in survival.

WHAT DO WE MEAN BY "STRESS"?

Stress is often defined as a threat, real or implied, to homeostasis, and homeostasis refers to the maintenance of a narrow range of vital physiological parameters necessary for survival. In common usage, stress usually refers to an event or succession of events that cause a response, often in the form of "distress" but also to a challenge that leads to a feeling of exhilaration, as in "good" stress. The term "stress" is full of ambiguities, however. It is often used to mean the event (stressor) or, sometimes, the response (stress response). Furthermore, it is frequently used in the negative sense of "distress," and sometimes it is used to describe a chronic state of imbalance in the response to stress. In this chapter, "stress" is used to refer to "stressors," that is, events that are threatening to an individual and that elicit physiological and behavioral responses.

The most commonly studied physiological systems that respond to stress are the hypothalmic-pituitary-adrenocortical (HPA) axis and the autonomic nervous system (ANS), particularly the sympathetic response of the adrenal medulla and sympathetic nerves. These systems respond in daily life according to stressful events, as well as to the diurnal cycle of rest and activity. Thus, these systems do more than respond to "stressors," even though they are frequently identified as "stress response systems." Behaviorally, the response to stress may consist of fight-or-flight reactions or involve health-related behaviors such as eating, alcohol consumption, smoking, and other forms of substance abuse. Another type of reaction to a potentially stressful situation is an increased state of vigilance, accompanied, at least in our species, by enhanced anxiety and worrying, particularly when the threat is ill defined or imaginary and when there is no clear alternative behavioral response that would end it. The behavioral responses to stress and states of anxiety are both capable of potentiating the production of the physiological mediators, for example, cortisol and catecholamines.

HOMEOSTASIS AND ALLOSTASIS

Allostasis is a term introduced by Sterling and Eyer (1988) to characterize how blood pressure and heart rate responses vary with experiences and time of day and also to describe changes in the setpoint of these parameters in hypertension. Sterling and Eyer used the change in setpoint as the primary example that distinguishes allostasis from homeostasis. Yet there is a much broader implication of what they discussed. In their paper, they state, "Allostasis emphasizes that the internal milieu varies to meet perceived and anticipated demand." This led us (McEwen and Stellar, 1993) to define allostasis more broadly than the idea of a changing setpoint, namely, as the process for actively maintaining homeostasis. The following discussion explains the rationale.

What is the internal milieu? As noted earlier, homeostasis, in a strict sense, applies to a limited number of systems, such as pH level, body temperature, and oxygen tension, that are essential for life and are, therefore, maintained over a narrow range as a result of their critical role in survival. These systems are not activated or varied to help the individual adapt to its environment. In contrast, "variation to meet perceived and anticipated demands" characterizes the state of the organism in a changing world and reflects the operation of most body systems in meeting environmental challenges, for example, through fluctuating hormones, heart rate and blood pressure, cytokines of the immune system, and other tissue mediators and hormones. Those mediators are most certainly not held constant, although their levels may usually operate within a range, and they participate in processes leading to adaptation as well as contributing to pathophysiology. The latter happens when they are produced insufficiently or in excess, that is, outside the normal range. We return to this last aspect later.

In our view, the systems that vary according to demand, such as the HPA axis and ANS, actually help maintain those systems that are truly homeostatic. Moreover, large variations in the HPA axis and ANS do not lead directly to death, as would large deviations in oxygen tension and pH. Therefore, we propose that allostasis is a much better term for physiological coping mechanisms than is homeostasis, which should be reserved for the parameters that are essential to maintain for survival. Therefore, allostasis is the process that keeps the organism alive and functioning, that is, maintaining homeostasis. Hence, we concur with Sterling and Eyer that allostasis should be defined as "maintaining stability through change," as, promoting adaptation and physiological coping, at least in the short run (McEwen, 1998).

We note, however, another view of homeostasis: that it can also mean the operation of coordinated physiological processes that maintain most of the steady states of the organism (Cannon, 1929). In this interpretation, homeostasis and allostasis might seem to mean almost the same thing. The reason they do not is that the notion of "steady state" is itself vague and does not distinguish between those systems essential for life and those that maintain the essential systems.

What are some examples of allostasis? Sterling and Eyer (1988) used variations in blood pressure as an example: in the morning, blood pressure rises when we get out of bed, and, to maintain consciousness, blood flow is maintained to the brain when we stand up. This type of allostasis helps to maintain oxygen tension in the brain. There are other examples: catecholamine and glucocorticoid elevations during physical activity mobilize and replenish, respectively, energy stores needed for the brain and body to function under challenge. These adaptations maintain essential metabolism and body temperature.

Examples of allostasis go beyond the immediate homeostatic maintenance of body temperature and pH to broader aspects of individual survival, for example, from pathogens or physical danger. In the immune system, acute stress-induced release of catecholamines and glucocorticoids facilitates the movement of immune cells to parts of the body, where they are needed to fight an infection or to produce other immune responses (Dhabhar and McEwen, 1999). Finally, in the brain, glucocorticoids and catecholamines act in concert to promote the formation of memories of events of potentially dangerous situations so that the individual can avoid them in the future (Roozendaal, 2000).

PROTECTIVE AND DAMAGING EFFECTS OF MEDIATORS
OF ALLOSTASIS

Because we believe that Sterling and Eyer (1988) identified and named a fundamental process of physiological coping that had been buried in the ambiguities of "homeostasis," it was not appropriate to use "allostasis" to refer to the pathologic aspects as well as the process of adaptation. We proposed the term "allostatic load" to mean the cost to the body of adapting repeatedly to demands placed upon it (McEwen and Stellar, 1993). We also proposed (McEwen, 1998) that, besides repeated stress, the inefficient management of allostatic systems, which has been called an "allostatic state" (Koob and LeMoal, 2001), is also responsible for an allostatic load (see Fig. 2.1). This is developed further, later in the chapter, after describing and contrasting the protective and damaging effects of the mediators of allostasis.

The most important aspect of the mediators associated with allostasis is that they have protective effects in the short run and yet can have damaging effects over longer time intervals if there are many adverse life events or if the hormonal secretion is dysregulated (McEwen, 1998). In contrast to Selye (1973), this view holds that mediators of allostasis have a spectrum of actions that depend on the time courses over which they are being produced and other events that are taking place at the same time. Later the chapter illustrates how the immediate effect of the secretion of mediators of allostasis such as glucocorticoids and catecholamines is largely protective and adaptive and then notes the damaging consequences that result from overproduction or dysregulation of the same mediators. Some of the examples given earlier to illustrate allostasis are repeated later.

For example, glucocorticoids, so named because of their ability to promote conversion of protein and lipids to usable carbohydrates, serve the body well in the short run by replenishing energy reserves after a period of activity, such as running away from a predator. Glucocorticoids also act on the brain to increase appetite for food and to increase locomotor activity and food-seeking behavior (Leibowitz and Hoebel, 1997), thus regulating behaviors that control energy intake and expenditure. This is useful when we have to run 2 miles, but it is not beneficial when we grab a bag of potato chips while writing a grant proposal or a paper. Inactivity and lack of energy expenditure creates a situation in which chronically elevated glucocorticoids can impede the action of insulin to promote glucose uptake. One of the results of this interaction is that insulin levels increase, and together insulin and glucocorticoid elevations promote the deposition of body fat. This combination of hormones also promotes the formation of atherosclerotic plaques in the coronary arteries (Brindley and Rolland, 1989).

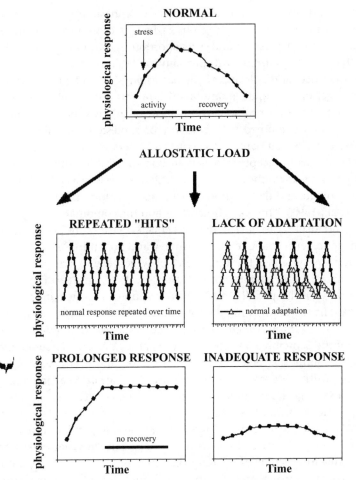

Figure 2.1: Four types of response patterns of allostatic mediators. The top panel illustrates the normal allostatic response, in which a response is initiated by a stressor, sustained for an appropriate interval, and then turned off. The remaining panels illustrate four conditions that lead to allostatic load: (1) repeated "hits" from multiple novel stressors, which may or may not change the response profiles of the mediators, leading to an allostatic state; (2) an example of an allostatic state involving lack of adaptation of the mediator to repeated presentations of the same situation; (3) an example of an allostatic state involving a prolonged response due to delayed shut down of the mediator in the aftermath of a stress or failure to show a normal diurnal rhythm; and (4) an example of an allostatic state involving the inadequate response of one mediator that leads to compensatory hyperactivity of other mediators (for example, inadequate secretion of glucocorticoid, resulting in increased levels of cytokines that are normally counterregulated by glucocorticoids). Figure drawn by Dr. Firdaus Dhabhar, Rockefeller University. Reprinted from McEwen (1998) by permission. Copyright © 2004 Massachusetts Medical Society. All rights reserved.

For the heart, we see a similar paradoxical biphasic role of allostasis mediators. As noted earlier, getting out of bed in the morning requires an increase in blood pressure and a reapportioning of blood flow to the head so that we can stand up without fainting (Sterling and Eyer, 1988). Blood pressure rises and falls during the day as physical and emotional demands change, providing adequate blood flow as needed. Yet repeatedly elevated blood pressure promotes generation of atherosclerotic plaques, particularly when combined with a supply of cholesterol and lipids and oxygen free radicals that damage the coronary artery walls (Manuck et al., 1995). Beta-adrenergic receptor blockers are known to inhibit this cascade of events and to slow down the atherosclerosis that is accelerated in dominant male cynomologus monkeys exposed to an unstable dominance hierarchy (Manuck et al., 1991). Thus catecholamines and the combination of glucocorticoids and insulin can have dangerous effects on the body, besides their important short-term adaptive roles (Brindley and Rolland, 1989).

The nervous system is the interpreter of which events are "stressful" and determines the behavioral and physiological responses to the stressor; it shows a similar paradoxical biphasic action of the mediators of allostatic load. In the brain, strong emotions frequently lead to "flashbulb" memories – for example, where we were and what we were doing when we heard of the September 11, 2001, terrorist attacks at the Pentagon and World Trade Center, or remembering the location and events associated with a very positive life event, like proposing marriage or receiving a promotion or award. Both catecholamines acting via beta-adrenergic receptors and glucocorticoid hormones acting via intracellular receptors play an important role in establishing these long-lasting memories, and a number of brain structures participate along with the ANS. The amygdala plays an important role in this type of memory (LeDoux, 1996), aided by the autonomic nervous system, which picks up a signal from circulating epinephrine (Cahill et al., 1994). The process is also aided by the hippocampus, which helps us remember "where we were and what we were doing" at the time the amygdala was turned on in such a powerful way (LeDoux 1996; Roozendaal, 2000).

Thus, epinephrine and glucocorticoids promote the memory of events and situations, which, in the future, may be dangerous, and this is an adaptive and beneficial function. The paradox for the brain comes when there is repeated stress over many days or when glucocorticoid levels remain high because of adrenal overactivity or poor shutoff of the stress response or the diurnal rhythm. Then there is atrophy of pyramidal neurons in the hippocampus and dentate gyrus (McEwen, 1999; Sousa et al., 2000) and inhibition of ongoing neurogenesis in the dentate gyrus (McEwen, 1999) as well as a possible loss of glial cells (Rajkowska et al., 1999). After prolonged periods of allostatic load, as in subordinate monkeys living in a dominance hierarchy, pyramidal neurons may actually die (Rozovsky et al., 2002). Through some or all of these processes, the hippocampus undergoes a shrinkage in size, with impairment of declarative, contextual, and spatial memory, and this can be picked up in the human brain by neuropsychological testing accompanied by magnetic resonance imaging in such conditions as recurrent depressive illness, Cushing's syndrome, posttraumatic stress disorder, mild cognitive impairment in aging, and schizophrenia (McEwen, 1999).

From Allostasis to Allostatic Load

Thus we see that protection and damage are the opposite, and seemingly unavoidable, extremes of the release of mediators of allostasis such as adrenal catecholamine and glucocorticoids, which provide some of the best examples to date of the organizing principles of allostasis and allostatic load. There are other physiological mediators that follow the same principles, and the concepts described here are not intended to be limited to the HPA axis or ANS. What are these mediators and how does their release progress from adaptation to damage, that is, from allostasis to allostatic load?

Table 2.1 provides a representative list of mediators and distinguishes between those that are primarily systemic in their production and those that are primarily produced locally within organs or tissues. However, it should also be noted that there is a remarkable degree to which many substances are produced both systemically and locally. For example, catecholamines and cytokines, which are produced locally, are also found in the bloodstream. This is an important factor when it comes to questions of measurement (as discussed later). Glucocorticoids and dehydroepiandrosterone (DHEA), which are primarily systemically released hormones of the adrenal cortex, are also produced locally – glucocorticoids by the developing thymus gland (Vacchio et al., 1994) and DHEA by the developing brain (Compagnone and Mellon, 1998). Yet it is the circulating levels of these two steroids that are considered to be the most relevant to many examples involving allostasis, allostatic states, and allostatic load.

The production and release of mediators of allostasis occurs according to two major patterns: (1) in response to challenge or demand, as in stressful experiences; and (2) as part of a housekeeping function that is governed by such things as the day-night, light-dark cycle. As described in detail later, not only stressors but also perturbation of the diurnal rhythm of glucocorticoid secretion by sleep deprivation and depressive illness provides some compelling examples of the long-term cost to the body that we refer to as allostatic load.

Table 2.1: Primary mediators of allostasis

Systemic Mediators
  Glucocorticoids
  Dehydroepiandrosterone
  Catecholamines (epinephrine, norepinephrine)
  Cytokines (e.g., IL-6, IL-1, TNF-alpha)
  Many systemic hormones (e.g., thyroid hormone, insulin, insulin-like growth
    factors, leptin)
  Many pituitary hormones (e.g., prolactin, corticotropin, growth hormone)
Tissue Mediators
  Corticotropin-releasing factor
  Excitatory amino acids
  Monoamines (e.g., serotonin, norepinephrine, epinephrine, histamine)
  Other neurotransmitters (e.g., gamma-aminobutyric acid [GABA], glycine)
  Other neuropeptides (e.g., neuropeptide Y, cholecystokinin, enkephalin,
    dynorphin, substance P)
  Many cytokines (e.g., TNF-alpha, IL-1, IL-6, IL-4, IL-10, IFN gamma)
  Some pituitary hormones (e.g., prolactin, proopiomelanocortin)
Genetic risk exacerbated by stressful events
  Hypertension (Gerin and Pickering, 1995): delayed decline of blood pressure
    after challenge in people with familial hypertension
  Diabetes (Lehman et al., 1991): increased incidence of type 1 diabetes after
    repeated stress in rats and family disruption in children (Hagglof et al., 1991)
Early development
  Prenatal stress, neonatal handling, maternal care and
    hypothalmic-pituitary-adrenocortical axis reactivity (Meaney et al., 1988);
    neonatal handling, which involves increased maternal licking and grooming of
    pups, reduces HPA reactivity and decreases brain aging (Liu et al., 1997)
  Abuse in childhood (Felitti et al., 1998): physical and sexual abuse increases
    poor health and maladaptive behavior in later life (DeBellis et al., 1994a,
    1994b; Taylor, 1999; Walker et al., 1999)
Diurnal rhythm and sleep
  Bone mineral density (Michelson et al., 1996): depression is associated with
    elevated diurnal HPA activity and increased loss of bone mineral density.
  Sleep loss and glucose (Leproult et al., 1997; Plat et al., 1999; Spiegel et al.,
    1999); sleep deprivaton leads to hyperglycemia (Van Cauter et al., 1992)
Lifestyle
  Diet (Kamara et al., 1998): fat-rich diet increases HPA activity
  Exercise (Dellwo and Beauchene, 1990; Perna and McDowell, 1995) increases
    HPA and sympathetic activity but decreases insulin resistance (Perseghin
    et al., 1996)
  Quantity of sleep (Spiegel et al., 1999), *see* sleep loss and glucose
  Smoking (Frederick et al., 1998; Verdecchia et al., 1995)
  Alcohol (Ogilvie and Rivier, 1997; Spencer and McEwen, 1990)
Repeated stressful experiences
  Atherosclerosis (Manuck et al., 1991): unstable dominance hierarchy
    accelerates atherosclerosis; this process can be prevented by beta-adrenergic
    blocking drugs
  Remodeling of hippocampal neurons and inhibition of neurogenesis by repeated
    stress (McEwen, 1999)
  Suppression of immunity by repeated stress (Dhabhar and McEwen, 1997)

How does the production of the mediators of allostasis give way to a wear and tear on the same systems that they are involved in protecting and promoting adaptation? The simplest conceptualization is chronic stress, that is, turning on the production of allostatic mediators such as cortisol and catecholamines repeatedly in response to novel, threatening events, as shown in the top left panel in Figure 2.1. Overexposure to mediators of allostasis results in the various types of pathophysiology and wear and tear described earlier. For instance, people who have had excessive stress in their lives, as measured by multiple periods of poverty-level income, show earlier aging, more depression, and an earlier decline of both physical and mental functioning (Lynch et al., 1997).

What about the effects of repeated stressful experiences on the response profiles of the mediators of allostasis? It is likely that repeated stress may change the temporal response profiles and lead, for example, to impaired shutoff or an elevated or reduced diurnal rhythm of cortisol or catecholamine production. In addition, these response profiles may be altered by genetic factors, early developmental influences, or the effects of lifestyle (see Table 2.1). Thus, the first step in many cases may be that the production of mediators becomes dysregulated and the setpoint for their regulation is changed, so that they are produced in elevated or reduced levels or according to an abnormal temporal pattern. If this perturbation becomes a chronic condition, it can be referred to as an allostatic state (Koob and LeMoal, 2001). Prolongation of the allostatic state can produce tissue pathophysiology, referred to as allostatic load.

In other words, there are circumstances in which the number of stressful events may not be excessive but in which the body fails to efficiently manage the response to challenges or maintain a normal diurnal rhythm; some examples of allostatic states are illustrated in the three remaining panels in Figure 2.1. The top right panel illustrates a failure to habituate to repeated stressors of the same kind. Measurement of cortisol in a repeated public-speaking challenge has revealed individuals who do not habituate, and these individuals, who lack self-confidence and self-esteem, may well be overexposing their bodies to stress hormones under many circumstances in daily life that do not overtly disturb other individuals (Kirschbaum et al., 1995).

The bottom panel of Figure 2.1 refers to a failure to turn off each stress response efficiently or to show a normal diurnal rhythm. For example, individuals with a genetic load, that is, two parents who are hypertensive, show prolonged elevation of blood pressure in the aftermath of a psychological stressor (Gerin and Pickering, 1995). Another example of the failure to shut off a response takes us into the realm of the housekeeping function of the mediators of allostasis, namely, the diurnal rhythm. Reduced amounts of sleep for a number of days results in elevated cortisol levels during the evening hours (Leproult et al., 1997; Spiegel et al., 1999). Sleep deprivation and elevated diurnal levels of cortisol are also features of major depression. One negative consequence of evening cortisol elevation is that it has a greater potential to cause a delayed hyperglycemic state than does cortisol elevation in the morning (Plat et al., 1999). In depressive illness, loss of bone mineral density has been reported that is linked to elevated diurnal glucocorticoid levels (Michelson et al., 1996). The loss of bone minerals and muscle protein are two of the recognized consequences of chronic elevation of glucocorticoids (Sapolsky et al., 1986).

The final example of an allostatic state comes from the notion that an acute response of a mediator of allostasis should be of sufficient magnitude to produce an adaptive response. If it is not, the systems that are affected by these mediators can themselves malfunction by overreacting. The bottom right panel of Figure 2.1 describes a situation in which the glucocorticoid response is inadequate to the needs of the individual genotype, resulting in excessive activity of other allostatic systems, such as the inflammatory cytokines, which are normally contained by elevated levels of cortisol and catecholamines. The Lewis rat illustrates a genetic contribution to this condition, having less corticosterone than the virtually syngenic Fischer rat. Lewis rats are vulnerable to inflammatory and autoimmune disturbances, not found in Fischer rats. In Lewis rats, these can be overcome by giving exogenous·glucocorticoids (Sternberg, 1997). Comparable human disorders in which lower-than-needed cortisol may play a role include fibromyalgia and chronic fatigue syndrome (Poteliakhoff, 1981; Ur et al., 1992; Crofford et al., 1994; Buske-Kirschbaum et al., 1997; Magarinos et al., 2001). This is discussed further, later in the chapter.

Thus the distinction between protection and damage, as far as hormonal mediators of allostasis are concerned, is related to the dynamics of the hormonal response. Similar considerations apply to other mediators such as excitatory amino acids (see discussion in the next section).

ALLOSTASIS AND ALLOSTATIC LOAD ARE ORGANIZING PRINCIPLES THAT APPLY TO MANY BIOLOGICAL MEDIATORS

Besides those associated with the HPA axis and ANS, there are many other biological mediators (see Table 2.1) and other examples of dysregulated adaptive systems that fit the definition of allostatic states and allostatic load. Two examples are considered here, one involving the brain and the other the immune system.

The increase in levels of extracellular glutamate in the hippocampus during restraint stress (Lowy et al., 1993) is involved in remodeling of dendrites of hippocampal CA3 pyramidal neurons. We know this because the remodeling is prevented by an N-methyl-D-aspartate (NMDA) receptor blocker and by phenytoin (Dilantin), a Na^+ channel blocker and antiepileptic drug (Magarinos and McEwen, 1995; Watanabe et al., 1992). There is a stress-induced elevation of extracellular glutamate, which can be measured by intrahippocampal microdialysis, that is attenuated by adrenalectomy, suggesting a dependence on adrenal steroids. This mechanism becomes less efficient as rats age, and this constitutes an example of an allostatic state in which the elevation of a mediator of allostasis fails to go back to baseline when the stressor is finished. When aging rats are subjected to restraint stress and microdialysis, there is an exacerbation of both the level of extracellular glutamate and a prolongation of the response after the stress is terminated (Lowy et al., 1995). This exacerbation leads to allostatic load at the tissue level because the elevated glutamate is likely to potentiate morphological and other effects that occur through activation of NMDA and other excitatory amino acid receptors. Aging rats show increases in Ca^{++} currents (Landfield and Eldridge, 1994; Porter et al., 1997) and changes in structure and function related to cytoskeleton and excitatory neurotransmission that are related to cognitive impairment (Nicolle et al., 1999; Sugaya et al., 1996, 1998).

The protective and damaging effects of stress on the immune system are illustrated by opposite effects of acute versus chronic stress on skin immunity: acute stress (2-hour restraint) has been shown to induce a significant enhancement of skin immunity (LeDoux, 1996). A stress-induced trafficking of leukocytes from the blood to the skin is one of the mediators of this immunoenhancement (Dhabhar and McEwen, 1999). Importantly, both the stress-induced changes in leukocyte trafficking and the stress-induced enhancement of immune function, measured as delayed-type hypersensitivity (DTH), have been shown to be dependent on adrenal stress hormones (Dhabhar et al., 1995, 1996; Dhabhar and McEwen, 1999). However, in contrast to acute stress, chronic stress significantly suppresses skin immunity and impairs leukocyte trafficking, and this may be one of the causes of the immunosuppression (Dhabhar and McEwen, 1997). Thus, in this case, allostatic load imposed by chronic stress has detrimental consequences for immune function, although it is not clear what type of allostatic state may be involved in suppressing DTH responses. One possibility is that repeated stress induces a down-regulation of immune cell trafficking and of the movement of immune cells into the tissues owing to a desensitization of the cell surface molecules and the cytokines and their receptors that normally mediate that acute-stress enhancement of cell trafficking and DTH (Dhabar and McEwen, unpublished speculations).

Infections and other types of pathogens trigger acute-phase responses and activate both innate and adaptive immune mechanisms. Normally, the course of an infection is characterized by a delicate balance between the inflammatory cytokines that activitate adaptive immune responses, the innate immune defenses and the actions of antiinflammatory factors such as glucocorticoids and antiinflammatory cytokines that keep the body's own defense from overreacting and causing damage or death (Miller et al., 2000; Munck et al., 1984). One type of allostatic state that does result in a measurable allostatic load is the dysregulation of immune system modulators that contributes, along with genetic risk factors, to autoimmune and inflammatory diseases such as multiple sclerosis, rheumatoid arthritis, and type 1 diabetes (Dowdell and Whitacre, 2000). (This is discussed later in the section on measurement). Another related type of allostatic state is the imbalance of immune system regulators that accompany chronic fatigue syndrome, fibromyalgia, and related types of hyperalgesia and chronic pain. The uncovering of these states exemplifies the type of meticulous research that is required to reveal mechanisms for a disorder that is debilitating to many people but that falls "between the cracks" of conventional medical diagnosis and treatment because of its subtle complexity and seeming inconsistency. This is discussed more fully later in relation to selection of which mediators of allostasis to measure.

ALLOSTASIS AND ALLOSTATIC LOAD INCLUDE CONTRIBUTIONS FROM GENES, DEVELOPMENT, DIURNAL RHYTHM, AND STRESS

Thus, allostasis represents the integrated output of interacting physiological systems, such as, but not confined to, the HPA axis and ANS. This is emphasized by the mediators listed in Table 2.1. An allostatic state reflects a response pattern in which these systems are overactive or dysregulated (or both). The overactivity and dysregulation may be manifested during the normal course of daily activities or as a result of stressful events, and it may be exacerbated by genetic risk factors such as diabetes or hypertension, as well as by early life events that make the physiological response systems particularly vulnerable to dysregulation or overactivity. Thus allostatic states and allostatic load include contributions of genetic risk factors, early developmental influences, the diurnal rhythm, lifestyle factors, and life stressors. Table 2.1 provides some examples that emphasize these different contributions.

THE MEASUREMENT OF ALLOSTASIS, ALLOSTATIC STATES, AND ALLOSTATIC LOAD

How can we measure allostasis and its consequences in terms of allostatic states and allostatic load, particularly when it comes to following the events that lead to disease over the life course in individual human subjects and groups of individuals? This is a major goal of the biologist in working with social scientists and epidemiologists in attempting to answer questions such as the relationship between working, living environments and socioeconomic conditions and health or disease. It is also one of the main reasons that the definition of terms should be made more precise. The distinction between allostatic states and allostatic load provides two types of end points that can be measured, at least in principle. On one hand, allostatic states refer to the response profiles of the mediators themselves. On the other hand, allostatic load focuses on the tissues and organs that show the cumulative effects of overexposure to the mediators of allostasis, either because of too much stress or because of various allostatic states (see Fig. 2.1). This section briefly considers the challenges and opportunities in measuring allostatic states and allostatic load.

For determining various allostatic states (e.g., Fig. 2.1) in human subjects, the choice of which mediators to measure depends, in large part, on where in the body one is able to measure them as noninvasively as possible. This can be done most easily by collecting urine or saliva; if necessary, blood; and, rarely and under special circumstances, cerebrospinal fluid. The choice is dictated by such factors as the size of the study, the cost of the assays, and the desire not to disrupt the lives of the subjects under study more than is absolutely necessary to ensure cooperation and minimize added stress and anxiety that can influence the secretion of the mediators being measured. This limits the choice to the circulating mediators such as glucocorticoids, DHEA, catecholamines, and certain cytokines (see Table 2.1). Salivary assays are particularly attractive, but then the question arises as to how to sample over time to get an adequate representation of a dynamic system because the levels of the mediators may fluctuate during the day and night. This is a topic unto itself and has been the subject of a number of methodological studies (see the Web site for MacArthur SES and Health Research Network: www.macses.ucsf.edu/). Portable monitoring of blood pressure and heart rate provide complementary information to the measurement of mediators in body fluids. The ease of such measurements also explains why the study of cardiovascular function as an end point of disease has progressed so far relative to other systems of the body that are sensitive to stress and show allostatic load.

Table 2.2: Some primary mediators and secondary outcomes

Primary mediators: assessment of allostatic states
  Elevated levels of inflammatory cytokines
  Elevated and flattened diurnal cortisol rhythms
  Elevated overnight urinary cortisol
  Low dehydroepiandrosterone: cortisol ratio
  Elevated levels of overnight urinary catecholamines
  Note: autonomic nervous system activity is also assessed indirectly by
    measuring blood pressure
  Abnormal insulin levels (also assessed indirectly as abnormal glucose levels)
Secondary outcomes: measures of allostatic load
  Brain: atrophy of brain regions, cognitive impairment
  Cardiovascular: atherosclerosis, left ventricular hypertrophy, clotting factors,
    homocysteine, oxidative stress markers
  Immune system: impaired wound healing, retarded immunization response,
    suppressed delayed-type hypersensitivity, chronic pain, and fatigue reflecting
    imbalance of immune system regulators in the central nervous system
  Metabolic: glycosylated hemoglobin, high-density: low-density lipoprotein,
    cholesterol, abdominal fat deposition, as measured by the waist:hip ratio,
    bone mineral density

Definitions: Primary mediators are circulating hormonal agents that produce a variety of effects on diverse target tissues throughout the body. These mediators interact with each other in producing primary and secondary effects and in regulating their own production. Secondary outcomes refer to biological parameters or functional states that are the products of the interactions of primary mediators (often more than one primary mediator) with tissue substrates. They reflect parameters that are themselves indicators of pathophysiological processes.

As far as assessing allostatic load, the fundamental issue is determining the extent to which the mediators of allostasis are a significant part of the causal chain leading to the secondary paethophysiological outcomes that represent allostatic load. Fortunately, for some of the most commonly measured mediators that can easily be measured in human subjects, there is considerable evidence that they are involved in diverse forms of allostatic load. Table 2.2 presents a list of some end points that can be used for cumulative assessment of allostatic load in various systems of the body. This section briefly describes the most easily measured and widely acting mediators of allostasis and indicates in a general way some of their connections to the pathophysiological secondary outcome measures of allostatic load. Four allostasis mediators are considered briefly: glucocorticoids, DHEA, catecholamines, and cytokines. They each have many effects on a variety of body systems, and their production and actions are interconnected.

Glucocorticoids

Glucocorticoids are among the most versatile of hormones, having important regulatory effects on the cardiovascular system, the regulation of fluid volume and response to hemorrhage, immunity and inflammation, metabolism, brain function, and reproduction (Sapolsky et al., 2000). Virtually every tissue in the body has intracellular glucocorticoid receptors, and some of the key effects are highlighted in the following list.

For *cardiovascular function,* the predominant effect of glucocorticoids is to permissively enhance cardiovascular function during times of acute stress, in part by enhancing sensitivity to catecholamines.

For control of *fluid volume,* glucocorticoids suppress the fluid volume increase that occurs in response to a hemorrhagic stressor, thus inhibiting potential damage from an overly exuberant response to damage.

In the case of *inflammation and immunity,* glucocorticoids are well known to suppress inflammation and the acute-phase response to an infection, keeping these responses under control to minimize damage that they might inflict if overactive. At the same time, glucocorticoids enhance initial mobilization of immune cells to sites of infection and shape the nature of the immune response, favoring, for example, humoral over cellular immunity. Finally, however, glucocorticoids ultimately contain all types of immune responses, which is why they are useful in the treatment of autoimmune disorders and in reducing rejection of organ transplants.

With regard to *metabolism,* glucocorticoids promote appetite and food-seeking behaviors and are classically described as promoting lipolysis, proteolysis, and gluconeogenesis (hence, their name). The chapter in this volume by Wingfield discusses glucocorticoids in relation to maintaining the balance between energy intake and expenditure in situations that force animals to enter an emergency life history stage. At the same time, they generally work in opposition to insulin, an anabolic hormone, except under the allostatic state of chronic glucocorticoid elevation in which case they act to promote hepatic glycogen deposition and lipogenesis (which leads to fat deposition and the expense of muscle protein loss), while raising insulin levels and impairing insulin actions on their receptors. The extreme consequence of this progression is type II diabetes, culminating in exhaustion of insulin production.

In the *central nervous system,* glucocorticoids inhibit glucose transport into brain cells and oppose the glucose-uptake-promoting effects of catecholamines that act by increasing cardiovascular activity and enhancing cerebral blood flow. As noted earlier, however, glucocorticoids increase appetite for food and work in opposition to the anorectic effects of stress-activated release of brain CRF, the neuropeptide that also directs

corticotropin release from the pituitary. Finally, as discussed earlier, glucocorticoids also biphasically modulate memory formation, with basal levels
enhancing formation of memories of emotionally charged events and stress
levels suppressing memory. Moreover, over long time periods involving repeated stress, glucocorticoids participate in a mechanism that causes atrophy of neural structures subserving memory.

With respect to *reproduction*, glucocorticoids participate in mechanisms
found in many species that inhibit reproduction, and this can be rational-
ized as a logical contribution to the response to stress that operates to delay
reproduction to a more auspicious time (Wingfield and Romero, 2000).

In summary, glucocorticoids fulfill the criterion of an allostatic mediator that can be measured noninvasively in human subjects and that gives
useful information about a wide range of normal and pathophysiological
conditions. As noted several times in this chapter, chronic elevations of glucocorticoids participate in a host of conditions that fall into the category of
allostatic load, such as hypertension, abdominal obesity, bone mineral loss,
loss of muscle mass, suppression of immune responses, memory impairment, and atrophy of brain structures such as the hippocampus. Chronic
insufficiency of glucocorticoids contributes to increased inflammatory and
autoimmune responses and may contribute to conditions of imbalance of
cytokines and pain mechanisms in fibromyalgia and chronic fatigue syndrome. Yet there are important interactions of glucocorticoids with other
primary mediators that are described in the following subsections.

DHEA

DHEA is an adrenal steroid that has a number of effects that can be described
as "functional antagonists" of glucocorticoid actions (Browne et al., 1993;
Kalimi et al., 1994; Wolf and Kirschbaum, 1999). Functional antagonism
refers to the fact that DHEA does not directly interact with the glucocorticoid receptor, and, in fact, there is no known receptor for DHEA in any
tissue. The antagonistic effects are most clearly demonstrated in terms of
the weight-inducing actions of elevated glucocorticoids in the Zucker rat
(Browne et al., 1993), as well as the actions of glucocorticoids to inhibit
memory and primed-burst potentiation (Fleshner et al., 1997), a form of
long-term potentiation. Bone mineral loss is correlated with reduced production of DHEA as a result of glucocorticoid therapy for autoimmune and
inflammatory disorders (Formiga et al 1997; Robinzon and Cutolo, 1999),
which has led to the suggestion that DHEA should be supplemented in individuals receiving glucocorticoid treatment for autoimmune and inflammatory disorders (Robinzon and Cutolo, 1999).

DHEA produces functional antagonism of glucocorticoid actions on a
number of immune parameters. It antagonizes the thymolytic actions of

glucocorticoids (Blauer et al., 1991), as well as the suppression of inflammatory cytokine production and cellular and humoral immune responses (Kalimi et al., 1994). DHEA is also thought to function as more than an antiglucocorticoid in preserving immune function – for example, after a thermal injury (Araneo and Daynes, 1995).

Glucocorticoid actions in relation to obesity may be related, at least in part, to the ability of glucocorticoids to increase glucose uptake in cultured cells (Nakashima et al., 1995). In principle, this action might work in opposition to glucocorticoid-induced insulin resistance. It is interesting to note that DHEA administration also antagonizes oxidative damage in brain, kidney and liver produced by acute hyperglycemia (Aragno et al., 1997).

In brain, interactions of DHEA with neurotransmitter systems such as serotonin, GABA, excitatory amino acids, and dopamine are implicated in the actions in the central nervous system, although, again, the cellular and molecular details of these interactions are not clear (Abadie et al., 1993). One of the reported actions of DHEA is to reduce depressive symptomatology, particularly in depressed elderly (Wolkowitz et al., 1997). In addition, DHEA, which declines with increasing age and has effects to enhance memory in aging rodents, has been implicated as a possible neuroprotective agent in aging, even though it does not appear to directly improve cognition in human subjects (Wolf and Kirschbaum, 1999). However, the data thus far collected do not strongly support this hypothesis, although it is still tempting to speculate that DHEA antagonizes and moderates the allostatic load produced by chronic elevation of glucocorticoids. In that connection, the mood-elevating actions attributed to DHEA may be related either to glucocorticoid antagonism or to the actions related to neurotransmitter systems such as serotonin.

In conclusion, despite the many caveats, and based on the functional antagonism of some glucocorticoid actions, the level of DHEA relative to glucocorticoids, both of which can be assessed in blood samples, is a potentially useful measure of the possible amelioration of pathophysiological glucocorticoid actions in relation to allostatic load. It is important to remember that low levels of DHEA are indicative of a higher potency of glucocorticoids.

Catecholamines
The other major, systemic mediator of allostasis besides glucocorticoids are the circulating catecholamines, which have effects that in some cases synergize with, and in other cases oppose, the actions of glucocorticoids. Glucocorticoids potentiate the actions of catecholamines while at the same time containing their release. In the adrenal medulla, glucocorticoids promote epinephrine synthesis by regulating the key enzyme, phenylethanolamine

N methyl transferase (PNMT; Goldstein and Pacak, 2000; Sapolsky et al., 2000). At the same time, catecholamines help maintain normal HPA function because adrenergic input to the adrenal cortex facilitates corticotropin-induced steroidogenesis (Young and Landsberg, 2000).

One of the most important distinctions is between norepinephrine release by the dispersed sympathetic nerves and epinephrine release by the adrenal medulla (Goldstein and Eisenhofer, 2000; Young and Landsberg, 2000). For example, in girls with chronic fatigue syndrome, morning resting epinephrine levels are elevated, whereas norepinephrine levels are not, compared with age-matched control subjects (Kavelaars et al., 2000). Whereas norepinephrine release is particularly important for the discrete regulation of blood-vessel constriction and blood flow and redistribution and influence over a host of organs such as the heart, spleen, and pancreas, epinephrine release is important for skeletal muscles that do not have extensive sympathetic innervation. One of the important actions of epinephrine in muscle is to retard the degradation of proteins, working against the catabolic actions of glucocorticoids (Young and Landsberg, 2000). Epinephrine release is also more closely related to emotional distress, whereas norepinephrine release is more related to physical exertion. Some of the other major effects of catecholamines are now considered in relation to the same processes that were discussed earlier for glucocorticoids (Goldstein and Eisenhofer, 2000; Young and Landsberg, 2000).

For *cardiovascular function,* the sympathetic nervous system maintains adequate cerebral blood flow in the transition from lying to standing and helps maintain the blood flow during upright physical activity. Adrenal medullary epinephrine is also released during daily physical activity, and this may be relevant to the normal metabolic activity of skeletal musculature, involving regulation of muscle glycogen stores and protecting muscle protein from degradation. In congestive heart failure, there is increased cardiac sympathetic activity in the face of declining sensitivity of the heart to respond to the catecholamines (Goldstein and Eisenhofer, 2000).

For control of *fluid volume,* both extreme underfilling and overfilling of the heart increase cardiac sympathetic activity, whereas hypernatremia causes inhibition of sympathetic activity and promotes natriuresis. Hyponatremia, on the other hand, increases sympathetic activity and promotes activity of the vasopressin and renin-angiotensin systems and aldosterone secretion that increases water and sodium retention.

In the case of *inflammation and immunity,* circulating catecholamines are involved in the mobilization and redistribution of immune cells in the body during stress and after a specific immune challenge, as is the case in delayed-type hypersensitivity (Dowdell and Whitacre, 2000; Spencer et al., 2000).

Catecholamines may also be involved in suppressing immune function in the spleen under stressful conditions (Shurin et al., 1994) because the spleen receives heavy sympathetic innervation (Bulloch, 2000) and is relatively protected from access to circulating endogenous glucocorticoids (Spencer et al., 2000). In general, enhancing sympathetic tone decreases both T-cell and NK cell functions but not the proliferation of splenic B cells (Dowdell and Whitacre, 2000). In contrast, chemical sympathectomy, although having varying results, does seem to increase the severity of autoimmune disorders (Dowdell and Whitacre, 2000).

As far as *metabolism,* catecholamines promote mobilization of fuel stores at a time of stress and act synergistically with glucocorticoids to increased glycogenolysis, gluconeogenesis, and lipolysis but exert opposing effects on protein catabolism, as noted earlier. One important aspect is regulation of *body temperature* (Goldstein and Eisenhofer, 2000). Epinephrine levels are also positively related to serum levels of HDL cholesterol and negatively related to triglycerides. However, perturbing the balance of activity of various mediators of metabolism and body weight regulation can lead to well-known metabolic disorders such as type 2 diabetes and obesity. For example, there is evidence that deficiency of leptin or leptin receptors decreases adrenal medullary epinephrine release, and, furthermore, that a developmental impairment in epinephrine production may contribute to the insulin-resistance syndrome. At the same time, increased sympathetic activation and norepinephrine release is elevated in hypertensive individuals and also related to higher levels of insulin, and there are indications that insulin further increases sympathetic activity in a vicious cycle (Arauz-Pacheco et al., 1996). Thus the origins of type 2 diabetes may be related on one hand to a deficient leptin system with decreased adrenal medullary activity and, on the other hand, to an activated sympathetic nervous system with elevated levels of norepinephrine and glucocorticoids, both of which contribute to insulin resistance and hyperglycemia (Bjorntorp et al., 1999).

In the *central nervous system,* catecholamines are associated with attention, vigilance, and arousal mechanisms (Goldstein and Pacak, 2000) and with the formation of memories related to strong emotions, as discussed earlier in this chapter.

In conclusion, epinephrine and norepinephrine are two somewhat independent indices of adrenal medullary and sympathetic neural activity that are related to changes in emotional state, physical activity, metabolism and body temperature, cardiac function, and fluid and electrolyte balance. Elevated levels of catecholamines, particularly during the night when sympathetic activity is normally decreased, are indicative of an allostatic state that may contribute to allostatic load. Separate measurements of norepinephrine and epinephrine and their metabolites in blood and urine are therefore useful indices of potentially abnormal allostatic states.

Cytokines

Cytokines are a diverse group of molecules that were first identified in relation to the acute-phase response and the subsequent activation of immune responses to a pathogen or other immunogenic agents. Inflammatory cytokines include IL-1, IL-2, IL-6, tumor necrosis factor (TNF), fibroblast growth factors, and interferons, whereas the antiinflammatory cytokines, so-named because they inhibit inflammatory cytokine production, include IL-4 and IL-10. These cytokines are produced locally by immune cells but are also produced by diverse organs such as the brain endothelial cells and the liver (Arkins et al., 2000; Obal and Krueger, 2000). As a result of their local production, cytokines often enter the circulation and can be detected in plasma samples. Sleep deprivation and psychological stress, such as public speaking, are reported to elevate inflammatory cytokine levels in blood (Altemus et al., 2001). Circulating levels of a number of inflammatory cytokines are elevated in relation to viral and other infections and contribute to the feeling of being sick, as well as the associated sleepiness, with both direct and indirect effects on the central nervous system (Arkins et al., 2000; Obal and Krueger, 2000). Circulating cytokines appear to have a limited access to the brain and also to induce a signal via the vagal system that induces the brain to express increased levels of its own inflammatory cytokines (Arkins et al., 2000; Obal and Krueger, 2000).

Inflammatory autoimmune diseases, such as multiple sclerosis, rheumatoid arthritis, and type 1 diabetes, reflect an allostatic state that consists of at least three principal causes: genetic risk factors such as those related to different major histocompatability complex alleles, factors that contribute to the development of tolerance to self-antigens or to antigens associated with food or other common agents, molecular mimicry between antigens on bacteria and self-antigens, and the hormonal milieu that regulates adaptive immune responses (Dowdell and Whitacre, 2000). Besides sex hormones, the relative insufficiency of glucocorticoids in the Lewis rat, compared with the Fischer rat, is often cited as an example of this latter influence and, as noted earlier, represents one type of allostatic state in which other agents that are normally counterregulated by glucocorticoids, such as inflammatory cytokines, may achieve the upper hand in a pathophysiological condition (Dowdell and Whitacre, 2000).

There is also evidence that cytokines are involved in the regulation of normal sleep, independently of their ability to induce fever (Moldofsky, 1995; Obal and Krueger, 2000). There is evidence for normal fluctuations in plasma levels of IL-1, IL-6, and TNF in relation to the diurnal cycle and sleep deprivation, whereas the story as far as circulating levels of interferons and fibroblast growth factors and normal sleep is less clear (Obal and Krueger, 2000). The antiinflammatory cytokines, IL-10 and IL-4, have been shown to inhibit sleep in rats, but there is a paucity of data on their involvement in normal sleep-wake cycles (Obal and Krueger, 2000).

Conditions such as allergies and exercise of an intensity that promotes muscle proteolysis tend to increase the production of cytokines that can be detected in blood (Moldofsky, 1995; Rohde et al., 1997; Borish et al., 1998; Cannon et al., 1999). Chronic fatigue syndrome (CFS) and fibromyalgia are related conditions that appear to represent an allostatic state in which there is dysregulation of cytokine production as well as a somewhat deficient level of circulating glucocorticoids that have counterregulatory effects on cytokine production (Dhabhar et al., 2000; Moldofsky, 1995). There is evidence that both mild exercise and allergies will exacerbate symptoms of CFS and fibromylagia (Peterson et al., 1994; Moldofsky, 1995; Borish et al., 1998; Cannon et al., 1999). In addition, viral infections are often reported as either precipitating or exacerbating CFS (Moldofsky, 1995).

Along with the symptoms of CFS and fibromyalgia, there are frequently reported, albeit somewhat inconsistent, signs of an allostatic state featuring elevated levels of certain inflammatory cytokines (Peterson et al., 1994; Moldofsky, 1995; Borish et al., 1998; Cannon et al., 1999) (Cannon et al., 1997; Mawle et al., 1997; Kavelaars et al., 2000). It is interesting to note that the balance of inflammatory to antiinflammatory cytokines may be affected in CFS sufferers, as shown by a deficient counterregulatory effect of both glucocorticoids and catecholamines on the balance between these two antagonistic types of cytokines in young women suffering from CFS (Kavelaars et al., 2000). Furthermore, as predicted from what we know of the linkage between peripheral and CNS cytokines, changes in circulating levels of inflammatory cytokines may signal increases in brain production of these same cytokines. Judging from recent work on a mouse model of CFS, these brain elevations may be more relevant to the symptoms of CFS and fibromyalgia than serum levels (Sheng et al., 1996). In this connection, hyperalgesia and perception of chronic pain has been associated with increased levels of TNF-alpha in the hippocampus and other brain regions in a rodent model in which the TNF-alpha appears to reduce the release of norepinephrine (Ignatowski et al., 1996; Ignatowski et al., 1997; Covey et al., 2000).

Cytokines of the proinflammatory type are associated with oxidative stress, and oxidative stress is linked to activation of genes via the transcription factor NfkB; type 2 diabetes is an example of an allostatic state in which NfKB is chronically activated (Bierhaus et al., 2001).

Besides the inhibitory actions of antiinflammatory cytokines, the proinflammatory response is also subject to regulation by two other primary agents: glucocorticoids and the parasympathetic nervous system. Parasympathetic inhibition of an allostatic state involving an out-of-control inflammatory response may be an effective treatment for septic shock (Borovikova et al., 2000). Glucocorticoids exert their effects directly on the production of inflammatory cytokines (Munck et al., 1984). They also inhibit the activity of the transcription factor NfkB, exerting direct and indirect effects at the level of transcriptional regulation (Wissink et al., 1998).

In conclusion, the measurement of circulating levels of inflammatory and antiinflammatory cytokines, as well as circulating glucocorticoids, DHEA, and catecholamines, provides clues as to the existence of allostatic states reflecting imbalance of various regulators of inflammation and adaptive immunity. Given the generally suppressive effects of glucocorticoids on inflammatory cytokine production (Spencer et al., 2000), cytokine levels should generally move in the opposite direction from glucocorticoid levels. One problem associated with the measurement of all of these markers is that their plasma levels may be very much related to the acute state of other processes related to such factors as recent exercise, allergies, sleep deprivation, and persistence of viral or other infections that exacerbate the existing allostatic state (Peterson et al., 1994; Moldofsky, 1995; Mawle et al., 1997; Borish et al., 1998; Cannon et al., 1999; Kavelaars et al., 2000). Thus challenge tests need to be designed to evaluate the responsiveness of each of the markers, whereas evaluations of "basal" or "diurnal" patterns of activity need to be carefully set up to detect possible allostatic states (Kirschbaum et al., 1995; Altemus et al., 2001; Stone et al., 2001).

Another, more general problem with evaluation of the levels of these mediators is the redundancy of their regulation and multiplicity of their interactions with each other, along with the fact that there are both positive and negative feedback loops for each of them. Thus, for example, as described earlier in this chapter, glucocorticoids enhance certain aspects of immune function acutely and yet can suppress immunity when present in high doses or over long time periods. Another example is that the transcriptional regulator NfkB has destructive and damaging effects on cells depending largely on the activity of other transcriptional regulators that are being regulated by other events going on in the cell (Lezoualc'h and Behl, 1998). Given these complex, reciprocal interactions and the fact that there is considerable redundancy in their control over each other, it seems best to treat them as a network in which perturbations in one variable can affect all of the others. This concept is discussed in the chapter by Singer, Ryff, and Seeman in this volume.

CONCLUSIONS

This chapter has attempted to clarify terminology related to traditional concepts of stress and homeostasis through use of the terms allostasis, allostatic states, and allostatic load. The goal is to facilitate the measurement of biological parameters that are related to the impact of the social and physical environment on health and risk for disease. This concluding section therefore addresses issues pertinent to both terminology and measurement.

Concerning terminology, I have used the term "stress" in this essay to refer to an event or events that are threatening to homeostasis or to individual integrity (both physiological and psychological). I have used allostasis in a broader sense than originally defined by Sterling and Eyer (1988) to refer to the active process of adaptation, that is, of reestablishing or maintaining homeostasis in the face of change. This recognizes that systems involved in adaptation may show sustained, elevated activity, now referred to as an allostatic state. Because the development of allostatic states and allostatic load may be related to lifestyle, sleep amount and patterns, as well as stressful experiences, allostasis is a broader term that includes "stress" as one of its components.

The term "allostatic load" is used here to refer to the cost to the body of being repeatedly forced to change or adapt as a result of lifestyle, altered day-night cycles, or stressors. Allostatic load results from many challenges or from inefficient management of the mediators. Just as stress is subsumed as a specific example of a stimulus for allostasis, chronic stress is subsumed as a specific subtype of allostatic load, because, in addition to stress, allostatic load also results from a chronically perturbed HPA axis due to genetic factors or early developmental events, or from a rich diet, smoking, drinking alcohol, and lack of exercise.

Thus, in contrast to homeostasis, allostasis refers to the process of adaptation of many bodily states simultaneously across various life circumstances through the expenditure of energy and production of mediators that produces adaptation in the short run but which can exacerbate disease over long time periods. As noted by Sterling and Eyer (1988) and later by Schulkin (Schulkin et al., 1994), the concept of allostasis also recognizes that it may be the result of anticipation of what the future may hold and that allostatic states and allostatic load may be produced as a result of anticipatory anxiety that is sustained over long time periods.

There are undoubtedly many readers who will want to continue using, or simply use out of habit, the traditional terms; they have a legitimate point. As a matter of semantics; one could always say that lifestyle factors, genetic and developmental contributions, and lack of exercise are stressful, using another definition of "stress," namely, that it refers to a state of unbalanced physiology. Thus, when stress is defined as a state of imbalance of any kind, it becomes much like "allostatic load." However, as noted in this chapter, the problem, then, is that the use of "stress" becomes ambiguous – does it refer to stressful events, to the response to those events, or to a chronic state of responding? Moreover, this use of "stress" does not distinguish between the process of adaptation (that we refer to as allostasis) and its consequences (that we prefer to call allostatic states and allostatic load).

As a solution to such ambiguity, allostasis, allostatic states, and allostatic load emphasize the process of adaptation to many situations involving change or challenge, whether they are caused by stressful events, by day-night cycles, or by continuing effects of lifestyle (diet, smoking, drinking, exercise or lack thereof), exacerbated by genetic risk factors and developmental contributions. Allostasis also clarifies an inherent ambiguity in the term "homeostasis" and distinguishes between the systems that are essential for life ("homeostasis") and those that maintain those systems in balance ("allostasis").

Furthermore, besides their inclusive breadth as organizing principles, allostasis, allostatic states and allostatic load also emphasize the intrinsic connections between the process of adaptation to a challenge that can also become damaging when prolonged or mismanaged. Because allostasis, allostatic states, and allostatic load are intrinsically mechanistic concepts, they challenge us to elucidate the mechanisms that distinguish between adaptation and the conditions that lead to disease. The breadth of the actions of the primary mediators of allostasis puts a different light on the General Adaptation Syndrome of Selye. Selye proposed that there was a stereotyped response to every stressful event. This has turned out not to be true (see Goldstein and Eisenhofer, 2000; Goldstein and Pacak, 2000). Instead, the mediators of allostasis are produced in different amounts in response to different signals and different situations that the organism encounters. However, as I have noted throughout this chapter, we now recognize that the same mediators produce a diverse number of effects depending on the specific differentiated characteristics of each organ system. Moreover, the mediators interact with each other, and the final physiologic outcome depends on the levels of the different mediators. Thus a new version of general adaptation syndrome might say that the primary mediators are almost universally involved in many aspects of adaptation (or allostasis), but that they contribute in varying combinations and to different degrees in orchestrating many forms of adaptation. They also participate in many forms of maladaptation when they are overused or improperly regulated.

Finally, no matter how appealing a theoretical framework may be, the practical application of the theory depends on making measurements that provide an empirical validation of the concepts and establish their usefulness in predicting changes in health over time. Unfortunately, ethical and financial limitations on the types of measurements that can be made on human subjects impose major problems for the biologist in finding markers that bear a meaningful relationship to pathophysiological outcomes. I have discussed in this chapter the selection of primary mediators for future studies. Clearly, compromises have to be made over what would be ideal. Yet considerable progress has been possible with the markers that were developed during the MacArthur Successful Aging Study, and these are described in another chapter in this volume by Singer, Ryff, and Seeman, along with the exciting prospects for future development of these measurements.

REFERENCES

Abadie, J. M., Wright, B., Correa, G., Browne, E. S., Porter, J. R., Svec, F. (1993). Effect of dehydroepiandrosterone on neurotransmitter levels and appetite regulation of the obese Zucker rat. *Diabetes* 42:662–9.

Adler, N. E., Boyce, T., Chesney, M. A., Cohen, S., Folkman, S., Kahn, R. L., Syme, L. S. (1994). Socioeconomic status and health: The challenge of the gradient. *Am Psychol* 49:15–24.

Adler, N. E., Marmot, M., McEwen, B. S., Stewart, J. E. (1999). Socioeconomic Status and Health in Industrial Nations: Social, psychological, and biological pathways. New York: *N Y Acad Sci* vol 896.

Altemus, M., Rao, B., Dhabhar, F. S., Ding, W., Granstein, R. D. (2001). Stress-induced changes in skin barrier function in healthy women. *J Invest Dermatol* 117:309–17.

Aragno, M., Brignardello, E., Tamagno, E., Gatto, V., Danni, O., Boccuzzi, G. (1997). Dehydroepiandrosterone administration prevents the oxidative damage induced by acute hyperglycemia in rats. *J Endocrinol* 155:233–40.

Araneo, B., Daynes, R. (1995). Dehydroepiandrosterone functions as more than an antiglucocorticoid in preserving immunocompetence after thermal injury. *Endocrinology* 136:393–401.

Arauz-Pacheco, C., Lender, D., Snell, P. G., Huet, B., Ramirez, L. C., Breen, L., Mora, P., Raskin, P. (1996). Relationship between insulin sensitivity, hyperinsulinemia, and insulin-mediated sympathetic activation in normotensive and hypertensive subjects. *Am J Hypertens* 9:1172–8.

Arkins, S., Johnson, R. W., Minshall, C., Dantzer, R., Kelley, K. W. (2000). Immunophysiology: The interactions of hormones, lymphohemopoietic cytokines, and the neuroimmune axis. In: *Coping with the Environment: Neural and Endocrine Mechanisms* (ed. B. S. McEwen), pp. 469–95. New York: Oxford University Press.

Bierhaus, A., Schiekofer, S., Schwaninger, M., Andrassy, M., Humpert, P. M., Chen, J., Hong M., Luther, T., Henle, T., Kloting, I., Morcos, M., Hofmann, M., Tritschler, H., Weigle, B., Kasper, M., Smith, M., Perry, G., Schmidt, A.-M., Stern, D. M., Haring, H.-U., Schleicher, E., Nawroth, P. P. (2001). Diabetes-associated sustained activation of the transcription factor nuclear factor-κB. *Diabetes* 50:2792–808.

Bjorntorp, P., Holm, G., Rosmond, R. (1999). Hypothalamic arousal, insulin resistance and type 2 diabetes mellitus. *Diabetic Med* 16:373–83.

Blauer, K. L., Poth, M., Rogers, W. M., Bernton, E. W. (1991). Dehydroepiandrosterone antagonized the suppressive effects of dexamethasone on lymphocyte proliferation. *Endocrinology* 129:3174–9.

Borish, L., Schmaling, K., DiClementi, J. D., Streib, J., Negri, J., Jones, J. F. (1998). Chronic fatigue syndrome: Identification of distinct subgroups on the basis of allergy and psychologic variables. *J Allergy Clin Immunol* 102:222–30.

Borovikova, L. V., Ivanova, S., Zhang, M., Yang, H., Botchkina, G. I., Watkins, L. R., Wang, H., Abumrad, N., Eaton, J. W., Tracey, K. J. (2000). Vagus nerve stimulation attenuates the systemic inflammatory response to endotoxin. *Nature* 405:458–62.

Brindley, D. N., Rolland, Y. (1989). Possible connections between stress, diabetes, obesity, hypertension and altered lipoprotein metabolism that may result in atherosclerosis. *Clin Sci* 77:453–61.

Browne, E. S., Porter, J. R., Correa, G., Abadie, J., Svec, F. (1993). Dehydroepiandrosterone regulation of the hepatic glucocorticoid receptor in the Zucker rat. The obesity research program. *J Steroid Biochem Molec Biol* 45:517–24.

Brunner, E. J., Marmot, M. G., Nanchahal, K., Shipley, M. J., Stansfeld, S. A., Juneja, M., Alberti, K. G. M. M. (1997). Social inequality in coronary risk: Central obesity and the metabolic syndrome. Evidence from the Whitehall II study. *Diabetologia* 40:1341–9.

Bulloch, K. (2000). Regional neural regulation of immunity: Anatomy and function. In: *Handbook of Physiology. Coping with the Environment: Neural and Endocrine Mechanisms* (ed. B. S. McEwen), pp. 353–79. New York: Oxford University Press.

Buske-Kirschbaum, A., Jobst, S., Wustmans, A., Kirschbaum, C., Rauth, W., Hellhammer, D. H. (1997). Attenuated free cortisol response to psychosocial stress in children with atopic dermatitis. *Psychosom Med* 59:419–26.

Cahill, L., Prins, B., Weber, M., McGaugh, J. L. (1994). Beta-adrenergic activation and memory for emotional events. *Nature* 371:702–4.

Cannon, J. G., Angel, J. B., Abad, L. W., Vannier, E., Mileno, M. D., Fagioli, L., Wolff, S. M., Komaroff, A. L. (1997). Interleukin-1, interleukin-1 receptor antagonist, and soluble interleukin-1 receptor type II secretion in chronic fatigue syndrome. *J Clin Immunol* 17:253–61.

Cannon, J. G., Angel, J. B., Ball, R. W., Abad, L. W., Fagioli, L., Komaroff, A. L. (1999). Acute phase responses and cytokine secretion in chronic fatigue syndrome. *J Clin Immunol* 19:414–21.

Cannon, W. (1929). The wisdom of the body. *Physiol Rev* 9:399–431.

Compagnone, N. A., Mellon, S. H. (1998). Dehydroepiandrosterone: A potential signalling molecule for neocortical organization during development. *Proc Natl Acad Sci U S A* 95:78–83.

Covey, W. C., Ignatowski, T. A., Knight, P. R., Spengler, R. N. (2000). Brain-derived TNF: Involvement in neuroplastic changes implicated in the conscious perception of persistent pain. *Brain Res* 859:113–22.

Crofford, L. J., Pillemer, S. R., Kalogeras, K., Cash, J. M., Michelson, D., Kling, M. A., Sternberg, E. M., Gold, P. W., Chrousos, G. P., Wilder, R. L. (1994). Hypothalamic-pituitary-adrenal axis perturbations in patients with fibromyalgia. *Arthritis Rheum* 37:1583–92.

DeBellis, M. D., Chrousos, G. P., Dorn, L. D., Burke, L., Helmers, K., Kling, M. A., Trickett, P. K., Putnam, F. W. (1994a). Hypothalamic-pituitary-adrenal axis dysregulation in sexually abused girls. *Clin Endocrinol Metab* 78:249–55.

DeBellis, M. D., Lefter, L., Trickett, P. K., Putnam, F. W. (1994b). Urinary catecholamine excretion in sexually abused girls. *J Am Acad Child Adolesc Psychiatry* 33:320–7.

Dellwo, M., Beauchene, R. E. (1990). The effect of exercise, diet restriction, and aging on the pituitary–adrenal axis in the rat. *Exp Geront* 25:553–62.

Dhabhar, F., McEwen, B. (1999). Enhancing versus suppressive effects of stress hormones on skin immune function. *Proc Natl Acad Sci U S A* 96:1059–64.

Dhabhar, F. S., McEwen, B. S. (1997). Acute stress enhances while chronic stress suppresses cell-mediated immunity in vivo: A potential role for leukocyte trafficking. *Brain Behav Immunol* 11:286–306.

Dhabhar, F. S., Miller, A. H., McEwen, B. S., Spencer, R. L. (1995). Effects of stress on immune cell distribution: Dynamics and hormonal mechanisms. *J Immunol* 154:5511–27.

Dhabhar, F. S., Miller, A. H., McEwen, B. S., Spencer, R. L. (1996). Stress-induced changes in blood leukocyte distribution: Role of adrenal steroid hormones. *J Immunol* 157:1638–44.

Dhabhar, F. S., Satoskar, A. R., Bluethmann, H., David, J. R., McEwen, B. S. (2000). Stress-induced enhancement of skin immune function: A role for γ interferon. *Proc Natl Acad Sci* 97:2846–51.

Dowdell, K., Whitacre, C. (2000). Regulation of inflammatory autoimmune diseases. In: *Coping with the Environment: Neural and Endocrine Mechanisms,* pp. 451–67. New York: Oxford University Press.

Felitti, V. J., Anda, R. F., Nordenberg, D., Williamson, D. F., Spitz, A. M., Edwards, V., Koss, M. P., Marks, J. S. (1998). Relationship of childhood abuse and household dysfunction to many of the leading causes of death in adults. The adverse childhood experiences (ACE) study. *Am J Prev Med* 14:245–58.

Fleshner, M., Pugh, C. R., Tremblay, D., Rudy, J. W. (1997). DHEA-S selectively impairs contextual-fear conditioning: Support for the antiglucocorticoid hypothesis. *Behav Neurosci* 111:512–17.

Formiga, F., Moga, I., Nolla, J. M., Navarro, M. A., Bonnin, R., Roig-Escofet, D. (1997). The association of dehydroepiandrosterone sulphate levels with bone mineral density in systemic lupus erythematosus. *Clin Exp Rheumatol* 15:387–92.

Frederick, S. L., Reus, V. I., Ginsberg, D., Hall, S. M., Munoz, R. F., Ellman, G. (1998). Cortisol and response to dexamethasone as predictors of withdrawal distress and abstinence success in smokers. *Biol Psychiatry* 43:525–30.

Gerin, W., Pickering, T. G. (1995). Association between delayed recovery of blood pressure after acute mental stress and parental history of hypertension. *J Hypertens* 13:603–10.

Goldstein, D. S., Eisenhofer, G. (2000). Sympathetic nervous system physiology and pathophysiology in coping with the environment. In: *Coping with the Environment: Neural and Endocrine Mechanisms,* pp. 21–43. New York: Oxford University Press.

Goldstein, D. S., Pacak, K. (2000). Catecholamines in the brain and responses to environmental challenges. In: *Coping with the Environment: Neural and Endocrine Mechanisms,* pp. 45–60. New York: Oxford University Press.

Hagglof, B., Bloom, L., Dahlquist, G., Lonnberg, G., Sahlin, B. (1991). The Swedish childhood diabetes study: Indications of severe psychological stress as a risk factor for type I (insulin-dependent) diabetes mellitus in childhood. *Diabetologia* 34:579–83.

Ignatowski, T. A., Chou, R. C., Spengler, R. N. (1996). Changes in noradrenergic sensitivity to tumor necrosis factor-α in brains of rats administered clonidine. *J Neuroimmunol* 70:55–63.

Ignatowski, T. A., Noble, B. K., Wright, J. R., Gorfien, J. L., Heffner, R. R., Spengler, R. N. (1997). Neuronal-associated tumor necrosis factor (TNF): Its role in noradrenergic functioning and modification of its expression following antidepressant drug administration. *J Neuroimmunol* 79:84–90.

Kalimi, M., Shafagoj, Y., Loria, R., Padgett, D., Regelson, W. (1994). Antiglucocorticoid effects of dehydroepiandrosterone (DHEA). *Mol Cell Biochem* 131:99–104.

Kamara, K., Eskay, R., Castonguay, T. (1998). High-fat diets and stress responsivity. *Physiol Behav* 64:1–6.

Kavelaars, A., Kuis, W., Knook, L., Sinnema, G., Heijnen, C. J. (2000). Disturbed neuroendocrine-immune interactions in chronic fatigue syndrome. *J Clin Endocrinol Metab* 85:692–6.

Kirschbaum, C., Prussner, J. C., Stone, A. A., Federenko, I., Gaab, J., Lintz, D., Schommer, N., Hellhammer, D. H. (1995). Persistent high cortisol responses to repeated psychological stress in a subpopulation of healthy men. *Psychosom Med* 57:468–74.

Koob, G. F., LeMoal, M. (2001). Drug addiction, dysregulation of reward, and allostasis. *Neuropsychopharmacology* 24:97–129.

Landfield, P. W., Eldridge, J. C. (1994). Evolving aspects of the glucocorticoid hypothesis of brain aging: Hormonal modulation of neuronal calcium homeostasis. *Neurobiol Aging* 15:579–88.

LeDoux, J. E. (1996). *The Emotional Brain.* New York: Simon & Schuster.

Lehman, C., Rodin, J., McEwen, B. S., Brinton, R. (1991). Impact of environmental stress on the expression of insulin-dependent diabetes mellitus. *Behav Neurosci* 105:241–45.

Leibowitz, S. F., Hoebel, B. G. (1997). Behavioral neuroscience of obesity. In: *Handbook of Obesity* (ed. G. A. Bray, C. Bouchard, W. P. T. James), pp. 313–58. New York: Marcel Dekker.

Leproult, R., Copinschi, G., Buxton, O., Van Cauter, E. (1997). Sleep loss results in an elevation of cortisol levels the next evening. *Sleep* 20:865–70.

Lezoualc'h, F., Behl, C. (1998). Transcription factor NF-κB: Friend or foe of neurons? *Mol Psychiatry* 3:15–20.

Liu, D., Diorio, J., Tannenbaum, B., Caldji, C., Francis, D., Freedman, A., Sharma, S., Pearson, D., Plotsky, P. M., Meaney, M. J. (1997). Maternal care, hippocampal glucocorticoid receptors, and hypothalamic-pituitary-adrenal responses to stress. *Science* 277:1659–62.

Lowy, M. T., Gault, L., Yamamoto, B. K. (1993). Adrenalectomy attenuates stress-induced elevations in extracellular glutamate concentrations in the hippocampus. *J Neurochem* 61:1957–60.

Lowy, M. T., Wittenberg, L., Yamamoto, B. K. (1995). Effect of acute stress on hippocampal glutamate levels and spectrin proteolysis in young and aged rats. *J Neurochem* 65:268–74.

Lynch, J. W., Kaplan, G. A., Shema, S. J. (1997). Cumulative impact of sustained economic hardship on physical, cognitive, psychological, and social functioning. *N Engl J Med* 337:1889–95.

Magarinos, A. M., Jain, K., Blount, E. D., Reagan, L., Smith, B. H., and McEwen, B. W. (2001). Peritoneal implantation of macroencapsulated porcine pancreatic islets in diabetic rats ameliorates severe hyperglycemia and prevents retraction and simplification of hippocampal dendrites. *Brain Res* 902: 282–7.

Magarinos, A. M., McEwen B. S. (1995). Stress-induced atrophy of apical dendrites of hippocampal CA3c neurons: involvement of glucocorticoid secretion and excitatory amino acid receptors. *Neuroscience* 69:89–98.

Manuck, S. B., Kaplan, J. R., Adams, M. R., Clarkson, T. B. (1995). Studies of psychosocial influences on coronary artery atherosclerosis in cynomolgus monkeys. *Health Psychol* 7:113–24.

Manuck, S. B., Kaplan, J. R., Muldoon, M. F., Adams, M. R., Clarkson, T. B. (1991). The behavioral exacerbation of atherosclerosis and its inhibition by propranolol. In: *Stress, Coping and Disease* (ed. P. M. McCabe, N. Schneiderman, T. M. Field, J. S. Skyler), pp. 51–72. Hove and London: Lawrence Erlbaum Associates.

Marmot, M. G., Smith, G. D., Stansfeld, S., Patel, C., North, F., Head, J., White, I., Brunner, E., Feeney, A. (1991). Health inequalities among British civil servants: the Whitehall II study. *Lancet* 337:1387–93.

Mawle, A. C., Nisenbaum, R., Dobbins, J. G., Gary H. E. Jr., Stewart, J. A., Reyes, M., Steele, L., Schmid, D. S., Reeves W. C. (1997). Immune responses associated with chronic fatigue syndrome: A case-control study. *J Infect Dis* 175:136–41.

McEwen, B. S. (1998). Protective and damaging effects of stress mediators. *New Eng J Med* 338:171–9.

McEwen, B. S. (1999). Stress and hippocampal plasticity. *Annu Rev Neurosci* 22:105–22.

McEwen, B. S. (2000). Allostasis and allostatic load: Implications for neuropsychopharmacology. *Neuropsychopharmacology* 22:108–24.

McEwen, B. S. Seeman, T. (1999). Protective and damaging effects of mediators of stress: Elaborating and testing the concepts of allostasis and allostatic load. *Ann N Y Acad Sci* 896:30–47.

McEwen, B. S., Stellar, E. (1993). Stress and the Individual: Mechanisms leading to disease. *Arch Int Med* 153:2093–101.

Meaney, M., Aitken, D., Berkel, H., Bhatnagar, S., Sapolsky, R. (1988). Effect of neonatal handling of age-related impairments associated with the hippocampus. *Science* 239:766–8.

Michelson, D., Stratakis, C., Hill, L., Reynolds, J., Galliven, E., Chrousos, G., Gold, P. (1996). Bone mineral density in women with depression. *N Engl J Med* 335:1176–81.

Miller, A. H., Pearce, B. D., Ruzek, M. C., Biron, C. A. (2000). Interactions between the hypothalamic-pituitary-adrenal axis and immune system during viral infection: Pathways for environmental effects on disease expression. In: *Coping with the Environment: Neural and Endocrine Mechanisms*, pp. 425–50. New York: Oxford University Press.

Moldofsky, H. (1995). Sleep, neuroimmune and neuroendocrine functions in fibromyalgia and chronic fatigue syndrome. *Adv Neuroimmunol* 5:39–56.

Munck, A., Guyre, P. M., Holbrook, N. (1984). Physiological functions of glucocorticoids in stress and their relation to pharmacological actions. *Endocr Rev* 5:25–44.

Nakashima, N., Haji, M., Sakai, Y., Ono, Y., Umeda, F., Nawata, H. (1995). Effect of dehydroepiandrosterone on glucose uptake in cultured human fibroblasts. *Metabolism* 44:543–8.

Nicolle, M. M., Colombo, P. J., Gallagher, M., McKinney, M. (1999). Metabotropic glutamate receptor-mediated hippocampal phosphoinositide turnover is blunted in spatial learning-impaired aged rats. *J Neurosci* 19:9604–10.

Obal, F. Jr., Krueger, J. M. (2000). Hormones, cytokines, and sleep. In: *Coping with the Environment: Neural and Endocrine Mechanisms*, pp. 331–49. New York: Oxford University Press.

Ogilvie, K. M., Rivier, C. (1997). Gender difference in hypothalamic-pituitary-adrenal axis response to alcohol in the rat: Activational role of gonadal steroids. *Brain Res* 766:19–28.

Perna, F. M., McDowell, S. L. (1995). Role of psychological stress in cortisol recovery from exhaustive exercise among elite athletes. *Int J Behav Med* 2:13–26.

Perseghin, G., Price, T. B., Petersen, K. F., Roden, M., Cline, G. W., Gerow, K., Rothman D. L., Shulman, G. I. (1996). Increased glucose transport-phosphorylation and muscle glycogen synthesis after exercise training in insulin-resistant subjects. *New Eng J Med* 335:1357–62.

Peterson, P. K., Sirr, S. A., Grammith, F. C., Schenck, C. H., Pheley, A. M., Hu, S., Chao, C. C. (1994). Effects of mild exercise on cytokines and cerebral blood flow in chronic fatigue syndrome patients. *Clin Diag Lab Immunol* 1:222–6.

Plat, L., Leproult, R., L'Hermite-Baleriaux, M., Fery, F., Mockel, J., Polonsky, K. S., Van Cauter, E. (1999). Metabolic effects of short-term elevations of plasma cortisol are more pronounced in the evening than in the morning. *J Clin Endocrin Metabol* 84:3082–92.

Porter, N. M., Thibault, O., Thibault, V., Chen, K. C., Landfield, P. W. (1997). Calcium channel density and hippocampal cell death with age in long-term culture. *J Neurosci* 17:5629–39.

Poteliakhoff, A. (1981). Adrenocortical activity and some clinical findings in acute and chronic fatigue. *J Psychosom Res* 25:91–5.

Rajkowska, G., Miguel-Hidalgo, J. J., Wei, J., Dilley, G., Pittman, S. D., Meltzer, H. Y., Overholser, J. C., Roth, B. L., Stockmeier, C. A. (1999). Morphometric evidence for neuronal and glial prefrontal cell pathology in major depression. *Biol Psychiatry* 45:1085–98.

Robinzon, B., Cutolo, M. (1999). Should dehydroepiandrosterone replacement therapy be provided with glucocorticoids? *Rheumatology* 38:488–95.

Rohde, T., MacLean, D. A., Richter, E. A., Kiens, B., Pedersen, B. K. (1997). Prolonged submaximal eccentric exercise is associated with increased levels of plasma IL-6. *Am J Physiol* 273:E85–91.

Roozendaal, B. (2000). Glucocorticoids and the regulation of memory consolidation. *Psychoneuroendocrinology* 25:213–38.

Rozovsky, I., Wei, M., Stone, D. J., Zanjani, H., Anderson, C. P., Morgan, T. E., Finch, C. E. (2002). Estradiol (E2) enhances neurite outgrowth by repressing glial fibrillary acidic protein expression and reorganizing laminin. *Endocrinology* 143:636–46.

Sapolsky, R., Krey, L., McEwen, B. S. (1986). The neuroendocrinology of stress and aging: The glucocorticoid cascade hypothesis. *Endocr Rev* 7:284–301.

Sapolsky, R. M., Romero, L. M., Munck, A. U. (2000). How do glucocorticoids influence stress responses? Integrating permissive, suppressive, stimulatory, and preparative actions. *Endocr Rev* 21:55–89.

Schulkin, J., McEwen, B. S., Gold, P. W. (1994). Allostasis, amygdala, and anticipatory angst. *Neurosci Biobehav Rev* 18:385–96.

Seeman, T. E., Singer, B. H., Rowe, J. W., Horwitz, R. I., McEwen, B. S. (1997). Price of adaptation – allostatic load and its health consequences: MacArthur studies of successful aging. *Arch Intern Med* 157:2259–68.

Selye, H. (1973). The evolution of the stress concept. *Am Scientist* 61:692–699.

Sheng, W. S., Hu, S., Lamkin, A., Peterson, P. K., Chao, C. C. (1996). Susceptibility to immunologically mediated fatigue in C57BL/6 versus Balb/c mice. *Clin Immunol Immunopath* 81:161–7.

Shurin, M. R., Zhou, D., Kusnecov, A., Rassinck, S., Rabin, B. S. (1994). Effect of one or more footshocks on spleen and blood lymphocyte proliferation in rats. *Brain, Behav Immun* 8:57–65.

Sousa, N., Lukoyanov, N. V., Madeira, M. D., Almeida, O. F. X., Paula-Barbosa, M. M. (2000). Reorganization of the morphology of hippocampal neurites and synapses after stress-induced damage correlates with behavioral improvement. *Neuroscience* 97:253–66.

Spencer, R., McEwen, B. S. (1990). Adaptation of the hypothalamic-pituitary-adrenal axis to chronic ethanol stress. *Neuroendocrinol* 52:481–89.

Spencer, R. L., Kalman, B. A, Dhabhar, F. S. (2000). Role of endogenous glucocorticoids in immune system function: Regulation and counterregulation. In: *Coping with the Environment: Neural and Endocrine Mechanisms*, pp. 381–423. New York: Oxford University Press.

Spiegel, K., Leproult, R., Van Cauter, E. (1999). Impact of sleep debt on metabolic and endocrine function. *Lancet* 354:1435–9.

Sterling, P., Eyer, J. (1988). Allostasis: A new paradigm to explain arousal pathology. In: *Handbook of Life Stress, Cognition and Health* (ed. S. Fisher, J. Reason), pp. 629–49. New York: John Wiley & Sons.

Sternberg, E. M. (1997). Neural-immune interactions in health and disease. *J Clin Invest* 100:2641–7.

Stone, A. A., Schwartz, J. E., Smyth, J., Kirschbaum, C., Cohen, S., Hellhammer, D., Grossman, S. (2001). Individual differences in the diurnal cycle of salivary free cortisol: a replication of flattened cycles for some individuals. *Psychoneuroendocrinology* 26:295–306.

Sugaya, K., Chouinard, M., Greene, R., Robbins, M., Personett, D., Kent, C., Gallagher, M., McKinney, M. (1996). Molecular indices of neuronal and glial plasticity in the hippocampal formation in a rodent model of age-induced spatial learning impairment. *J Neurosci* 16:3427–43.

Sugaya, K., Greene, R., Personett, D., Robbins, M., Kent, C., Bryan, D., Skiba, E., Gallagher, M., McKinney, M. (1998). Septo-hippocampal cholinergic and neurotrophin markers in age-induced cognitive decline. *Neurobiol Aging* 19:351–61.

Taylor S. E. (1999). The lifelong legacy of childhood abuse. *Am J Med* 107:399–400.

Ur, E., White, P. D., Grossman, A. (1992). Hypothesis: Cytokines may be activated to cause depressive illness and chronic fatigue syndrome. *Eur Arch of Psychiatry Clin Neurosci* 241:317–22.

Vacchio, M. S., Papadopoulos, V., Ashwell, J. D. (1994). Steroid production in the thymus: Implications for thymocyte selection. *J Exp Med* 179:1835–46.

Van Cauter, E., Shapiro, E. T., Tillil, H., Polonsky, K. S. (1992). Circadian modulation of glucose and insulin responses to meals: Relationship to cortisol rhythm. *Am J Physiol* 262:E467–75.

Verdecchia, P., Schillaci, G., Borgioni, C., Ciucci, A., Zampi, I., Battistelli, M., Gattobigio, R., Sacchi, N., Porcellati, C. (1995). Cigarette smoking, ambulatory blood pressure and cardiac hypertrophy in essential hypertension. *J Hypertens* 13:1209–15.

Walker, E. A., Gelfand, A., Katon, W. J., Koss, M. P., Von Korff, M., Bernstein, D., Russo, J. (1999). Adult health status of women with histories of childhood abuse and neglect. *Am J Med* 107:332–9.

Watanabe, Y., Gould, E., Cameron, H. A., Daniels, D. C., McEwen, B. S. (1992). Phenytoin prevents stress- and corticosterone-induced atrophy of CA3 pyramidal neurons. *Hippocampus* 2:431–6.

Wingfield, J. C., Romero, L. M. (2000). Adrenocortical responses to stress and their modulation in free-living vertebrates. In: *Coping with the Environment: Neural and Endocrine Mechanisms,* pp. 211–34. New York: Oxford University Press.

Wissink, S., van Heerde, E. C., van der Burg, B., van der Saag, P. T. (1998). A dual mechanism mediates repression of NF- B activity by glucocorticoids. *Mol Endocrinol* 12:355–63.

Wolf, O. T., Kirschbaum, C. (1999). Actions of dehydroepiandrosterone and its sulfate in the central nervous system: Effects on cognition and emotion in animals and humans. *Brain Res Rev* 30:264–98.

Wolkowitz, O. M., Reus, V. I., Roberts, E., Manfredi, F., Chan, T., Raum, W. J., Ormiston, S., Johnson, R., Canick, J., Brizendine, L., Weingartner, H. (1997). Dehydroepiandrosterone (DHEA) treatment of depression. *Biol Psychiatry* 41:311–18.

Young, J. B., Landsberg, L. (2000). Synthesis, storage and secretion of adrenal medullary hormones: Physiology and pathophysiology. In: *Handbook of Physiology. Section 7. The Endocrine System,* pp. 3–19. New York: Oxford University Press.

3 Merging of the Homeostat Theory with the Concept of Allostatic Load

David S. Goldstein

CANNON AND "HOMEOSTASIS"

Walter B. Cannon, extending Claude Bernard's concept of the *milieu intérieur* (Fig. 3.1) taught that coordinated body processes work toward the goal of an ideal set of steady states – "homeostasis." (Cannon, 1929a, 1929b, 1939). In emergencies, rapid activation of homeostatic systems – especially of what Cannon called the "sympathico-adrenal system" – would preserve the internal environment, by producing compensatory and anticipatory adjustments that would enhance the likelihood of survival.

It is by now clear that activities of daily life, such as meal ingestion, speaking, changing posture, and movement – that is, not only emergencies – are associated with continual alterations in sympathetic nervous system and adrenomedullary hormonal system outflows, maintaining appropriate blood flow to the brain, body temperature, delivery of metabolic fuel to body organs, and so forth. Each of these activities is associated with a somewhat different set of "normal" apparent steady states, directed by the brain and determined by coordinated actions of a variety of effector systems. This principle leads directly to the concept of "allostasis," as discussed later.

Cannon wrote that not only do activities of physiological systems change to maintain homeostasis but that these alterations also influence behaviors of the organism that in turn contribute to that maintenance. For instance, he noted that animals deprived of salt develop "salt hunger," and hypoglycemia evokes a hunger for calories (Cannon, 1939). This concept of physiology-induced alterations in motivational states and behavior was subsequently expanded considerably in the research of Curt Richter in the subsequent decades.

SELYE AND "STRESS"

Hans Selye introduced and popularized stress as a medical scientific idea. According to Selye's theory, stress is the nonspecific response of the body to

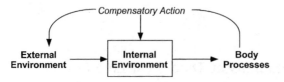

Figure 3.1: Bernard's concept of the *milieu intérieur,* in which the internal environment is maintained by continuous adjustments in bodily processes.

any demand upon it (Selye, 1974). Responses to stressors would have specific and nonspecific components, and Selye referred to only the nonspecific component as "stress." After removal of specific responses from consideration, a nonspecific syndrome would remain. Although nonspecific with respect to the inciting agents, the stress response itself was viewed to consist of a stereotyped pathological pattern, with enlargement of the adrenal glands, involution of the thymus gland (associated with atrophy of lymph nodes and inhibition of inflammatory responses), and peptic bleeding or ulceration.

More than a half century elapsed before Selye's doctrine of nonspecificity underwent experimental testing (Pacak et al., 1998). If values for dependent neuroendocrine measures (e.g., plasma corticotropin, epinephrine) increased as a function of the intensity of any stressor, with part of the response specific for the stressor and part nonspecific "stress," then if a_n and a_x were constants relating the intensity of stressor "x" to the nonspecific and specific components of the response, and if b_n and b_y were constants

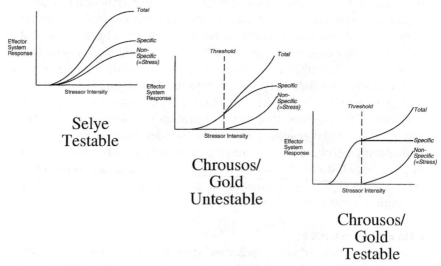

Figure 3.2: Testability of models of stress that relate effector system responses to stressor intensity.

relating the intensity of stressor "y" to the nonspecific and specific components of the response, the ratio $a_n : b_n$ would be the same for both (indeed all) stressors. The findings of Pacak et al. failed to support this prediction (Fig. 3.2).

Chrousos and Gold (1992) modified the doctrine of nonspecificity by proposing that above a threshold intensity, any stressor would elicit the "stress syndrome" (Fig. 2.3). If above the threshold intensity the specific response continued to increase, while the nonspecific response emerged, the total response would be the sum of two components, both of which could vary. This situation would result in simultaneous algebraic equations that could not be solved, obviating testing of the doctrine of nonspecificity. If the specific component of the response were assumed to plateau when the nonspecific response began to emerge, and if there were two intensities of two stressors (x and X for one stressor and y and Y for the other), both above the threshold intensity, then for effector systems A and B, $A_y/A_y = Y/y = B_y/A_y$. The findings of Pacak et al. (1998) also failed to support this idea.

By now researchers have largely abandoned Selye's stress theory. More modern theories have viewed stress as a threat to homeostasis, where the response has a degree of specificity, depending among other things on the organism's perception of the stressor and perceived ability to cope with it (Goldstein, 2001). This brief chapter pursues the notion of a homeostatic definition of stress.

A HOMEOSTATIC DEFINITION OF STRESS

In a home temperature control system (Fig. 3.3, left), the thermostat plays a central role by sensing discrepancy between the setpoint, determined by a regulator, and the temperature, which produces differential bending of metal bands in the thermostat. This type of system is a classical example of regulation by negative feedback. Home temperature control systems always include multiple effectors. The redundancy comes at relatively little cost, compared with four advantages. The multiplicity extends the range of control of external temperatures where the internal temperature can be maintained; when a single effector fails to function, others are activated compensatorily, helping maintain the temperature at about the set level; one can pattern the use of the effectors as appropriate to maximize economy and efficiency; and multiple effectors minimize wear and tear on the components.

Analogously, stress occurs when the organism senses a disruption or a threat of disruption of homeostasis. Central to the present theory is that the body possesses numerous homeostatic comparators, called "homeostats." Each homeostat compares information with a setpoint for responding,

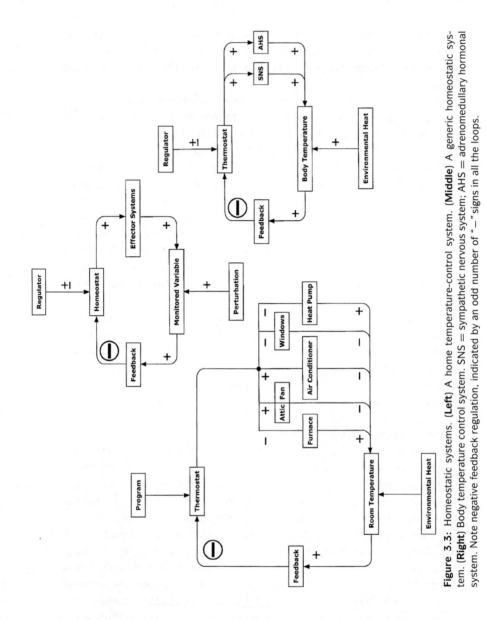

Figure 3.3: Homeostatic systems. **(Left)** A home temperature-control system. **(Middle)** A generic homeostatic system. **(Right)** Body temperature control system. SNS = sympathetic nervous system; AHS = adrenomedullary hormonal system. Note negative feedback regulation, indicated by an odd number of "–" signs in all the loops.

determined by a regulator (Fig. 3.3, middle). Homeostatic systems typically use multiple effectors to change values for the controlled variable. The loop is closed by monitoring changes in the levels of the controlled variable, via one or more monitored variables.

A tremendous array of homeostatic systems detect perturbations of monitored variables. In particular, in line with the home heating analogy, afferent information to the brain about cutaneous and blood temperature determines activities of cholinergic and noradrenergic nerve fibers in the skin that regulate sweating and vasomotor tone (Frank et al., 1997).

PRINCIPLES OF HOMEOSTAT OPERATION

Homeostatic systems operate according to a few principles, which, despite their simplicity, can explain complex physiological phenomena and help to resolve persistently controversial issues in the area of stress and disease.

Homeostatic systems always include regulation by negative feedback. Increases in values of the monitored variable result in changes in effector activity that oppose and thereby "buffer" changes in that variable. This feedback regulation can be modulated at several levels and therefore can be complex.

Induction of a positive feedback loop in a homeostatic system evokes instability. An example would be renin-angiotensin-aldosterone system activation in congestive heart failure. Activation of this system increases sodium retention and vascular tone, leading to increased cardiac preload and afterload that worsen the congestive heart failure. Therefore, treatment with an angiotensin converting enzyme inhibitor or angiotensin II receptor blocker can successfully treat congestive heart failure (Kluger et al., 1982).

Another example may be fainting reactions. Fainting is preceded by high circulating epinephrine levels (Mosqueda-Garcia et al., 1997). This elicits skeletal muscle vasodilation, and total peripheral resistance to blood flow falls. If there were enough "shunting" of blood to the skeletal muscle, then cerebral blood flow would fall. When the state of cerebral hypoperfusion reached consciousness, the person would not feel "right." This would evoke more adrenomedullary secretion of epinephrine, and the consequent neurocirculatory positive feedback loop would lead to critical brainstem hypoperfusion and loss of consciousness within minutes.

People who faint repeatedly do not feel completely normal between episodes. Patients who are susceptible to neurocardiogenic syncope often complain of chronic fatigue, headache, chest pain, orthostatic intolerance, difficulty concentrating, and heat intolerance, any of which can be debilitating. In essence this may reflect consequences of long-term allostatic load, as discussed subsequently.

Induction of a positive feedback loop "nested" in a larger system that includes negative feedback can lead to a new steady-state group of settings and values for monitored variables, rather than "explosion" of the system. For example, a distressing situation might elicit fear, resulting in release of norepinephrine in the brain and epinephrine in the periphery, both of which could augment vigilance behavior and heighten the experience of distress, resulting in greater fear (Aston-Jones et al., 1994). The organism could enter an "escape mode," with a different set of homeostatic regulatory settings; however, there is a risk of the positive feedback look leading to a behavioral "explosion" (i.e., panic) or to a pathophysiologic "explosion," (i.e., pulmonary edema). The notion of induction of a nested positive feedback loop can also provide a model for developmental changes in adolescence, in which stability would actually be abnormal, but there is a greater chance for both psychological and physiogical disorders to emerge.

Homeostatic systems generally use more than one effector, for the same reasons as home temperature control systems do so. Effector redundancy extends the ranges of control of monitored variables. It enables compensatory activation of alternative effectors, assuming no change in homeostat settings (Fig. 3.4). Examples of compensatory activation in physiology include augmentation of sympathoneural responsiveness by adrenalectomy, hypophysectomy, or thyroidectomy (Udelsman et al., 1987; Goldstein et al., 1993; Fukuhara et al., 1996). Finally, effector redundancy introduces the potential for patterned effector responses. Patterning of neuroendocrine, physiological, and behavioral effectors increases the likelihood of adaptiveness to the particular challenge to homeostasis, providing another basis for natural selection to favor the evolution of systems with multiple effectors. Finally, multiple effectors decrease effects of allostatic load on each of the components of the homeostatic system.

Different homeostats can regulate the activity of the same effector system (Fig. 3.4). For instance, the osmostat and volustat share the vasopressin effector (Quillen and Cowley, 1983). Blockade of afferent information to or interference with the function of a homeostat increases the variability of levels of the monitored variable. Thus, baroreceptor deafferentiation increases the variability of blood pressure, as does bilateral destruction of the nucleus of the solitary tract, the likely brainstem site of the arterial barostat (Nathan and Reis, 1977).

Even a simple homeostatic reflex reflects stress, when a perceived discrepancy between a setpoint for a monitored variable and information about the actual level of that variable elicits compensatory responses to decrease the discrepancy. Thus, one way of looking at stress is as a condition where expectations, whether genetically programmed, established by prior learning, or deduced from circumstances, do not match the current or

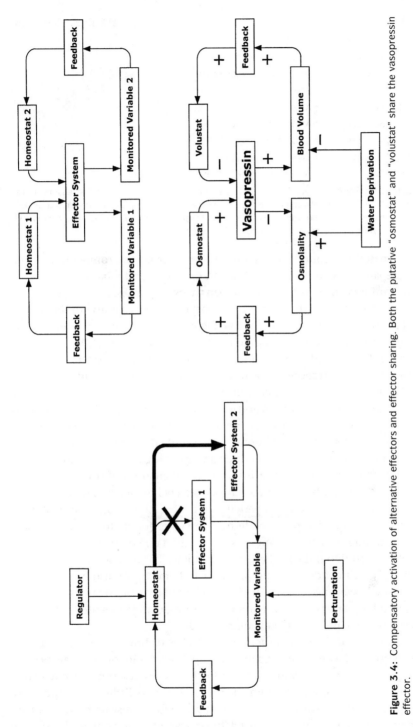

Figure 3.4: Compensatory activation of alternative effectors and effector sharing. Both the putative "osmostat" and "volustat" share the vasopressin effector.

Stressor	Neuroendocrine Pattern
Orthostasis	↑ SNS = ↑ AHS = ↓ X > ↑ HPA
Cold	↑ SNS > ↑ AHS > ↓ X > ↑ HPA
Glucoprivation	↑ AHS > ↑ HPA > ↓ X > ↑ SNS
Immobile Fear	↑ AHS > ↑ HPA > ↓ X > ↑ SNS
Fight / Rage	↑ SNS > ↑ AHS = ↓ X > ↑ HPA
Flight / Escape	↑ SNS = ↑ AHS = ↓ X > ↑ HPA
Faint / "Give Up"	↑ AHS = ↑ X = ↓ SNS > ↑ HPA

Figure 3.5: Some neuroendocrine patterns to physical or psychological stressors. In reality, the distinction between these types of stressors is blurred. SNS = sympathetic nervous system; AHS = adrenomedullary hormonal system; X = vagus nerve (cranial nerve X); HPA = hypothalamo-pituitary-adrenocortical axis.

anticipated perceptions of the internal or external environment, and this discrepancy between what is observed or sensed and what is expected or programmed elicits patterned, compensatory responses.

In contrast with the doctrine of nonspecificity, according to the homeostatic theory of stress, activities of effector systems are coordinated in relatively specific patterns, including neuroendocrine patterns. These patterns serve different needs. For instance, sympathetic nervous system activation predominates in response to orthostasis, moderate exercise, and exposure to cold, whereas adrenomedullary hormonal system activation predominates in response to glucoprivation and emotional distress (Fig. 3.5). Activation of the hypothalamo-pituitary-adrenocortical system seems especially prominent in distressing stressors sensed as novel (de Boer et al., 1989).

For each stress, neuroendocrine and physiological changes are coupled with behavioral changes. For instance, the regulation of total body water in humans depends on an interplay between behavior (the search for water and drinking), an internal experience or feeling (thirst), and the elicitation of a neurohumoral response pattern (in this case, dominated by vasopressin, the antidiuretic hormone, and to a lesser extent angiotensin, a potent stimulator of drinking). Evoked changes in homeostat function often produce not only neuroendocrine and physiological effects but also behavioral responses; however, because of traditional boundaries among physiology, endocrinology, and psychology, interactions producing integrated patterns of response remain incompletely understood.

In terms of the body's thermostat, studies of humans exposed to cold or with mild core hypothermia have provided support for the notion of primitive specificity of neuroendocrine stress responses. Cold exposure increases plasma norepinephrine levels, with little if any increases in plasma epinephrine levels (Frank et al., 1997), consistent with sympathetic neuronal activation and relatively less adrenomedullary hormonal activation. Mild core hypothermia increases antecubital venous levels of norepinephrine but not epinephrine. Both norepinephrine and epine- phrine levels in arterial plasma increase in this setting, but the norepine- phrine response is greater. These findings make sense, in that one can main- tain body temperature effectively by sympathetically mediated cutaneous vasoconstriction, piloerection, and shivering. When these mechanisms give way, and core temperature falls, then high circulating epinephrine levels in- crease the generation of calories, associated with the experience of distress, which motivates escape and avoidance and augments norepinephrine re- lease from sympathetic nerve terminals for a given amount of nerve traffic.

Distress is aversive to the organism, but the homeostat theory does not assume an equivalence of noxiousness (i.e., negatively reinforcing proper- ties) with production of pathological changes. Because the homeostat the- ory does not assume pathologic effects of distress, the theory does not as- sume that distress causes disease. Selye characterized distress as unpleasant or harmful (Selye, 1974), without separating these two very different char- acteristics. He never incorporated the relationship between distress and dis- ease explicitly in his theory. As noted earlier, Selye's theory emphasized the nonspecificity of the stress response, whereas according to the homostatic theory, the experience of distress responses depends on the character, in- tensity, and meaning of the stressor as perceived by the organism and on the organism's perceived ability to cope with it. Distress responses, like all stress responses, have a "purpose," mitigating effects of a stressor in some way. This applies not only to neuroendocrine aspects of those responses (such as the glucose counterregulatory actions of pituitary-adrenocortical and adrenomedullary stimulation during insulin-induced hypoglycemia) but also to psychological aspects (such as conditioned aversive and instru- mental avoidance learning). Distress responses evolved and probably con- tinue to be expressed even in higher organisms, including humans who ac- tually are only rarely exposed to truly "fight-or-flight" agonistic encounters, because of the importance of those responses in instinctive communication. Selye's theory did not consider the communication aspect of distress.

"ALLOSTASIS" AND "ALLOSTATIC LOAD"

Levels of physiological activity required to reestablish or maintain home- ostasis differ, depending on continually changing conditions in which the organism finds itself (e.g., running vs. standing vs. lying down). "Allosta- sis," a term used by Sterling and Eyer in 1988 (McEwen, 1998b), refers to levels of activity required for the individual to "maintain stability through change" – that is, to adapt (McEwen, 1998a; Schulkin et al., 1998; McEwen, 2000). In terms of the homeostat theory, allostasis refers to the set of apparent steady states maintained by multiple effectors. In the analogy of the home temperature control system, one can regulate temperature at different levels through the appropriate use of effectors. Among individuals, levels of glucose, blood pressure, body temperature, metabolism, and so forth are normally held stable at different levels, with different patterns of effector activation.

Homeostat resetting redefines the conditions required to maintain homeostasis. Regulation around an altered apparent steady state is the essence of allostasis. This would be analogous to a different thermostatic setting in winter compared with summer. A neuroendocrine example would be hyperglycemia that occurs due to exercise. Even in anticipation of the need for metabolic fuel, by activation of "central command" the blood glucose level increases to a new steady state value. Resetting alters activities of multiple effector systems required to maintain allostasis, at least for short durations. During stress, short-term changes in homeostatic settings generally enhance the long-term well-being and survival of the organism. Responses during exercise provide an obvious example. When superimposed on a substrate of pathology, however, homeostatic resetting can cause harm. For instance, in the setting of ischemic heart disease, global or patterned increases in sympathetic outflows increase cardiac work, with the resulting imbalance between oxygen supply and demand potentially precipitating angina pectoris, myocardial infarction, or sudden death.

"Allostatic load" (McEwen and Stellar, 1993) refers to effects of prolonged continuous or intermittent activation of effectors involved in allostasis. In the analogy of the home temperature control system, allostatic load would increase if a window or door were left open. In this situation, one or more effectors would be activated frequently or even continuously. An even more extreme example would be having the air conditioner and the furnace on at the same time, as is often the case in an overheated apartment in the spring when there is a warm day before the boilers have been shut down. Continued use of the furnace and air conditioner in opposition to one another, an example of an inefficient "allostatic state," consumes fuel and contributes to wear and tear on both pieces of equipment. Long-term allostatic load – the wear-and-tear cost of adaptation – provides a conceptual basis for studying long-term health consequences of stress.

MEDICAL AND PSYCHOLOGICAL CONSEQUENCES OF STRESS AND ALLOSTASIS

The homeostatic theory of stress and the concept of allostasis can help understand chronic as well as acute medical consequences of stress. Chronic activation of effectors in allostatic states promotes wear and tear, or allostatic load. For instance, chronic elevations in sympathetic neuronal and hypothalamic-pituitary-adrenocortical outflows might worsen insulin resistance or accelerate cardiovascular hypertrophy. Chronic activation of hypothalamic-pituitary-adrenocortical activation and release of endogenous excitatory amino acids in the brain can lead to remodeling of neurons in the hippocampus and impairment of cognitive function, processes that may participate in psychiatric illnesses such as major depression (McEwen, 2000).

Another application of the homeostatic idea to medical consequences of stress is the perceived ability to cope. As noted earlier, an organism experiences distress upon sensing that the effector responses will not be sufficient to restore or maintain allostasis. In contrast with distress, stress does not imply a conscious experience. For instance, even heavily sedated humans have substantial adrenomedullary stimulation in response to acute glucoprivation. Indeed, the greater extent the adrenomedullary response to the same stressor in alert compared with sedated humans might provide a measure of the distress. Distress instinctively elicits observable signs and pituitary-adrenocortical and adrenomedullary activation (Goldstein, 1995, 2001). Via these neuroendocrine changes, distress could worsen pathophysiologic processes. For instance, because of adrenomedullary activation, in a patient with coronary artery stenosis distress could elicit cardiovascular stimulation and produce an excess of myocardial oxygen consumption over supply, precipitating myocardial infarction or lethal ventricular arrhythmias. Moreover, long-term distress could augment both the risk of a mood disorder and the risk of worsening coronary disease.

Long-term physical or mental consequences of stress would depend on long-term effects of allostatic load (Fig. 3.6). Prolonged, intensive activation of effector systems could exaggerate effects of intrinsic defects in any of them, just as increased air pressure in a tire could expand and eventually "blow out" a weakened area. It is not difficult to imagine that repeated or long-term stress or distress could lead to a medical or psychiatric "blowout."

Maintenance of allostatic states requires energy. This requirement is perhaps clearest in the allostasis of core temperature. In mammals, maintenance of a constant core temperature accounts for a substantial proportion of total body energy expenditure at rest. One may hypothesize that reducing allostatic load exerts beneficial health effects, just as one may hypothesize that excessive allostatic load exerts deleterious health effects. In the analogy of the home temperature control system, maintaining a temperature of 60° Fahrenheit in the summer would require a great expenditure of energy and involve cooling systems being on continuously, whereas in the winter, maintaining the same temperature would be energy efficient. One

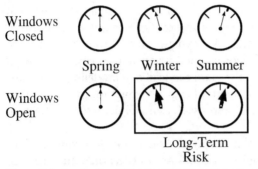

Figure 3.6: Application of the thermostat analogy to understand the concepts of allostasis, allostatic load, and long-term consequences of allostatic load. In the spring, home temperature is easily controlled. In the winter, one turns down the thermostat, so that the system operates with a different goal temperature (allostasis). In the summer, one turns up the thermostat. If the windows are left open, then in the summer and winter, maintaining allostasis requires increased work of the effector systems, and in the long term, this would increase the risk of system breakdown.

can imagine that the likelihood of system breakdown would depend on the extent of long-term energy use by the effector systems.

Chronic effector system activation might alter the efficiency of the homeostatic system itself. For instance, chronic sympathetic nervous stimulation of the cardiovascular system could promote cardiovascular hypertrophy, "splinting" arterial baroreceptors in stiff blood vessel walls, which in turn would contribute to systolic hypertension and the risk of heart failure, kidney failure, and stroke.

Moreover, an inappropriately large adrenomedullary response to a stressor might exaggerate the experience of emotional distress (Schachter and Singer, 1962). Exaggerated distress responses might increase the risk of worsening an independent pathologic process, such as in panic-induced angina pectoris (Mansour et al., 1998; Wilkinson et al., 1998).

SUMMARY: MERGING OF THE HOMEOSTAT THEORY WITH THE CONCEPT OF ALLOSTASIS

In summary, this chapter reflects a merging of the homeostat theory of stress with the concept of allostatic load. Until this conceptual merging, the homeostat theory did not lead easily to testable predictions about long-term effects of stress and distress, and the concept of allostatic load did not incorporate determinants of that load as sensed discrepancies between afferent information and setpoints for responding, leading to patterned alterations in activities of multiple effectors. Merging of the homeostat theory of stress with the notions of allostasis and allostatic load can provide a basis for explaining and predicting physical and psychiatric effects of acute and chronic stress.

Stress is an interdisciplinary topic, and understanding the health consequences of stress requires an integrative approach. Research and ideas about stress must move beyond considering only one effector system, such as the hypothalamo-pituitary-adrenocortical system, and only one monitored variable, such as serum glucose levels, to incorporate multiple effectors and multiple homeostatic systems that are regulated in parallel. They must also move beyond the notion of a single set of ideal values for monitored variables – homeostasis – to incorporate dynamic changes in homeostatic settings – allostasis, with effects that accumulate or emerge over time. Merging of the homeostatic definitions of stress and distress with the concept of allostasis should provide a better understanding of the roles of stress and distress in chronic diseases and also provide a conceptual basis for the further development of scientific integrative medicine.

REFERENCES

Aston-Jones, G., Rajkowski, J., Kubiak, P., Alexinsky, T. (1994). Locus coeruleus neurons in monkey are selectively activated by attended cues in a vigilance task. *J Neurosci* 14:4467–80.

Cannon, W. B. (1929a). *Bodily Changes in Pain, Hunger, Fear and Rage*. New York: D. Appleton.

Cannon, W. B. (1929b). Organization for physiological homeostasis. *Physiol Rev* 9:399–431.

Cannon, W. B. (1939). *The Wisdom of the Body*. New York: W. W. Norton.

Chrousos, G. P., Gold, P. W. (1992). The concepts of stress and stress system disorders. Overview of physical and behavioral homeostasis. *JAMA* 267:1244–52.

de Boer, S. F., Van der Gugten, J., Slangen, J. L. (1989). Plasma catecholamine and corticosterone responses to predictable and unpredictable noise stress in rats. *Physiol Behav* 45:789–95.

Frank, S. M., Higgins, M. S., Fleisher, L. A., Sitzmann, J. V., Raff, H., Breslow, M. J. (1997). Adrenergic, respiratory, and cardiovascular effects of core cooling in humans. *Am J Physiol* 272:R557–62.

Fukuhara, K., Kvetnansky, R., Cizza, G., Pacak, K., Ohara, H., Goldstein, D. S., Kopin, I. J. (1996). Interrelations between sympathoadrenal system and hypothalamo-pituitary-adrenocortical/thyroid systems in rats exposed to cold stress. *J Neuroendocrinol* 8:533–41.

Goldstein, D. S. (1995). *Stress, Catecholamines, and Cardiovascular Disease*. New York: Oxford University Press.

Goldstein, D. S. (2001). *The Autonomic Nervous System in Health and Disease*. New York: Marcel Dekker.

Goldstein, D. S., Garty, M., Bagdy, G., Szemeredi, K., Sternberg, E. M., Listwak, S., Deka-Starosta, A., Hoffman, A., Chang, P. C., Stull, R., Gold, P. W., Kopin, I. J. (1993). Role of CRH in glucopenia-induced adrenomedullary activation in rats. *J Neuroendocrinol* 5:475–86.

Kluger, J., Cody, R. J., Laragh, J. H. (1982). The contributions of sympathetic tone and the renin-angiotensin system to severe chronic congestive heart failure: response to specific inhibitors (prazosin and captopril). *Am J Cardiol* 49:1667–74.

Mansour, V. M., Wilkinson, D. J., Jennings, G. L., Schwarz, R. G., Thompson, J. M., Esler, M. D. (1998). Panic disorder: Coronary spasm as a basis for cardiac risk? *Med J Australia* 168:390–2.

McEwen, B. S. (1998a). Stress, adaptation, and disease. Allostasis and allostatic load. *Ann N Y Acad Sci* 840:33–44.

McEwen, B. S. (1998b). Stress, adaptation, and disease. Allostasis and allostatic load. *Ann N Y Acad Sci* 840:33–44.

McEwen, B. S. (2000). Allostasis and allostatic load: Implications for neuropsychopharmacology. *Neuropsychopharmacology* 22:108–24.

McEwen, B. S., Stellar, E. (1993). Stress and the individual. Mechanisms leading to disease. *Arch Intern Med* 153:2093–101.

Mosqueda-Garcia, R., Furlan, R., Fernandez-Violante, R., Desai, T., Snell, M., Jarai, Z., Ananthram, V., Robertson, R. M., Robertson, D. (1997). Sympathetic and baroreceptor reflex function in neurally mediated syncope evoked by tilt. *J Clin Invest* 99:2736–44.

Nathan, M. A., Reis, D. J. (1977). Chronic labile hypertension produced by lesions of the nucleus tractus solitarii in the cat. *Circ Res* 40:72–81.

Pacak, K., Palkovits, M., Yadid, G., Kvetnansky, R., Kopin, I. J., Goldstein, D. S. (1998). Heterogeneous neuroendocrine responses to various stressors: A test of Selye's doctrine of non-specificity. *Am J Physiol* 275:R1247–55.

Quillen, E. W. Jr., Cowley, A. W. Jr. (1983). Influence of volume changes on osmolality-vasopressin relationships in conscious dogs. *Am J Physiol* 244:H73–9.

Schachter, S., Singer, J. (1962). Cognitive, social, and physiological determinants of emotional state. *Psychol Rev* 69:379–99.

Schulkin, J., Gold, P. W., McEwen, B. S. (1998). Induction of corticotropin-releasing hormone gene expression by glucocorticoids: implication for understanding the states of fear and anxiety and allostatic load. *Psychoneuroendocrinology* 23:219–43.

Selye, H. (1974). *Stress without Distress*. New York: New American Library.

Udelsman, R., Goldstein, D. S., Loriaux, D. L., Chrousos, G. P. (1987). Catecholamine-glucocorticoid interactions during surgical stress. *J Surg Res* 43:539–45.

Wilkinson, D. J. C., Thompson, J. M., Lambert, G. W., Jennings, G. L., Schwarz, R. G., Jeffreys, D., Turner, A. G., Esler, M. D. (1998). Sympathetic activity in patients with panic disorder at rest, under laboratory mental stress, and during panic attacks. *Arch Gen Psychiatry* 55:511–20.

4 Operationalizing Allostatic Load

Burton Singer, Carol D. Ryff, and Teresa Seeman

INTRODUCTION

In elderly populations, comorbidity in the form of multiple co-occurring chronic conditions is the norm rather than the exception. For example, in the United States, 61% of women and 47% of men aged 70 to 79 report two or more chronic conditions. These figures rise to 70% of women and 53% of men aged 80 to 89 with 2+ chronic conditions (Jaur and Stoddard, 1999). No single form of comorbidity occurs with high frequency, but rather a multiplicity of diverse combinations are observed (e.g., osteoarthritis and diabetes, colon cancer and coronary heart disease, depression and osteoporosis). This diversity underscores the need for an early warning system of biomarkers that can signal early signs of dysregulation across multiple biological systems.

One response to this challenge was the introduction of the concept of allostatic load (McEwen and Stellar, 1993; McEwen, 1998; McEwen and Seeman, 1999) as a measure of the cumulative biological burden exacted on the body through attempts to adapt to life's demands. The ability to adapt successfully to challenges has been referred to as allostasis (Sterling and Eyer, 1988; McEwen, 2002; Schulkin, 2003). This notion emphasizes the biological imperative that, to survive, "an organism must vary parameters of its internal milieu and match them appropriately to environmental demands" (Sterling and Eyer, 1988). When the adaptive responses to challenge lie chronically outside of normal operating ranges, wear and tear on regulatory systems occurs, and allostatic load accumulates.

Dysregulation in one or more of the biological systems involved in allostasis is reflected in characterizations of the multiple pathways to pathophysiology and a diversity of chronic conditions. Thus, any operationalization of the concept of allostatic load should signal pending onset of diverse kinds of disability and disease. In addition, allostatic load, measured at

113

any age or via longitudinal assessments, should represent the biological signature of cumulative antecedent challenges. It should serve as an early warning indicator of downstream comorbidity and also reflect co-occurring antecedent risk factors. The purposes of this chapter are to (1) review the extant operationalizations of allostatic load and their rationale, (2) relate indicators of allostatic load to the known mechanisms of the underlying biological systems that they purport to reflect, and (3) describe promising research directions that would yield more nuanced operationalizations of allostatic load than heretofore.

In the next section, we review the indicators of allostatic load and associated scoring schemes for this concept that are currently in use. We also present evidence that allostatic load reflects antecedent psychosocial challenges, particularly in the realm of social relationships. We review the predictive capabilities of these scoring schemes for four categories of downstream health consequences: mortality, incident cardiovascular disease, and declines in physical and cognitive functioning. In the section "Additional Measures and Study Designs," we discuss the augmentation of current panels of biomarkers that would substantially refine our ability to monitor allostatic load. In "Indicators of Allostatic Load and Systemic Dynamics," we describe the linkage between current indicators of allostatic load and the mechanisms of operation of several underlying biological systems. Particular attention is given to metabolic pathways and multiple routes to insulin resistance. "New Possibilities for Measuring Allostatic Load" introduces metabolic profiling based on ^{1}H-NMR (nuclear magnetic resonance) spectroscopy, a new technology that could substantially improve the measurement of allostatic load and provide much closer linkage to fine-grained representations of metabolic networks. The profiles would represent metabolic system contributions to allostatic load. Analogous profiling for the immune, neuroendocrine, and autonomic nervous systems, for example, would represent the basis for much more fine-grained scoring of allostatic load than can be achieved with extant technology. Some of these possibilities are described in the final section of this chapter.

APPROACHES TO SCORING ALLOSTATIC LOAD

Three analytical strategies have been used to date to calculate allostatic load scores for individuals: (1) summation of the number of biomarkers with levels at or above (respectively below) "elevated-risk" levels; (2) weighted summation of standardized biomarker scores, with weights derived from canonical correlation analyses; and (3) recursive partitioning classification of persons into empirically determined high, intermediate, or low categories of allostatic load. We describe and evaluate each of these scoring schemes, using the restricted set of biomarkers available in the MacArthur Study of Successful Aging. We also indicate the potential, or lack thereof, of these scoring schemes for quantifying the concept of allostatic load with more comprehensive sets of biomarkers.

Elevated-Risk Zones

Because high scores on any measure of allostatic load are designed to indicate that an individual is at increased risk for a diversity of deleterious downstream outcomes, it is useful to examine the distribution of biomarker measurements in a general population and identify the extreme quartiles (or quintiles) as candidate cut points for defining elevated-risk zones. It is also important to emphasize that cut points defining the boundary of an elevated-risk zone are not intended to be levels that would be regarded as clinically significant for disease diagnosis. Thus, there is a tension between setting such boundaries at levels that correspond to an early warning of negative later-life consequences but not so extreme as to be regarded by physicians as clinically significant levels. Our use of upper (respectively lower) quartile cut points is motivated by the objective of empirically identifying zones that must contain persons at increased risk, while at the same time avoiding clinically significant cutoff levels. We used 10 biomarkers ascertained in the MacArthur Study of Successful Aging, a longitudinal study of relatively high-functioning men and women aged 70–9 years at the time of baseline data collection in 1988–9 (Berkman et al., 1993; Seeman et al., 1997a). The biomarkers, the systems they reflect, and their risk-zone boundaries – corresponding to upper (respectively lower) quartiles of the distribution of measured values – are listed in Table 4.1, Six of the biomarkers are secondary mediators reflecting primarily components of the metabolic syndrome (syndrome X). The four non–syndrome X markers are primary mediators that reflect functioning of the hypothalmic-pituitary-adrenocortical (HPA) axis and the sympathetic nervous system. Using resting levels of all biomarkers is predicated on a hypothesis that their values reflect cumulative wear and tear in the associated systems.

With these criteria at hand, we define allostatic load score for an individual to be the number of biomarkers in the list in Table 4.1 for which the individual satisfies the stated inequality. Contributions from different biological systems are all weighted equally in this scoring of allostatic load. Higher scores are interpreted as reflecting more extensive wear and tear on the systems being directly measured and possibly on other systems (e.g., the immune system) that communicate with those we are measuring but for which there was no assessment in the MacArthur Aging Study.

Table 4.1: Risk-zone and system designation for individual biomarkers

Risk zone	System
Highest quartile	
Systolic blood pressure (>= 148 mm Hg)	Metabolic
Diastolic blood pressure (>= 83 mm Hg)	Metabolic
Waist–hip ratio (>= 0.94)	Metabolic
Ratio total cholesterol/HDL (>= 5.9)	Metabolic
Glycosylated hemoglobin (>= 7.1%)	Metabolic
Urinary cortisol (>= 25.7 ug/g creatinine)	HPA axis
Urinary norepinephrine (>= 48 ug/g creatinine)	Sympathetic nervous system
Urinary epinephrine (>= 5 ug/g creatinine)	Sympathetic nervous system
Lowest quartile	
HDL cholesterol (<= 37 mg/dl)	Metabolic
DHEA-S (<= 350 ng/ml)	HPA axis

Note: HDL = high-density lipoprotein; DHEA-S = dehydroepiandrosterone sulfate.

Support for this hypothesis derives from two complementary bodies of evidence. First, in elderly populations (age 70+), increasing allostatic load score is associated with increased mortality, incident cardiovascular disease (CVD), and decline in cognitive and physical functioning 7 years beyond baseline assessments (Seeman et al., 2001). In addition, elevated levels of individual biomarkers are associated with multiple chronic conditions. For example, hypercortisolism is associated with co-occurrence of depression and osteoporosis. Indeed, depression is a major, and largely unrecognized, risk factor for osteoporosis (Cizza et al., 2001). Second, persons with cumulative negative social relationship histories have higher allostatic load scores in later life than persons with cumulative positive social relationship histories (Ryff et al., 2001). In the short term (i.e., in response to acute challenges), negative relationships are associated with increased HPA axis reactivity, immune suppression, and increased sympathetic nervous system (SNS) activity (Kiecolt-Glaser et al., 1994, 1998). However, persistent negative relationship challenges necessitate repeated adaptation of these systems, and this is what, over a long period of time, gives rise to the wear and tear that results in the biomarkers of Table 4.1 attaining elevated-risk-zone levels and, ultimately, negative health consequences (Kiecolt-Glaser, 1999; Kiecolt-Glaser et al., 2002). In an interesting study of men in the Normative Aging Study, high levels of hostility were associated with elevated allostatic load based on a similar panel of biomarkers to those in Table 4.1 (Kubzansky et al., 2000).

We illustrate these associations using relationship and biomarker assessments on a demographically representative subsample ($n = 106$) of respondents in the Wisconsin Longitudinal Study (WLS). The WLS is a long-term survey of a random sample of 10,317 men and women who graduated from Wisconsin high schools in 1957. Survey data were collected from the original respondents in 1957, 1975, and 1992–3. Biomarkers, together with an additional round of survey information and supplementary questionnaires on relationships, were ascertained on the previously mentioned subsample in 1997–8. Childhood measures, a parental bonding scale (Parker et al., 1979), assessed the extent to which an individual had caring, supportive, and affectionate relationships with each of his or her parents. There were separate Father Caring and Mother Caring scales. The experience of uncaring and even abusive interactions with one or both parents is anticipated to be a defining feature of a negative social relationship pathway. It is also hypothesized to contribute to wear and tear on multiple biological systems, the effects of which would become manifest in later life.

Regarding midlife relationships, four aspects of connection to a spouse or significant other are hypothesized to contribute to cumulative relationship profiles that should, in turn, be associated with later-life biological indicators of wear and tear on the body. These are emotional, sexual, intellectual, and recreational modes of intimacy. They were assessed using the PAIR (Personal Assessment of Intimacy in Relationships) Inventory (Schaefer and Olson, 1981). The emotional and sexual subscales were included because of their focus on the most intimate forms of connection between two people. The intellectual and recreational subscales emphasize mutually enjoyed experience, companionship, and the scope of shared communication. The PAIR seeks to identify the degree to which each partner feels intimate in each of the four relational areas.

Formal specification of relationship pathways required the following steps:

(1) We cross-classified individuals as positive (+) or negative (−) on the Father Caring and Mother Caring components of the Parental Bonding Scale according to whether their score on the respective scale is above or below the median. Each person thus would have a (Father Caring, Mother Caring) pair of valences which can be (−,−), (−,+), (+,−), or (+,+).

(2) Scores on the emotional (E) and sexual (S) subscales of the PAIR inventory were combined into an E + S score, representing probes of the most personal and intimate aspects of spousal ties. Analogously, the intellectual (I) and recreational (R) scores were combined into an I + R score. Then persons were scored as positive (+) or negative (−) on each of the combined scales according as they were above or below the median on the respective

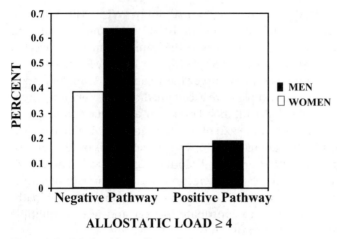

Figure 4.1: Relationship profiles and allostatic load.

combination. Each individual had a pair of (E + S, I + R) valences, which could also be any one of the four options (−,−), (−,+), (+,−), or (+,+).

(3) A person was defined to be on the **negative relational pathway** if she or he scored (−,−) on the (Father Caring, Mother Caring) pair or on the E + S AND I + R pair in adulthood (or both). An individual was defined to be on a **positive relational pathway** if she or he had at least one + on the Father Caring/Mother Caring pair AND at least one + on E + S or I + R. Thus, the positive path requires some positive relational experience with one or both parents in childhood and at least one of the combined forms of intimacy in adulthood. This pathway again underscores the cumulative nature of positive emotional experience with significant others in childhood and adulthood.

Figure 4.1 shows the associations between elevated-risk-zone allostatic load scores at ages 59–60 and antecedent cumulative relational histories in the WLS.

Complementing the associations shown in Figure 4.1 are the relationships between allostatic load scores at given ages and downstream consequences such as mortality, incident cardiovascular disease, and decline in physical and cognitive functioning. Figure 4.2 shows the association between baseline allostatic load scores and four categories of health outcomes on the same population (MacArthur Aging Study) 7 years later.

A comparison of these associations with corresponding predictions from the Wisconsin Longitudinal Study (WLS) cannot be carried out at the

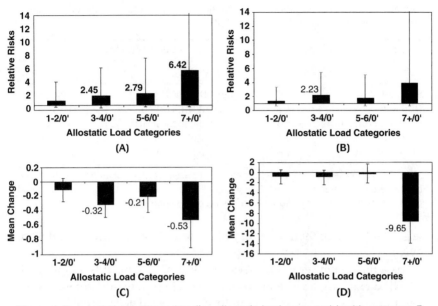

Figure 4.2: Association between baseline allostatic load score and health outcomes 7 years later. **(A)** Mortality risks by allostatic load. **(B)** Incident cardiovascular disease risks by allostatic load. **(C)** Changes in physical functioning and allostatic load. **(D)** Changes in cognitive functioning and allostatic load.

present time because baseline biomarker data collection on that population was only completed in 1998 (4 years ago at the time of this writing). Unfortunately, there is, at the present time, a paucity of longitudinal follow-up data on nearly all human populations for which this set of biomarker data is available. This situation will improve over the next several years as follow-up data from the Taiwan Aging Study (Weinstein and Willis, 2001), the WLS (Sewall et al., 2001), and the Wisconsin Community Relocation Study (Smider et al., 1996) become available.

The rich antecedent life history data in the WLS does not have a counterpart in the MacArthur Aging Study. Thus, in showing the chronological progression from early life social relationships to their counterparts in adulthood and biomarker assessmentat ages 59–60, followed by biomarker assessments at ages 70–9 and then downstream health outcomes 7 years later, we are using the WLS and the MacArthur Aging populations as a synthetic cohort. The disparities in measures across studies as exhibited here in prospective follow-up data post biomarker assessments and in antecedent life history data, are presently pervasive. Some evidence that these data gaps will be closing in the near future can be seen in several ongoing investigations

Table 4.2: Co-occurring biomarkers in elevated-risk zones of the Wisconsin longitudinal study (Ages 59–60; $n = 106$)

Most prevalent pairs		Most prevalent triples	
Pair	Frequency	Triple	Frequency
SBP and DBP	19/106	SBP, DBP, WHR	12/106
DBP and WHR	19/106	DBP, WHR, cortisol	10/106
DBP and cortisol	18/106	Total cholesterol/HDL, HDL, cortisol	9/106
WHR and cortisol	17/106	DBP, WHR, HDL	8/106

Note: Of 106 cases, 72 have 2+ biomarkers in elevated-risk zones and 44 have 3+ biomarkers in elevated-risk zones. SBP = systolic blood pressure; DBP = diastolic blood pressure; WHR = waist-hip ratio; HDL = high-density lipoprotein.

(e.g., Hertzman, 1999; Power and Mathews, 1998; Wadsworth and Kuh, 1997).

As a final point regarding elevated-risk-zone scoring, it is useful to ask which pairs and triples of biomarkers co-occur in elevated-risk zones. Our interest in multisystem contributions to allostatic load derives from epidemiological evidence that comorbidity in the form of multiple co-occurring chronic conditions is the norm rather than the exception in elderly populations (Jaur and Stoddard, 1999). Table 4.2 shows these biomarker frequencies for the age 59–60 population in the WLS.

Aside from the most prevalent pairs and triples, a key point to observe is the great diversity of combinations of biomarkers that co-occur in elevated-risk zones (i.e., 72/106 persons have 2+ biomarkers in elevated-risk zones). This diversity has a counterpart in comorbidity data in that no single form of comorbidity occurs with high frequency. Rather, a multiplicity of diverse conditions are observed (e.g., osteoarthritis and diabetes, colon cancer and coronary heart disease, depression, and osteoporosis). This is precisely why a scoring scheme for allostatic load must be able to reflect possible dysregulation across multiple biological systems.

Canonical Weights
An obvious limitation of the elevated-risk zone scoring scheme is the fact that it does not take full advantage of the measured values of each biomarker. Instead it uses a cutoff (quartiles) for specifying the boundary of a risk zone for each biomarker. In addition, there is no clear argument supporting the contention that all biomarkers, even among the limited set we have used, should be equally weighted for prediction of diverse forms of downstream outcomes. The surprising aspect of the predictions shown

in Figure 4.2 is that a single allostatic-load-scoring algorithm is reasonably effective for identifying elevated risk of such varied outcomes as mortality, incident cardiovascular disease, and decline in physical and cognitive functioning. The elevated-risk-zone scoring scheme for allostatic load is also reflective of antecedent challenges. This is exemplified by Figure 4.1, showing a substantially higher proportion of individuals on a negative relationship pathway at high allostatic load in comparison to the percentage at high allostatic load on a positive relationship pathway.

To move beyond the elevated-risk-zone scoring algorithm, we carried out a canonical correlation analysis (Karlamangla et al., 2002) using changes in physical and cognitive functioning measures as health outcomes and the same 10 biomarkers indicated earlier as predictor variables. The objective of such an analysis is to determine which linear combination of biomarkers is maximally correlated with which linear combination of functional change scores. The maximal correlation achievable between such linear combinations is called the canonical correlation of the set of biomarkers with the given set of functional change-scores. The weights in the best linear combinations for predicting decline in physical and cognitive functioning are called canonical weights. The linear combination of biomarker scores, using canonical weights, is then used as a scoring scheme for allostatic load. To ensure that all biomarkers are on a common scale, the canonical weights are applied to standardized biomarker scores (i.e the standardized score, z_i, for the ith individual on the kth biomarker is defined as

$$z_i^{(k)} = [y_i^{(k)} - (\text{mean of } y^{(k)} \text{ values})]/(\text{standard deviation of } y^{(k)} \text{ values}),$$

where $y_i^{(k)}$ = measured value of the kth biomarker for individual i. Then allostatic load score is defined as

$$\text{AL} = w_1 z^{(1)} + w_2 z^{(2)} + \cdots + w_{10} z^{(10)},$$

where w_k is the canonical weight associated with the kth biomarker, and $z^{(k)}$ is the standardized score for the kth biomarker. For the ith individual's standardized numerical score on the kth biomarker, the variable $z^{(k)}$ takes on the value $z_i^{(k)}$. Second, the minus sign on the weight for diastolic blood pressure indicates that low values, reflective of hypotension, are what place this elderly (aged 70–9) population at risk. Low diastolic blood pressure (DBP) would never be picked up by the elevated-risk-zone scoring scheme described previously because it only focuses on high DBP as an elevated-risk condition. Table 4.3 lists the canonical weights and the 25th and 75th percentiles of 200 bootstrap estimates of the weights based on data from the MacArthur Aging Study (see Karlamangla et al., 2002, for full technical details).

Table 4.3: Canonical weights for allostatic load scoring

Biomarkers	Weights	(25th, 75th) percentiles of 200 bootstrap estimates[a]
Urinary epinephrine	0.60	[0.49, 0.71]
Urinary norepinephrine	0.12	[0.00, 0.24]
Urinary cortisol	0.195	[0.07, 0.32]
Dehydroepiandrosterone-sulfate	0.175	[−0.35, 0.00]
Waist-hip ratio	0.185	[0.06, 0.31]
Glycosylated hemoglobin	0.295	[0.16, 0.43]
HDL cholesterol	—	—
Total cholesterol/HDL ratio	—	—
Systolic blood pressure	0.18	[0.02, 0.34]
Diastolic blood pressure	−0.355	[−0.48, −0.23]

[a] Technical details about bootstrap estimates are given in Karlamangla et al., 2002. HDL = high-density lipo-protein.

Note: The canonical correlation between the biomarkers and functional decline measures is 0.505 with bootstrap interval (0.48, 0.53).

Several points should be emphasized regarding this system of weights. First, at least for purposes of predicting declines in physical and cognitive functioning, the cholesterol measures do not contribute to allostatic load, as in the elevated-risk-zone scoring of the previous section. Second, the canonical correlation analysis suggests that moderate and higher levels of DBP are associated with better downstream physical and cognitive functioning.

A useful test of this operationalization of allostatic load is to ask whether it can be used to distinguish between cumulative social relationship histories that have different degrees of adversity relative to advantage from childhood thru adulthood. Figure 4.3, based on the social relationship histories in the WLS, shows this kind of discrimination.

It is interesting to observe that if the canonical correlation analysis is carried out using only the four primary mediators – epinephrine, norepinephrine, cortisol, and dehydroepiandrosterone-sulfate (DHEA-S) – and the same measures of decline in physical and cognitive functioning as in the previous analysis, the overall canonical correlation is 0.43 with bootstrap interval (0.40, 0.46) (Karlamangla et al., 2002). This suggests that these primary mediators alone serve as a good early warning system for subsequent functional decline. An important next step in the development of this kind of scoring scheme for biomarkers would be to carry out canonical correlation analyses with mortality and incident cardiovascular disease as outcomes. This exercise would reveal which biomarkers are of greatest importance for predicting which health outcomes. The elevated-risk-zone

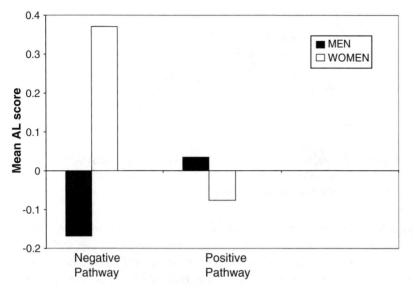

Figure 4.3: Canonical allostatic load (AL) score and relationship pathways.

algorithm is too crude for such distinctions because it automatically in-cludes all biomarkers with equal weight.

Recursive Partitioning

A more sensitive methodology for scoring allostatic load than either of the strategies we have described is recursive partitioning (RP; Breiman et al., 1984; Zhang and Singer, 1999). The basic idea of RP is to begin with a set of candidate predictor variables (e.g., the biomarkers used in the previous discussion) and a well-defined set of outcomes (e.g., mortality, decline in physical and cognitive functioning, etc.). The first step in developing an allostatic load scoring scheme is for the RP algorithms to search among the predictor variables and cut points along each of them to identify the best single predictor variable – with accompanying cut point – such that indi-viduals on one side of the cut point are predicted to end up in one of the outcome categories and individuals on the other side of it are predicted to be in an alternative outcome category. This literally *partitions* the original population into two groups: those scoring above the cut point on the desig-nated variable and those scoring below it. Figure 4.4 shows the result of this step using the MacArthur Aging Study population with the 10 biomarkers listed in Table 3.1 and the outcome categories being dead or alive at 7 years beyond baseline biomarker data collection.

DBP was identified as the best single predictor variable, but also that the high-risk condition is, DBP < 60 mm Hg, that is, a condition associated

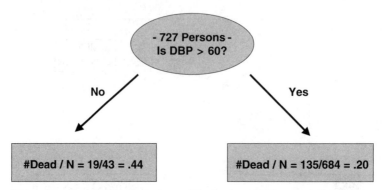

Figure 4.4: First stage in recursive partitioning tree. DBP = diastolic blood pressure.

with hypotension. Thus, low DBP is a high-risk condition for decline in physical and cognitive functioning, as demonstrated earlier in the canonical correlation analysis, and also for mortality within 7 years, as indicated here. We reiterate that this condition is masked in the original elevated-risk-zone methodology, because that scoring scheme focused entirely on high blood pressure as a risky condition. In addition, there was no systematic strategy associated with the elevated-risk-zone methodology that could have clearly identified a useful cut point for defining "low" DBP. The designation of such cut points as part of the RP algorithms represents a major advantage of this technology over either of the techniques described previously.

After the first partitioning, the next step in developing an allostatic load scoring scheme is to partition each of the subpopulations separately (i.e., those with DBP < 60 mm Hg and those with DBP > 60 mm Hg) by identifying which predictor variable and with what accompanying cut point would lead to the best mortality prediction for the given subpopulation. There is also an option of retaining one or both of the terminal nodes identified in the initial partitioning and thereby prohibiting further refinement of the prediction rules. In the present application, the subpopulation with DBP < 60 mm Hg was placed in a terminal node with a high mortality rate (44%). The subpopulation with DBP > 60 mm Hg was further partitioned into those with DBP > 80 mm Hg and those with DBP < 80 mm Hg. Those with DBP > 80 mm Hg were placed in a terminal node with intermediate mortality rate (27%), and those with DBP < 80 mm Hg were further partitioned into subpopulations with high-density lipoprotein (HDL) cholesterol > 36 mg/dL and those with HDL cholesterol < 36 mg/dL. Figure 4.5 shows this partitioning along with subsequent assignment of the HDL partitioned populations to terminal nodes.

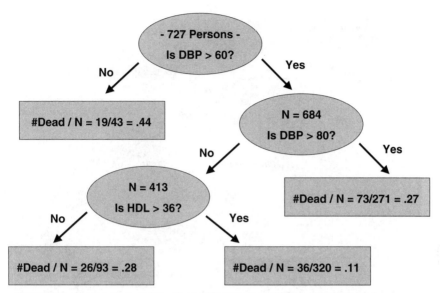

Figure 4.5: Second stage in recursive partitioning tree. DBP = diastolic blood pressure.

If we were stop the possible partitioning at this point, we would have identified three allostatic load categories, described by the following Boolean statements:

High allostatic load: {DBP < 60 mm Hg}
                                            [Mortality rate within 7 years = 44%]
Intermediate allostatic load: {(60 ≤ DBP ≤ 80) AND (HDL ≤ 36)}
or {DBP > 80}   [Mortality rate within 7 years = 27%]
Low allostatic load: {(60 ≤ DBP ≤ 80) AND (HDL > 36)}
                                            [Mortality rate within 7 years = 11%]

Using only two of the original set of biomarkers, diastolic blood pressure and HDL cholesterol, already identifies three categories of allostatic load and distinctly different associated mortality rates. However, if we continue the partitioning, without overfitting the data, we obtain the following more elaborate prediction tree with corresponding Boolean statements defining High, Intermediate, and Low levels of allostatic load (Fig. 4.6). A Boolean statement summary of membership in terminal nodes according to a classification of high, intermediate, and low levels of allostatic load load is given in Figure 4.7.

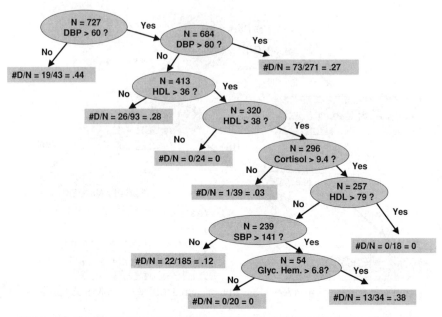

Figure 4.6: Final recursive partitioning tree. DBP = diastolic blood pressure; HDL = high-density lipoprotein; SBP = systolic blood pressure; Gly. Hem. = glycosylated hemoglobin.

High AL – [DBP < = 60] OR {[60 < DBP < = 80] AND [cort > 9.4] AND [SBP > 141]

AND [Glyc. Hem. > 6.1]}

(Proportion Dead within 7 years = .42)

Intermediate AL – {[60 < DBP < = 80] AND [HDL < = 36]} OR [DBP > 80]

(Proportion Dead within 7 years = .27)

Low AL -- [60 < DBP < = 80] AND {[36 < HDL < = 38]

OR ([HDL > 38] AND [cort < = 9.4]

OR ([38 < HDL < = 79] AND [cort > 9.4] AND

[SBP > 141] AND [Glyc. Hem. < = 6.8])

OR ([cort > 9.4] AND [HDL > 79])

OR ([38 < HDL < = 79] AND [SBP < = 141]

(Proportion Dead within 7 years = .08)

Figure 4.7: Allostatic load (AL) categories specified by Boolean statements. Abbreviations as in Figure 4.6.

Table 4.4: Proportions in allostatic load (AL) categories × Age

Category	Age 70–9 (Mac Aging; $N = 727$)	Age 59–60 (WLS; $N = 106$)
Low AL	.394	.415
Intermediate AL	.500	.538
High AL	.106	.047

Note: Mac Aging = MacArthur Study of Successful Aging; WLS = Wisconsin Longitudinal Study.

This specification of allostatic load categories uses only 5 of the original 10 biomarkers, although all of them were a priori candidates to enter into the RP-based rules for predicting downstream mortality. We would antici-pate that different combinations of biomarkers with correspondingly differ-ent Boolean statements would be associated with declines in physical and cognitive functioning or incident cardiovascular disease. The development of such RP-derived categories lies in the future.

The RP scoring rule shows an interesting age dependence (Table 4.4). An analogous age dependence showing higher fractions of older populations at higher allostatic load has also been demonstrated using the elevated-risk-score algorithm (Seeman et al., 2002).

In the aged 70–79 population, 10.6% of the people are in the high allo-static load category in comparison with only 4.7% in that category among the aged 59–60 population. A slightly higher percentage (41.5%) of the younger population is in the low allostatic load category relative to the corresponding percentage (39.4%) in the older population.

The RP allostatic load categories in Figure 4.7 also reflect antecedent life history profiles. To illustrate this point, life history profiles were constructed using data from the WLS. Individual lives were represented by information from five domains: (1) family background and early life experiences; (2) occupational experiences from first job through employment status at age 59–60; (3) adult family life; (4) mental health, psychological outlooks, and beliefs in adulthood; and (5) physical health in adulthood. Persons who had predominantly negative experiences in three of the first four domains were defined to have essentially negative profiles. Persons who had very positive childhood experiences (domain 1) but who had decidedly nega-tive experiences on at least two of domains (2), (3), and (4) were defined to have poor adult profiles subsequent to positive childhood profiles. For details about the life history profile construction, see Low and colleagues (2002). These profiles are more nuanced than the relationship pathways discussed earlier in that they use information from multiple life domains,

Table 4.5: Proportions in allostatic load (AL) categories × life history profile

	Allostatic load category		
Profile	Low AL	Intermediate AL	High AL
Essentially negative	.30	.70	0
Positive childhood but			
negative adulthood	.60	.30	.10

relationships being one component. We hypothesize that persons with persistently negative experiential profiles should be more concentrated in the intermediate-to-high allostatic load categories at age 59–60 relative to persons who had very positive childhoods despite their negative adulthood experiences. Table 4.5 shows the distribution of persons across allostatic load categories as a function of membership in either of these life history groups.

Table 4.5 suggests that there is a long-term protective effect from a positive childhood, even in the face of considerable adversity in adulthood. The protective effect manifests itself in the much larger percentage of people in the low allostatic load category (60%) relative to the corresponding percentage (30%) for those who have persistently negative profiles. However, there is a vulnerable subset (10%) in the group in which individuals start out positive and then have negative adulthoods in that they end up in the high allostatic load category at age 59–60.

Gender Differences

Continuing the discussion with the same set of biomarkers, it is important to observe that there are gender differences in the distributions of each of these measures. This raises the question of whether allostatic load scoring schemes should be set up on a gender-specific basis, or whether there are gender differences in the distribution of allostatic load using the scoring methods described above. Across several U.S. populations, mean levels of the primary mediators, cortisol, epinephrine, and norepinephrine, are higher for women than for men, and DHEA-S mean levels are lower for women relative to those for men (Goldman et al., 2004). For the six measures associated with metabolic syndrome (syndrome X), men have higher mean levels than women (Goldman et al., 2004). This would suggest that the primary mediators dominate women's contributions to allostatic load scores by any of the methods described earlier, while metabolic syndrome biomarkers dominate men's contributions to allostatic load scores. This is

Table 4.6: Distribution of allostatic load (AL) score × gender

	AL – Score		
	0–1	2	3+
Men ($n = 57$)	.26	.23	.51
Women ($n = 49$)	.39	.30	.31

indeed the case for the WLS and MacArthur Aging populations; however, it would be useful to have comparable data for a much more extensive set of populations before we could make statements of general principle.

Using the elevated-risk-zone scoring scheme for allostatic load, Table 4.6 shows the distribution of allostatic load scores by gender, ascertained in the WLS ($N = 106$).

The men have a larger percentage (51%) with 3+ biomarkers in elevated-risk zones than women (31%). However, among women with allostatic load score of 3+, more have primary mediators in the elevated-risk zone than do men with allostatic load scores of 3+. The contributions to allostatic load scores for men are more heavily weighted toward biomarkers associated with the metabolic syndrome (syndrome X).

A limitation of the allostatic load scoring scheme used for the calculations in Table 4.6 is that the cut points defining the boundaries of the elevated-risk zones are quartiles for a general population with comparable numbers of men and women. This facilitates cross gender comparisons with allostatic load calculated the same way for everyone. It does not allow for the fact that risk-zone boundaries can be quite different for men and women, however. For example, the lower quartile of the waist-hip ratio distribution for men far exceeds the upper quartile for women across several U.S. populations (Goldman et al., 2004). This suggests that it would be useful to respecify allostatic load scoring by each of the methods we have described so that they are gender specific.

In parallel with such a revision of allostatic load scoring should be gender-specific construction of life history profiles. There is a substantial literature (e.g., Kudeilka et al., 2000) documenting gender differences in subjective aspects of human stress response and its physiological correlates. The focus of this literature is short-term response – reactivity studies – to specific challenges. In contrast to this local-in-time focus, allostatic load is designed to reflect longer-term cumulative effects of multiple challenges (i.e., wear and tear on multiple biological systems). Thus, we hypothesize that life history profiles that incorporate information about multiple kinds of challenges, in which gender differences in reactivity have also been demonstrated, should be associated with gender-specific consequences in allostatic load scores. For example, among persons aged 30–65, it has been shown (Orth-Gomer et al., 2000) that elevated cortisol, norepinephrine, and epinephrine are consequences of marital stress in women but not in men. In addition, women in this age range have a worse prognosis than men following acute myocardial infarctions. Among working women, recurrent coronary events were associated with marital stress and not work stress. Work stress, on the other hand, has been associated more with increased DBP and systolic blood pressure (SBP) and coronary heart disease in men relative to women. These findings indicate that different experiences should be regarded as "adverse" on pathway specifications for women relative to men. They also indicate that different biomarkers contribute to allostatic load for men and women and that they do so for different kinds of challenge. Thorough exploration of this general topic lies in the future. It represents an important aspect of the research agenda associated with the concept of allostatic load.

ADDITIONAL MEASURES AND STUDY DESIGNS

The previous section focused on operationalizations of allostatic load in human populations, in which ten biomarkers that happened to be a convenience sample in the MacArthur Aging Study were the basis for allostatic load scoring. In addition, empirical work to date has used allostatic load scores based on assessments at a single point in time. We view this development as simply the initial stage in a research program with two principal objectives: (1) refinement of allostatic load scoring using biomarkers that reflect possible dysregulation across a more diverse set of biological systems than those represented in the panel in Table 4.1 and (2) characterization of allostatic load histories over the life course and their association with a taxonomy of psychosocial life history profiles.

Selective Addition of Biomarkers

Relative to our first objective it is important to incorporate immune measures that reflect general levels of proinflammatory activity in an individual. A multiplicity of cytokines could be used for this purpose, but there would also be the option of simultaneously assessing multiple arms of the immune system. The many possible options for assessing immune function highlight a basic tension involved in operationalizing allostatic load. In particular, characterization of the dynamics of the immune system – and the multiple ways in which it could malfunction – suggest that a large battery of immune measures be incorporated in an allostatic load scoring scheme.

On the other hand, because our larger objective is to use biological indicators that reflect possible malfunction across multiple systems (e.g., HPA axis, sympathetic and parasympathetic nervous system, immune system), it is important to have a limited set of biomarkers in place for each system, but use those that are *sensitive to broadly based dysregulation* while not being overly burdensome when employed in large population studies.

As a case in point consider the incorporation of IL-6, meaning the study of the protein and determination of levels of the soluble receptor sIL-6r, as in a panel of biomarkers to be integrated in allostatic load scoring scheme. A rationale for this choice would be that IL-6 is often associated with two other inflammatory cytokines, interleukin-1 (IL-1) and tumor necrosis factor. Furthermore, elevated levels of IL-6 are found in the context of trauma, infection, fever, and stressful experience. The central point is that it reflects multiple forms of immune system activation. In addition, multiple epidemiological studies have shown that IL-6 levels tend to increase with increasing age, and that they are especially high in elderly persons with disability and overall poor health. In the context of mental illness, many studies have shown that depressed persons have elevated levels of IL-6 and sIL-6r. Thus, elevated levels of IL-6 reflect diverse forms of disability which, in turn, can be consequences of diverse types of comorbid conditions. The lack of disease specificity is, from our perspective, a strength rather than a weakness. Allostatic load scoring emphasizes biomarkers that are predictive of impairment in broad categories of functioning.

Our discussion thus far has focused entirely on resting levels of biomarkers. The working hypothesis has been that excessively elevated (or low) values reflect cumulative wear and tear that has taken place over substantial periods of time prior to the assessments used in allostatic load scoring. Challenge studies and ambulatory measurements represent useful sources of data to refine operationalizations of allostatic load. In the case of challenge studies, it would be necessary to identify patterns of reactivity that reflect dysregulation in one or more biological systems. In addition, such patterns of reactivity would, at a minimum, have to be associated with downstream chronic conditions, disability, and mortality. It would also be necessary to demonstrate that they reflect particular kinds of psychosocial life history profiles. Such analyses would serve to establish particular reactivity patterns (i.e., information based on local-in-time assessments) as related to much longer-in-time downstream consequences and antecedent histories.

To illustrate this idea, consider cortisol response to a driving simulation challenge (Seeman et al., 1995.) and to sets of cognitive challenges (Miller and Lachman, 2002). In the former case, there are three categories of response curves: (1) a rapid rise in cortisol accompanying the challenge and a return to baseline levels within 30 minutes following the challenge; (2) a rapid rise in cortisol accompanying the challenge, but persistent elevated levels for 2+ hours following it; and (3) no response at all. One would hypothesize that a category (2) response pattern is associated with dysregulation in the HPA axis and that this condition should be associated with later-life incident cardiovascular disease of some kind (Sapolsky, 1993, 1996). Furthermore, persons with a category (2) response pattern would be hypothesized to have antecedent psychosocial profiles with significant adverse conditions.

Empirical tests of these longer-in-time associations lie in the future. They would be essential for incorporating response pattern (2), for example, into a Boolean statement that described high allostatic load. In particular, one could conceive of an augmentation to the Boolean statement for high or intermediate allostatic load score in Figure 4.7 by the phrase, category (2) cortisol response to driving challenge. Analogously, one would anticipate that category (1) cortisol responses would be part of a low allostatic load Boolean statement. Nonresponse to a single challenge could either imply blunted sensitivity – a negative condition – or simply that high-speed driving is not perceived by the individual to be a challenge. These alternatives cannot be differentiated within a single challenge study. Thus multiple challenge studies could be introduced as part of a protocol for assessing allostatic load. The open questions at the present time are these: which challenges would be most useful; and which biomarkers should be assessed to ascertain patterns of reactivity?

Finally, we mention that an expanded panel of biomarkers should include multiple salivary cortisol assessments throughout the day. This would facilitate our ability to associate high trough levels of glucocorticoids with the development of the metabolic syndrome. The elevation of trough glucocorticoids, consequential to mild but chronic stressors, biases organisms toward storage of calories as fat. This phenomenon is not discernible from mean levels of cortisol, because elevated troughs are usually accompanied by a reduction in peak concentrations of glucocorticoids, leaving daily mean values invariant (Dallman et al., 2000).

Longitudinal Studies

A major limitation of the human population-based studies in which multiple biomarkers have been assessed is that even when they are rich in longitudinal data at the psychosocial level (Wadsworth and Kuh, 1997; Power and Mathews, 1998; Sewell et al., 2002), they have only a single round of biological assessments. Conversely, longitudinal studies such as the Framingham

Study (Dawber, 1980; Kannel, 2000) that are rich in multiple biomarker assessments over time tend to have limited psychosocial data. This has severely limited our capacity to study the dynamics of allostatic load in human populations in relation to the process of cumulative challenge that is taking place at the psychosocial level. Nevertheless, there is currently considerable interest in facilitating population-based longitudinal studies that will simultaneously ascertain multiple biomarkers and detailed psychosocial data (Finch et al., 2001). Because blood, urine, and saliva are the easiest and most cost-effective biofluids to collect in large population samples, the challenge for improving operationalizations of allostatic load resides in using technologies that will facilitate the assessment of nuanced biological characterizations of responses to challenges and the experience of living in a diverse range of environments.

INDICATORS OF ALLOSTATIC LOAD AND SYSTEM DYNAMICS

A criticism of allostatic load measurement to date is that the biomarkers and the accompanying scoring schemes have not been linked to the mechanisms of operation of the underlying systems that they purport to represent. In this section, we present a mechanistic rationale for the inclusion of the measures in Table 4.1 in operationalizations of allostatic load.

There are several levels of description at which mechanism and system dynamics can be discussed. In the context of metabolism, a useful place to start is the metabolic pathways chart (see http://biol.net/Pathways.htm). Focusing on primary metabolism, one considers those chemical reactions that transduce energy; synthesize necessary monomers, cofactors, and other small molecules; and assemble the molecules of cellular functions (i.e., proteins, nucleic acids, carbohydrates, and lipids). Then, shifting to secondary metabolism, there are a vast number of reactions that involve metabolites such as cholesterol and processes such as the excretion of nitrogen. Directly linked to six of the allostatic load biomarkers in Table 4.1 (i.e., those listed as associated with metabolic function) is insulin, a small protein that plays a central role in the control of intermediary metabolism. Insulin has major effects on both carbohydrate and lipid metabolism and significantly influences protein and mineral metabolism. Metabolic pathways involving insulin and its dynamics – described at the level of chemical reactions – are incompletely known, despite the vast and intricate extant metabolic pathway charts. These knowledge gaps define the limits to which we can delineate the mechanisms of insulin action.

Insulin secretion and its role in regulating intermediary metabolism contributes to allostatic load through the phenomenon of insulin resistance. Insulin resistance occurs when the normal amount of insulin secreted by the pancreas is not able to bind to receptors on cells that facilitate the passage of glucose into the cell. Some of the known mechanistic pathways leading to insulin resistance have, as a starting point, enlarged adipose tissue mass (Kahn and Flier, 2000). This is reflected in one of the biomarkers in Table 4.1, namely, waist-hip ratio (WHR). High WHR values do not, by themselves, guarantee insulin resistance; however, they do represent one component of an early warning system of possible dysregulation in the metabolic pathways regulated by insulin secretion. There is also evidence for insulin resistance, initiated by other means, contributing to obesity and, hence, high WHR. Regardless of the direction of causality, the linkages between obesity and insulin resistance provide an important rationale for including WHR as a contributor to allostatic load.

People who are insulin resistant typically have an imbalance in their blood lipids. They have increased levels of triglycerides and decreased HDL cholesterol. In addition they often have elevated levels of low-density lipoprotein (LDL), thereby implying high levels of the ratio of total/HDL cholesterol. In the presence of insulin resistance, blood glucose levels are elevated. An excellent marker of long-term serum glucose levels is glycosylated hemoglobin. Finally, persons with insulin resistance typically have high blood pressure. Thus, the cluster of biomarkers (WHR, HDL cholesterol, total/HDL cholesterol ratio, glycosylated hemoglobin, SBP and DBP) are all associated with the phenomenon of insulin resistance, which, in turn, is implicated in dysregulation of multiple forms of intermediary metabolism. Although the complete pathway specifications going from the inability of insulin to bind to receptors on cells that facilitate glucose passage to elevated-risk-zone levels of these six biomarkers are still unknown, sufficient detail is already in hand to support these measures as indicators of allostatic load. The central point is that the extant evidence on the structure and dynamics of metabolic networks at the level of chemical reactions (i.e., mechanism at this level of detail) is linked (via what must still be regarded as strong associational evidence) with low levels of HDL cholesterol and high levels of the other five markers.

Despite the known gaps in metabolic pathway specificity related to insulin secretion and the consequences of insulin resistance (see Kahn and Flier [2000] for specifics regarding such gaps in the obesity-insulin resistance links), there is already progress in providing more refined and detailed characterizations of metabolic processes in simpler systems such as *Escheria coli*. In particular, the annotated genome sequence for *E.coli* has been used with biochemical and physiological information to reconstruct a complete metabolic network (Edwards et al., 2001). These authors also tested the hypothesis that *E. coli* uses its metabolism to grow at a maximal rate. They demonstrated that the *E. coli* metabolic network is optimized to maximize growth under a range of experimental conditions. Indeed, they obtained a quantitative genotype-phenotype relationship for metabolism in bacterial cells. An important limitation of their study is that it only characterizes metabolic flux in a steady state. The determination of metabolic concentrations and global system dynamics still lies in the future. Comparable specification of metabolic mechanism for the pathways associated with insulin secretion in humans is an even more challenging enterprise. However, the *E. coli* example serves to clarify the kinds of additional studies that are necessary to specify metabolic mechanism at a much more refined level than the extant chemical reaction pathway representations. The implications of such detailed specificity for operationalizing allostatic load are as follows: (1) it would provide a basis for identifying the causal mechanisms connecting insulin resistance with elevated-risk-zone levels of the six metabolic system markers used in previous studies of allostatic load and (2) it would suggest more refined levels of measurement to assess dysregulation at critical points along metabolic pathways.

Although the discussion, thus far, has focused on linking the six metabolic system markers in our original list (Table 4.1) to a more mechanistic description of the action of insulin and its dysregulation, it is important to emphasize the linkage among the HPA axis, metabolism, and our cortisol and DHEA-S measures (McEwen, 2000). In this regard, we note that glucocorticoids promote the conversion of protein and lipids to usable carbohydrates. They facilitate the replenishing of energy reserves after a period of activity. This directly links the normal action of glucocorticoids with the metabolic pathways connected to insulin secretion. Glucocorticoids also act on the brain to increase appetite for food and to increase locomotor activity and food seeking behavior (Leibowitz and Hobel, 1997). Thus glucocorticoids regulate behavior that controls energy input and expenditure. In contrast to this kind of normal functioning, chronically elevated glucocorticoids (e.g., as exemplified by high levels of 12-hour urinary cortisol [Table 4.1]) may be a consequence of persistent stressful experience (e.g., negative relationship histories, economic hardship, persistent perceived discrimination), poor sleep, or rich diet. Chronically elevated glucocorticoids also serve to promote insulin resistance. The combination of elevated insulin and glucocorticoids promotes the deposition of body fat, and this combination promotes the formation of athersclerotic plaques in the coronary arteries (Brindley and Rolland, 1989). This cascade of events supports the utilization of 12-hour urinary cortisol levels as an indicator of the HPA axis contribution to an operationalization of allostatic load.

DHEA-S is known to produce functional antagonism to the action of glucocorticoids, and particularly to their actions on a number of immune parameters. The precise mechanism of DHEA-S actions is unknown; what is known is that it antagonizes the thymolytic actions of glucocorticoids (Blauer et al., 1991), as well as the suppression of inflammatory cytokine production and cellular and immune responses (Rohleder et al., 2001). DHEA-S is also thought to function as more than an antiglucocorticoid in preserving immune function, for example, after a thermal injury (Araneo and Daynes, 1995). Thus, low levels of DHEA-S can be associated with multiple forms of dysregulation, thereby motivating our inclusion of it in a panel of biomarkers that can reflect allostatic load.

There are also established relationships between glucocorticoids and cognitive functioning. At an associational level, this was pointed out earlier in the section on approach to allostatic load scoring where high allostatic load in elevated-risk-zone scoring was predictive of decline in cognitive functioning 7 years beyond the baseline allostatic load assessments (Fig. 4.2, panel 3). In addition, high canonical weight allostatic load scores were associated with declines in cognitive functioning (Karlamangla et al., 2002). More mechanistic evidence for these connections are rooted in the discovery of adrenal steroid receptors in the hippocampus. For example, in studies of holocaust survivors the brain shows signs of atrophy as a result of elevated glucocorticoids and severe traumatic stress (Sapolsky, 1992). Declines of hippocampal-related cognitive functions, such as spatial and episodic memory, occur in human subjects who show increases in HPA-axis activity assessed over periods of 4–5 years (Lupien et al., 1994; Seeman et al., 1997b). Recent evidence also indicates that the most severely impaired persons have a significantly smaller hippocampal volume compared with the least impaired individuals (Lupien et al., 1996). The occurrence of hippocampal atrophy is not attributable to glucocorticoids alone. Other factors play a role, including endogenous excitatory amino acid neurotransmitters, change in dentate gyrus neuron number, and atrophy of dendritic processes. For a detailed review of these mechanisms, see McEwen (2000, 2002).

Finally, we mention some of the rationale for our including the catecholamines, epinephrine and norepinephrine, in our original panel of allostatic load biomarkers. The catecholamines and their concomitant effects on other physiological functions, such as blood pressure, heart rate, and lipolysis, suggest that they are good indicators of the stress to which individuals are exposed. Epinephrine output is mainly influenced by mental stress, whereas norepinephrine is more sensitive to physical activity and body posture. These bodily effects are also linked to increased health risks. Long-lasting elevated catecholamine levels are considered to contribute to the development of athersclerosis and predispose to myocardial ischemia (Krantz and Manuck, 1984; Rozanski et al., 1988). Elevated catecholamine levels also make the blood more prone to arterial obstruction and myocardial infarction. The role of the catecholamines in hypertension has also been investigated (e.g., Nelesen and Dimsdale, 1994). In the study of psychosocial aspects of musculoskeletal disorders (e.g., Moon and Sauter, 1996), it is generally assumed that psychological stress plays an important role by influencing various bodily functions including muscle tension and, thus, form a link to neck, shoulder, and back pain problems. Jobs with a high prevalence of musculoskeletal disorders, such as repetitive assembly-line work, are characterized by highly elevated sympathetic arousal and slow unwinding after work (Johanson et al., 1976; Melin et al., 1999).

NEW POSSIBILITIES FOR MEASURING ALLOSTATIC LOAD

An obvious limitation of the extant allostatic load measurement and scoring schemes is that they are based on a small number of indicators of dysregulation in complex dynamical systems, where breakdown can occur at many sites. In particular, they are not capable of monitoring the diverse possibilities for dysregulation that are present in metabolic, immune, neuroendocrine, and autonomic nervous system networks. Moving to more elaborate and fine-grained measurement and allostatic load scoring requires essentially different technologies from those that have been used to date. An illustration of new possibilities is provided by the recent development of high-throughput methods for assessing metabolic systems (Nicholson et al., 2002). We briefly summarize this technology and indicate, by example, how it can be used to score the metabolic contribution to allostatic load in a much more nuanced manner than heretofore.

Metabonomics

Currently, there is great interest in characterizing pathways to diverse diseases by relating gene expression to phenotypic outcomes. Several technologies are being developed to achieve this end, namely, genomics and transcriptomics, which examine genetic complement and gene expression, respectively; proteomics, which involves the analysis of protein synthesis and cell signaling; metabolomics, which investigates metabolic regulation and fluxes in individual cells or cell types; and metabonomics – the determination of systemic biochemical profiles and regulation of function in whole organisms by analyzing biofluids and tissues. Metabonomics deals with detecting, identifying, quantitating, and cataloguing the history of time-related metabolic changes in an integrated biological system. Multidimensional metabolic trajectories can then be related to the biological

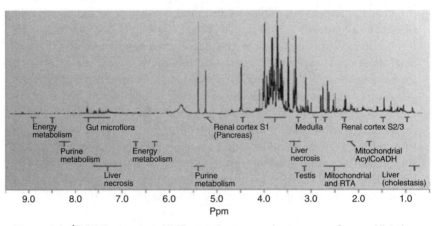

Figure 4.8: ^{1}H-NMR spectrum. NMR = nuclear magnetic resonance. Source: Nicholson et al. (2002).

events in an ongoing pathophysiological process (Lindon et al., 2000, 2001; Holmes et al., 2001).

Biofluids such as urine and plasma can be readily collected in large human population studies. Thus ^{1}H-NMR spectroscopy can be used with biofluid samples to obtain a spectroscopic fingerprint in which the spectral intensity distribution is determined by the relative concentrations of solutes, and in some cases by their intermolecular interactions (Nicholson et al., 1989). The essential concept behind NMR spectroscopy is the fact that some atomic nuclei possess a nonzero magnetic moment. This property is quantized and leads to discrete energy states in a magnetic field. Nuclei such as ^{1}H, ^{13}C, ^{15}N, and ^{31}P can undergo transitions between these states when radio-frequency pulses of appropriate energy are applied. The exact frequency of a transition depends on the type of nucleus and its electronic environment in a molecule. For example, ^{1}H nuclei in a molecule give NMR peaks at frequencies (chemical shifts) that are characteristic of their chemical environment. Figure 4.8 shows a 600-MHz ^{1}H-NMR spectrum of rat urine. The figure indicates the spectral biomarker windows that are diagnostic for a subset of diverse pathophysiological conditions. The essential point is that one NMR spectrum can carry information on a wide range of metabolites released under very diverse conditions (potentially hundreds of signatures of dysregulation in a single spectral measurement).

High-frequency ^{1}H-NMR is particularly useful for our purposes, because nearly all metabolic intermediates have unique ^{1}H-NMR signatures.

A one-dimensional spectrum of urine typically contains many thousands of sharp lines from hundreds, or potentially thousands, of metabolites. ^{1}H-NMR spectra of urine are dominated by low-molecular-weight compounds, whereas plasma contains both low- and high-molecular weight components, which give a wide range of single line widths: protein and lipoprotein signals dominate, with small-molecule fingerprints superimposed on them (Nicholson et al., 1995).

A scoring scheme for allostatic load, using ^{1}H-NMR spectra, can be developed according to the following strategy. First, identify a population with well-documented psychosocial and medical histories and with minimal, if any, chronic conditions. Using, for example, 12-hour integrated urine samples (as in the protocol for the MacArthur Aging Study; Seeman et al., 1997), ^{1}H-NMR spectra should be generated for, say, several hundred such persons. Then identify a population of persons who have cumulative adverse psychosocial profiles, such as the negative relationship pathways discussed earlier – and who have one or more chronic conditions, possibly of diverse character. Then using the ^{1}H-NMR spectra from both groups, we could apply the recursive partitioning methodology (discussed earlier in the section on measurement) to produce a description of the features of the spectra that distinguish the two populations. A Boolean statement describing such features, in terms of chemical shifts and amplitudes in the spectra, would represent an initial, coarse-level representation of elevated allostatic load. This representation could be refined by further classifying the persons with adverse histories according to increasing severity of experience and chronic conditions and identifying the features of their ^{1}H-NMR spectra that correspond to distinct levels of severity.

A recent study of coronary artery disease (Brindley et al., 2002) illustrates the potential of ^{1}H-NMR spectroscopy, using serum samples, to serve as both an effective noninvasive technology for disease diagnosis and also as a tool for discriminating among populations on the basis of metabolic profiles. Twelve spectral regions (see online supplement to Brindley et al., 2002), associated with different lipids, clearly separated persons with triple vessel disease from those with normal coronary arteries. In addition, the spectral classification also served to discriminate among subgroups with different disease severity. From the perspective of disease diagnosis, the NMR-based metabonomic methodology achieved comparable performance when compared with the standard invasive angiography. The linkage between this novel case study and operationalization of allostatic load is the fact that the ^{1}H-NMR spectra clearly discriminated among distinct populations. That the populations were defined by disease categories and the absence of disease, as opposed to psychosocial experience categories, is unimportant. The central

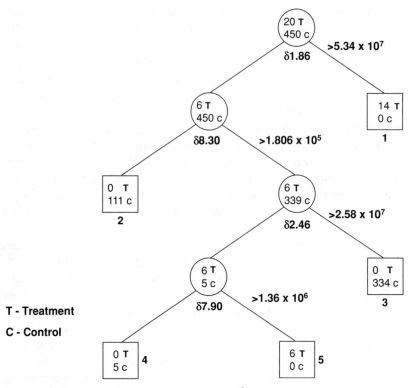

Figure 4.9: Recursive partitioning classification of ^{1}H-NMR resonance. NMR = nuclear magnetic spectra.

idea is that the metabolic profiles define distinct signatures between populations within which there is still considerable variability.

To illustrate these ideas in a simple setting, consider a population of 470 rats, 20 of which have been treated with a liver toxin and 450 control rats. Figure 4.9 shows a classification tree that results from applying the RP methodology to the 470 ^{1}H-NMR spectra (Ku et al., 2002).

The first split (top of the tree, Fig. 4.9) identifies 14 treated and 0 control animals with amplitude $> 5.34 \times 10^7$ at chemical shift 1.86, whereas a mixture of 6 treated and 450 control animals do not satisfy this condition. This suggests that this single criterion identifies some, but not all, of the treated animals. The heterogeneous population with 6 treated and 450 control animals is then further partitioned with the objective of identifying other chemical shifts and spectral amplitudes that separate the two groups. The successive further partitioning and identification leads to the full set of terminal nodes on the tree. At each stage, chemical shifts and

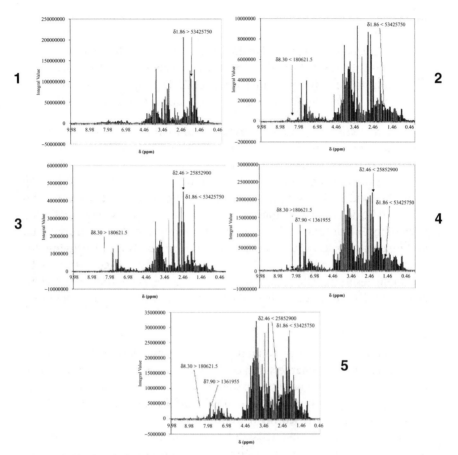

Figure 4.10: Terminal node spectra.

amplitude cut points are determined to minimize the impurity in terminal nodes – that is, they seek the cleanest separation of treated from control animals.

Figure 4.10 shows the spectra of a treated animal at each of terminal nodes 1 and 5 and of control animals at terminal nodes 2, 3, and 4. These spectra clarify the substantial within- and between-group variation that must be dealt with in identifying the discriminating features of a population of spectra.

In the context of scoring allostatic load, the following Boolean statement characterizing terminal nodes for rats treated with liver toxin, would specify the features of ^{1}H-NMR spectra that represent a metabolic contribution to allostatic load (in this case it would be a contribution derived from a liver toxin):

> *Criteria for Elevated Allostatic Load* – [ampl. > 5.34×10^7 @
> chem.shift 1.86] *or* {[ampl. < 5.34×10^7 @ chem. shift 1.86] AND
> [ampl. > 1.81×10^5 @ chem. shift 8.30] AND [ampl. < 2.59×10^7 @
> chem. shift 2.46] AND [ampl. > 1.36×10^6 @ chem. shift 7.90].

Different criteria would be consequential to different challenges. However, the logical union of such criteria would represent the metabolic contribution to allostatic load from a diversity of challenges.

This example is designed to illustrate the use of ^{1}H-NMR for scoring allostatic load in the simplest possible setting. Much richer representations can be derived by RP analyses of spectra from urine samples collected over time. In addition, it is important to emphasize that environmental and pharmacological challenges are not the only basis for identifying distinct metabolic profiles. Genetic variation is also associated with specific metabolic responses (Holmes et al., 2001). This opens the way toward characterizing the metabolic profiles of (genetic $\times$ environment) interactions.

DISCUSSION

We began by reviewing three methods for scoring allostatic load, using a limited set (10) of biomarkers that could reflect dysregulation across several biological systems. It is essential to summarize the strengths and weaknesses of these algorithms to identify the most fruitful next steps that should be pursued in refining the scoring of allostatic load.

With regard to elevated-risk-zone scoring, the positive features are the following: (1) the boundaries of the risk zones respect the requirement of having early warning, but not clinically significant, cut points on indicators of possible dysregulation; (2) the same allostatic load scoring algorithm predicts multiple downstream consequences: mortality, incident cardiovascular disease, and decline in physical and cognitive functioning; (3) the same allostatic load score can be attained via different combinations of biomarkers, thereby reflecting the diversity of forms of dysregulation that occur in large populations; (4) allostatic load scores reflect the cumulative impact of multiple biomarkers, even when no single biomarker, by itself, is strongly associated with one or more downstream outcomes; and (5) allostatic load scores reflect variation in cumulative antecedent adversity, particularly with regard to social relationships. Weaknesses of elevated-risk zone scoring are as follows: (1′) low levels on biomarkers such as DBP and cortisol are not incorporated in risk zones, thereby ignoring conditions such as hypotension and hypocortisolism; (2′) all biomarkers are weighted equally in scoring allostatic load, despite the fact that they contribute differentially to different downstream outcomes; and (3′) the current scoring algorithm only uses cross-sectional information on biomarkers.

Concerning weakness (1′), simply setting a lower cut point at a quartile or quintile for the full population is not a sufficiently sensitive strategy for specifying risk of hypotension or hypocortisolism. Cut-point determination requires a nonlinear regression strategy focused on particular outcomes. In this regard, the RP methodology, used to specify levels of allostatic load for mortality prediction, identified 60 mm Hg as a cut point for DBP – that is, DBP <60 mm Hg represents an elevated-risk condition. Different outcomes, for example, decline in physical or cognitive functioning, would yield different RP-determined categories of allostatic load from those in Figure 4.7, some of which might be associated with low levels on one or more biomarkers. This kind of analysis lies in the future and represents an important next step in refining allostatic load scoring algorithms. This simple comparison of RP with current elevated-risk-zone scoring suggests that RP is a considerably more sensitive tool for operationalizing allostatic load. Our experience to date supports this contention.

With regard to weakness (2′), the problem of differential weighting of biomarkers when focusing on particular outcomes is resolved in the canonical weight scoring algorithm (Table 4.3), for which decline in physical and cognitive functioning represented the set of outcomes of interest. Here epinephrine, DBP, and glycosylated hemoglobin had the highest (in magnitude) weights, whereas cholesterol was not included in the scoring. With different outcomes, we would anticipate different weights. An important feature of canonical weight scoring, analogous to RP, is that it also identifies low DBP as a high-risk condition (note the negative sign on the weight for DBP). A limitation of canonical weight scoring relative to RP is also revealed in this analysis. We learn that low DBP is an elevated-risk condition, but only RP identifies a level below which people are in danger.

For purposes of predicting downstream health outcomes, it would be useful to compare the capabilities of elevated-risk-zone scoring, canonical weighting, and RP on a broad range of outcomes. A comparative analysis of this kind should be high on the research agenda for increasing our understanding of allostatic load scoring. In addition, all three methods should be used with a broader range of biomarkers, including, for example, IL-6, serum albumin, and fibrinogen, which are available in extant data from the MacArthur Aging Study.

Finally, regarding longitudinal assessments of biomarkers and the scoring of allostatic load (3′) an initial operationalization, focused on cardiovascular risk, has recently been put forth by Karlamangla et al. (2004). Using measurements obtained in the Coronary Artery Risk Development in Young Adults (CARDIA) Study, seven biomarkers were assessed on the same individuals over five waves of data collection. These were SBP and DBP, fasting glucose, fasting insulin, waist-hip ratio, LDL cholesterol, and total-HDL cholesterol ratio. The CARDIA study is based on 5,115 men and women aged 18–30 years, recruited from two racial groups (white and non-Hispanic black) in four urban areas, with data collection taking place in 1985–6, 1987, 1990, 1992, and 1995. Thus, each individual has five assessments for the seven biomarkers. Using the biomarker histories as the unit of analysis, Karlamangla and colleagues (2004) showed that for each biomarker there were three nonoverlapping sets of curves, describing the full population and representing low, medium, and high levels of cardiovascular risk. It is striking that, for all persons, membership in a given risk category for a particular biomarker in 1985 implied membership in the same risk category 10 years later. This suggested that a useful way to quantify cumulative risk was to assign an individual a score of 0 if he or she was in the lowest risk category of histories on a given marker, a score of 1 for membership in the intermediate risk category, and a score of 2 for membership in the highest risk categoriy of histories. Then, for each person, a risk accumulation score (an operationalization of allostatic load for cardiovascular risk) was defined to be the sum of the scores across all seven biomarkers. Here allostatic load scores can range from 0 to 14.

The distribution of allostatic load scores was examined to test the hypothesis that lower socioeconomic status would be associated with greater longitudinal accumulation of cardiovascular risk. Women who had graduated college by 1995 had lower allostatic load scores than women who were less educated. White men who were college graduates did better than those who had high school or lower education. There was no significant relationship between education and allostatic load score for black men. These analyses represent an initial step toward relating longitudinal allostatic load scores to psychosocial experience – in this case, social stratification defined by levels of education.

Whether the persistence over time found in cardiovascular-related biomarkers holds for persons in the same age range using immune, neuroendocrine, or sympathetic nervous system measures requires investigation in other studies. In addition, it may be that such persistence breaks down at older ages, thereby requiring different strategies for scoring allostatic load (i.e., age-dependent scoring algorithms). It is also imperative to have future longitudinal assessments of biomarkers on individuals who have richly characterized psychosocial histories. This would be the basis for studying the dynamic interplay between cumulative life challenges and the dynamics of multiple biological systems.

Going beyond the operationalizations of allostatic load that involve relatively few biomarkers, our discussion of metabonomics indicated how a high-throughput technology based on ^{1}H-NMR spectroscopy could assess possible dysregulation at multiple sites in metabolic networks. Indeed, metabolic profiles, derived from spectra that can reveal the presence of hundreds of metabolites in biofluids (urine, serum, saliva) have the potential to transform our methodology for scoring allostatic load. It will be essential in future studies to carry out metabonomic analyses on populations that are well characterized in terms of longitudinal assessments of life challenges. Preferably, metabolic profiling should be based on biofluid samples collected over time, thereby identifying the dynamic trajectories of metabolite release that reflect a broad array of environmental challenges. The development of comparable technologies for assessing the fine-grained dynamics of the immune and neuroendocrine systems would be highly desirable. Ultimately this would lead to a multidimensional analogue of the metabolic profiles, one in which we would have the scoring of allostatic load based on measures that are more closely linked to the dynamics of multiple intercommunicating biological systems (Sternberg, 2000). Such developments lie in the future.

The concept of allostasis, or the ability of biological systems to adapt to challenges and maintain internal viability, has been a central notion for characterizing normal operating conditions of whole organisms. The disruption of allostasis, resulting from cumulative wear and tear on biological systems, has been referred to as allostatic load. Operationalization of this idea, the focus of this chapter, is a subject still in its infancy. Indeed, an exciting challenge for future research is refinement of operationalizations of both allostasis and allostatic load. These concepts are of fundamental importance for a much more nuanced representation of the functioning of whole organisms than heretofore.

REFERENCES

Araneo, B., Daynes, R. (1995). Dehydroepiandrosterone functions as more than an anti-glucocorticoid in preserving immunocompetence after thermal injury. *Endocrinology* 136:393–401.

Berkman, L. F., Seeman, T. E., Albert, M., Blazer, D., Kahn, R., Mohs, R., Finch. C., Schneider, E., Cotman, C., McClearn, G., Nesselroade, J., Featherman, D., Garmezy, N., McKhann, G., Brim, G., Prager, D., Rowe, J. W. (1993). High, usual and impaired functioning in community-dwelling older men and women: Finds from the MacArthur Foundation Research Network on Successful Aging. *J Clin Epidemiol* 46:1129–40.

Blauer, K. L., Poth, M., Rogers, W. M., Benton, E. W. (1991). Dehydroepiandrosterone antagonizes the suppressive effects of dexamethasone on lymphocyte proliferation. *Endocrinology* 129(6):3174–9.

Breiman, L., Friedman, J., Olshen, R., Stone, C. J. (1984). *Classification and Regression Trees*. Belmont, CA: Wadsworth.

Brindley, D. N., Rolland, Y. (1989). Possible connections between stress, diabetes, obesity, hypertension and altered lipoprotein metabolism that may result in atherosclerosis. *Clin Sci* 77:453–61.

Brindley, J. T., Antti, H., Holmes, E., Tranter, G., Nicholson, J. K., Bethell, H. W. L., Clarke, S., Schofield, P. M., McKilligan, E., Mosedale, D. E., Granger, D. J. (2002). Rapid and noninvasive diagnosis of the presence and severity of coronary heart disease using ^{1}H-NMR- based metabonomics. *Nat Med* 8:1439–44.

Cizza, G., Ravn, P., Chrousos, Gold, P. W. (2001). Depression: A major unrecognized risk factor for osteoporosis. *Trends Endocrinol Metab* 12:198–203.

Dallman, M. F., Akana, S. F., Bhatnagar, S., Bell, M. E., Strack, A. M. (2000). Bottomed out: metabolic significance of the circadian trough in glucocorticoid concentrations. *Int J Obes Relat Metab Disord* 24(Suppl. 2):S40–6.

Dawber, T. (1980). *The Framingham Study*. Cambridge: Harvard University Press.

Edwards, J. S., Ibarra, R. U., Palsson, B. O. (2001). *In silico* predictions of *Escheria coli* metabolic capabilities are consistent with experimental data. *Nat Biotech* 19:125–30.

Finch, C. E., Vaupel, J. W., Kinsella, K. (2001). *Cells and Surveys*. Washington, DC: National Academy Press.

Goldman, N., Weinstein, M., Glei, D., Singer, B., Seeman, T. (2004). Sex Differentials in Biological Risk Factors: Estimates from Population-based Surveys. *J. Women's Health* (in press).

Hertzman, C. (1999). The biological embedding of early experience and its effects on health in adulthood. *Ann N Y Acad Sci* 896:85–95.

Holmes, E., Nicholson, J. K., Tranter, G. (2001). Metabonomic classification of genetic variations in toxicological and metabolic responses using probabilistic neural networks. *Chem Res Toxicol* 48:182–91.

Jaur, L., Stoddard, S. (1999). *Chartbook on Women and Disability in the U.S.* Washington, DC: U.S. National Institute on Disability and Rehabilitation Research.

Johanson, G., Aronsson, G., Lindstrom, B. O. (1976). Social psychological and neuroendocrine stress reactions in highly mechanized work. *Ergonomics* 21:583–99.

Kahn, B. B., Flier, J. S. (2000). Obesity and insulin resistance. *J Clin Invest* 106:473–81.

Kannel, W. B. (2000). Fifty years of framingham study contributions to understanding hypertension. *J Hum Hyperten* 14:83–90.

Karlamangla, A., Singer, B., Rowe, J., McEwen, B., Seeman, T. (2002). Allostatic load as a predictor of future health: MacArthur Studies of Successful Aging. *J Clin Epidemiol* 55:696–710.

Karlamanga, A., Singer, B., Williams, D., Schwartz, J., Matthews, K., Kiefe, C., Seeman, T. (2004). Impact of socio-economic status on longituditol accumulation of cardiovascular risk in young adults. The CARDIA Study. *Social Sci Med* (in press).

Kiecolt-Glaser, J. K., Malarkey, W., Cacioppo, J. T., Glaser, R. (1994). Stressful personal relationships: Endocrine and immune function. In: *Handbook of Human Stress and Immunity* (ed. R. Glaser, J. K. Kiecolt-Glaser), pp. 321–39. San Diego: Academic Press.

Kiecolt-Glaser, J. K., Glaser, R., Cacioppo, J. T., Malarkey, W. B. (1998). Marital stress: Immunologic, neuroendocrine, and autonomic correlates. *Ann N Y Acad Sci* 840:649–55.

Kiecolt-Glaser, J. K. (1999). Stress, personal relationships, immune function: Health implications. Norman Cousins Memorial Lecture 1998. *Brain Behav Immun* 13:61–72.

Kiecolt-Glaser, J. K., McGuire, L., Robles, T. F., Glaser, R. (2002). Emotions, morbidity, and mortality: New perspectives from psychoneuroimmunology. *Ann Rev Psychol* 53:83–107.

Krantz, D. S., Manuck, S. B. (1984). Acute psychophysiologic reactivity and risk of cardiovascular disease: A review and methodologic critique. *Psychol Bull* 96:435–64.

Ku, J., Holmes, E., Nicholson, J. K., Pelczer, I., Singer, B. H. (2002). *Recursive partitioning classification of ^{1}H-NMR spectra*. Technical report, Office of Population Research, Princeton University.

Kubzansky, L. D., Kawachi, I., Sparrow, D. (2000). Socioeconomic status, hostility and risk factor clustering in the Normative Aging Study: Any help from the concept of allostatic load? *Ann Behav Med* 21:330–8.

Kudeilka, B., Hellhammer, D., Kirschbaum, C. (2000). Sex differences in human stress responses. In: *Encyclopedia of Stress, vol. 3* (ed. G. Fink), pp. 424–9. San Diego: Academic Press.

Leibowitz, S. F., Hobel, B. G. (1997). Behavioral neuroscience of obesity. In: *Handbook of Obesity* (ed. G. A. Gray, C. Bouchard, W. P. T. James), pp. 313–58. New York: Marcel Dekker.

Lindon, J. C., Nicholson, J. K., Holmes, E., Everett, J. R. (2000). Metabonomics: metabolic processes studied by NMR spectroscopy of biofluids. *Concepts Magn Reson* 12: 289–320.

Lindon, J. C., Holmes, E., Nicholson, J. K. (2001). Pattern recognition methods and applications in biomedical magnetic resonance. *Prog Nucl Magn Res Spectrosc* 39:1–40.

Low, C., Ryff, C. D., Hale, L. E., Singer, B. (2002). *Life histories and allostatic load*. Technical Report, Institute on Aging, University of Wisconsin.

Lupien, S., DeLeon, M. J., DeSanti, S., Convit, A., Tannenbaum, B. M., Nair, N. P. V., McEwen, B. S., Hauger, R. L., Meaney, M. J. (1996). Longitudinal increase in cortisol during human aging predicts hippocampal atrophy and memory deficits. *Abstr Soc Neurosci* 22(#740.1):1889.

Lupien, S., Lecours, A. R., Lussier, I., Schwarz, G., Nair, N. P. V., Meaney, M. J. (1994). Basal cortisol levels and cognitive deficits in human aging. *J Neurosci* 14:2893–903.

McEwen, B. S. (1998). Protective and Damaging Effects of stress mediators. *New Engl J Med* 338:171–9.

McEwen, B. S. (2000). The neurobiology of stress: From serendipity to clinical relevance. *Brain Res* 886:172–89.

McEwen, B. (2002). Sex, stress, and the hippocampus: Allostasis, allostatic load and the aging process. *Neurobiol of Aging* 23:921–39.

McEwen, B. S., Seeman, T. E. (1999). Protective and damaging effects of mediators of stress: Elaborating and testing the concepts of allostasis and allostatic load. *Ann N Y Acad Sci* 896:30–47.

McEwen, B. S., Stellar, E. (1993). Stress and the individual: Mechanisms leading to disease. *Arch Int Med* 153:2093–3101.

Melin, B., Lundberg, U., Soderlund, J., Granquist, M. (1999). Psychological and physiological stress reactions of male and female assembly workers: A comparison between two different forms of work organization. *J Organ Behav* (20):47–61.

Miller, L. M. S., Lachman, M. E. (2002). *Stress reactivity and cognitive performance in adulthood*. Technical report, Department of Psychology Brandeis University.

Moon, S. D., Sauter, S. L. (Eds.). (1996). *Psychosocial Aspects of Musculoskeletal Disorders in Office Work*. London: Taylor and Francis.

Nelesen, R. A., Dimsdale, J. E. (1994) Hypertension and adrenergic functioning. In: *Adrenergic Dysfunction and Psychobiology* (ed. O. G. Cameron), pp. 257–76. Washington, DC: American Psychiatric Press.

Nicholson, J. K., Connelly, J., Lindon, J. C., Holmes, E. (2002). Metabonomics: A platform for studying drug toxicity and gene function. *Nat Rev Drug Discovery* 1:153–61.

Nicholson, J. K., Foxall, P., Spraul, M., Farrant, R. D., Lindon, J. C. (1995). 750 MHz ^{1}H and ^{1}H – ^{13}C NMR spectrosc of human blood plasma. *Anal Chem* 67:793–811.

Nicholson, J. K., Wilson, I. D. (1989). High resolution proton NMR spectroscopy of biological fluids. *Prog Nucl Magn Res Spectrosc* 21:449–501.

Orth-Gomer, K., Wamala, S. P., Horsten, M., Schenck-Gustafsson, K., Schneiderman, N., Mittelman, M. (2000). Marital Stress Worsens Prognosis in Women with Coronary Heart Disease: The Stockholm Female Coronary Risk Study. *JAMA* 284:3008–14.

Parker, G., Tupling, H., Brown, L. B. (1979). A parental bonding instrument. *Br J Med Psych* 52:1–10.

Power, C., Matthews, S. (1998). Accumulation of health risks across social groups in a national longitudinal study. In: *Human Biology and Social Inequality* (ed. S. S. Strickland, P. S. Shetty), pp. 36–57. Cambridge: Cambridge University Press.

Rozanski, A., Bairey, C. N., Krantz, D. S., Friedman, J., Resser, K. J., Moreil, M., Hilton-Chalton, S., Hestrin, L., Bientendorf, J., Berman, D. S. (1988). Mental stress and the induction of silent myocardial ischemia in patients with coronary artery disease. *New Engl J Med* 318:1005–11.

Rohleder, N., Schommer, N. C., Hellhammer, D. H., Engel, R., Kirschbaurs, C. (2001). Sex differences in glucocorticoid sensitivity of proinflammatory cytokine production after psychosocial stress. *Psycho Somatic Med* 63(6):966–72.

Ryff, C. D., Singer, B. H., Wing, E., Love, G. D. (2001). Elective affinities and uninvited agonies: Mapping emotion with significant others onto health. In: *Emotion, Social Relationships, and Health* (ed. C. D. Ryff, B. H. Singer), pp. 133–75. New York: Oxford University Press.

Sapolsky, R. (1992). *Stress, the Aging Brain and the Mechanisms of Neuron Death*. Cambridge: MIT Press.

Sapolsky, R. (1993). *Why Zebras Don't Get Ulcers*. New York: W. H. Freeman.

Sapolsky, R. (1996). Stress, glucocorticoids, and damage to the nervous system: The current state of confusion. *Stress* 1:1.

Schaefer, M. T., Olson, D. H. (1981). Assessing intimacy: The PAIR Inventory. *J Marital Fam Ther* 7:47–60.

Schulkin, J. (2003). *Rethinking Homeostasis*, Cambridge, MA: MIT Press.

Seeman, T. E., Berkman, L. F., Gulanski, B. I., Robbins, R. J., Greenspan, S. L., Charpentier, P. A., Rowe, J. W. (1995). Self-esteem and neuroendocrine response to challenge: MacArthur Studies of Successful Aging. *J Psychosom Res* 39:69–84.

Seeman, T. E., Singer, B. H., Rowe, J. W., Horwitz, R. I., McEwen, B. S. (1997a). The price of adaptation – allostatic load and its health consequences: MacArthur Studies of Successful Aging. *Arch Int Med* 157:2259–68.

Seeman, T. E., McEwen, B. S., Singer, B. H., Albert, M. S., Rowe, J. W. (1997b). Increase in urinary cortisol excretion and memory declines: MacArthur Studies of Successful Aging. *J Clin Endocr Metab* 82:2458–65.

Seeman, T. E., McEwen, B. S., Rowe, J. W., Singer, B. H. (2001). Allostatic load as a marker of cumulative biological risk: MacArthur Studies of Successful Aging. *Proc Nat Acad Sci U S A* 98:470–5.

Seeman, T., Singer, B., Ryff, C. D., Dienberg Love, G., Levy-Storms, L. (2002). Social relationships, gender, and allostatic load across two age-cohorts. *Psychosom Med* 64:395–406.

Sewall, W. H., Hauser, R. M., Springer, K. W., Hauser, T. S. (2001). *As we age: The Wisconsin Longitudinal Study, 1957–2001*. CDE Working Paper No. 2001–09. Madison: Center for Demography and Ecology, University of Wisconsin.

Smider, M. A., Essex, M. J., Ryff, C. D. (1996). Adaptation to community relocation: The interactive influence of psychological resources and contextual factors. *Psycholo Aging* 11:362–71.

Sterling, P., Eyer, J. (1988). Allostasis: A new paradigm to explain arousal pathology. In: *Handbook of Life Stress, Cognition, and Health* (ed. S. Fisher, J. Reason), pp. 631–51. New York: John Wiley & Sons.

Sternberg, E. M. (2000). *The Balance Within: The Science Connecting Health and Emotions*. New York: W. H. Freeman.

Wadsworth, M., Kuh, D. (1997). Childhood influences on adult health: A review of recent work from the British 1946 national birth cohort study, the MRC national survey of health and development. *Paediatr Perinat Epidemiol* 11:2–20.

Weinstein, M., Willis, R. J. (2001). Stretching social surveys to include bioindicators: Possibilities for the Health and Retirement Study, Experience from the Taiwan Study of the Elderly. In: *Cells and Surveys* (ed. C. E. Finch, J. W., Vaupel, K. Kinsella), pp. 250–75. Washington, DC: National Academy Press.

Zhang, H., Singer, B. (1999). *Recursive Partitioning in the Health Sciences*. New York: Springer-Verlag.

5 Drug Addiction and Allostasis

George F. Koob and Michel Le Moal

INTRODUCTION

Allostasis is a concept developed originally by neurobiologist Peter Sterling and epidemiologist James Eyer to explain the physiological basis for changes in patterns of human morbidity and mortality. They observed that the baby boom generation of individuals born after World War II had reached an age when major causes of death were renal, cerebral, and cardiovascular disease and that the single largest contributor to these diseases was hypertension. These researchers could find no explanation for these findings and no explanation for why hypertension was most prevalent where social disruption was greatest (Sterling and Eyer, 1988). They argued that the only possible link between the sociopsychological and physiological phenomena is in the brain. Using this backdrop, they argued for a form of brain-body regulation different from *homeostasis,* that of *allostasis.*

Homeostasis can be defined as "preserving constancy in the internal environment." Allostasis can be defined as stability through change. Claude Bernard is credited with being the first to suggest that the internal milieu of the body is critically important for establishing and maintaining stable states within the body (this concept was later termed homeostasis by Walter Cannon in 1926). Bernard argued as early as 1859–60 that two environments affect an organism: a *general milieu* that is the outside world of inanimate and animate objects, and an *internal milieu,* where the elements of a living body find an optimal climate for operation. Originally described largely for the circulatory system, Bernard later argued for involvement of the lymphatic systems as well (Cannon, 1929, 1932, 1935).

Bernard further argued that not only was an organism required to respond to outside stimuli to regulate energy by bodily adjustments, but that

This is publication number 14154-NP from The Scripps Research Institute. The authors would like to thank Mike Arends for his help with manuscript preparation.

Table 5.1: Homeostasis versus Allostasis

Homeostasis	Allostasis
Normal setpoint	Changing setpoint
Physiologic equilibrium	Compensated equilibrium
No anticipation of demand	Anticipation of demand
No adjustment based on history	Adjustment based on history
Adjustment carries no price	Adjustment and accomodation carry a price
No pathology	Leads to pathology

the body also maintained the internal milieu remarkably constant in the face of such challenges making the organism at some level independent of the exterior challenges. Finally, he argued that all the vital mechanisms of the body have but one goal of maintaining the conditions of the life of the internal milieu constant (Cannon, 1929, 1932, 1935).

Using the arousal-stress continuum as their physiological framework, Sterling and Eyer argued that allostasis has several unique characteristics (see Table 5.1). Allostasis is far more complex than homeostasis. For example, in homeostasis the bodily system returns physiological parameters to a specific "setpoint." In allostasis, there is a continuous reevaluation of the organism's need and continuous readjustments to new setpoints, depending on demand. However, if demand continues unabated the setpoint remains adjusted out of the homeostatic range. Another difference between allostasis and homeostasis is that the former can anticipate altered need and the system can make adjustments in advance. In homeostasis, increased need triggers a signal for negative feedback mechanisms. In this type of situation, the break with homeostasis may become large before a correction is made, particularly if resources are not available. Homeostatic systems do not change according to experience, whereas allostatic systems can be hypothesized to use past experience to anticipate demand (Sterling and Eyer, 1988).

Allostasis involves a feedforward mechanism rather than the negative feedback mechanisms of homeostasis (see Fig. 5.1). An example of such a feed-forward system is illustrated in the interaction between corticotropin-releasing factor (CRF) and norepinephrine in the brainstem and basal forebrain. A feed-forward mechanism has many advantages because in homeostasis when increased need produces a signal, negative feedback can correct the need, but the time required may be great and the resources may not be available. However, in allostasis there is continuous reevaluation of need and continuous readjustment of all parameters toward new setpoints. Thus, there is a fine matching of resources to needs. Yet it is precisely this ability to mobilize resources quickly and use feed-forward mechanisms

Feed-Forward CRF-NE-CRF
Stress System

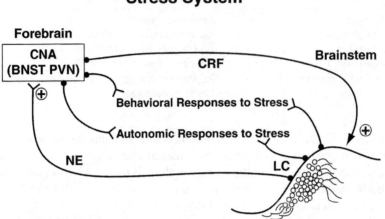

Figure 5.1: Diagram describing a feed-forward system whereby corticotropin-releasing factor (CRF) could activate brainstem noradrenergic activity, which in turn activates forebrain CRF activity, effectively closing the loop. Such a hypothesized mechanism could explain the potentiation of stress responses with repeated exposure that could lead to psychopathology. Note that behavioral and autonomic responses could be activated in both the brainstem and the forebrain as partial exits from the closed part of the loop. CNA = central nucleus of the amygdala; BNST = bed nucleus of the stria terminalis; PVN = paraventricular nucleus; NE = norepinephrine; LC = locus coeruleus. [Reprinted by permission of Elsevier Science from "Corticotropin-releasing factor, norepinephrine and stress," by G. F. Koob **Biol. Psychiatry** 46:1167–80. Copyright 1999 by the Society of Biological Psychiatry.]

that leads to an allostatic state and an ultimate cost of an allostatic mechanism that is known as allostatic load (McEwen, 1998). An *allostatic state* can be defined as a state of chronic deviation of the regulatory system from its normal (homeostatic) operating level. *Allostatic load* can be defined as the long-term cost of allostasis that accumulates over time and reflects the accumulation of damage that can lead to pathological states. Allostatic load is the consequence of repeated deviations from homeostasis that take on the form of changed setpoints that require more and more energy to defend. Such allostatic adjustments have been hypothesized to move to the level of allostatic load by four situations (McEwen, 2000). First, the challenge may be frequent, usually environmental. Second, the allostatic change may not habituate to repeated challenge. Third, there may be an inability to shut off the allostatic processes. Fourth, there may be inadequate allostatic responses that trigger compensatory responses in other systems.

The failure of allostatic change to habituate or not to shut off is inherent in a feed-forward system that is in place for rapid, anticipated challenge to homeostasis. Yet the very physiological mechanism that allows rapid response to environmental challenge becomes the engine of pathology if adequate time or resources are not available to shut off the response. Thus, for example, chronically elevated blood pressure is "appropriate" in an allostatic model to meet environmental demand of chronic arousal but is "certainly not healthy" (Sterling and Eyer, 1988).

As is evident in an article from the journal *Neuropsychopharmacology* (Koob and Le Moal, 2001), allostatic mechanisms are hypothesized to be involved in maintaining a functioning brain reward system. However, repeated challenges, as in the case of drugs of abuse or environmental challenges, lead to the brain's attempts to maintain stability, but at a cost. The deviation from normal brain reward thresholds is termed an "allostatic state" and represents a chronic elevation of reward setpoint fueled by not only dysregulation of reward circuits but also activation of brain and hormonal stress responses.

Drug addiction is a chronic relapsing disorder characterized by compulsive drug taking behavior with impairment in social and occupational functioning. From a psychiatric perspective, drug addiction has aspects of both impulse control disorders and compulsive disorders (Fig. 5.2). Impulse control disorders are characterized by an increasing sense of tension or arousal before committing an impulsive act, pleasure, gratification or relief at the time of committing the act, and there may or may not be regret, self reproach or guilt following the act (American Psychiatric Association, 1994). In contrast, compulsive disorders are characterized by anxiety and stress before committing a compulsive repetitive behavior, and relief from the stress by performing the compulsive behavior. In drug addiction, individuals are hypothesized to move from an impulsive disorder to a compulsive disorder where there is a shift from positive reinforcement driving the motivated behavior to negative reinforcement driving the motivated behavior. Drug addiction has been conceptualized as a disorder that progresses from impulsivity to compulsivity in a collapsed cycle of addiction comprised of three stages: preoccupation-anticipation, binge-intoxication, and withdrawal-negative affect (Koob and Le Moal, 1997). Different theoretical perspectives ranging from experimental psychology, social psychology and neurobiology can be superimposed on these three stages which are conceptualized as feeding into each other, becoming more intense, and ultimately leading to the pathological state known as addiction (Koob and Le Moal, 1997). Thus, addiction is conceptualized as a cycle of spiraling dysregulation of brain reward systems that progressively increases resulting in the compulsive use of drugs.

A critical aspect of this conceptualization of addiction is that the behavioral changes are paralleled by shifts in functional activity of specific

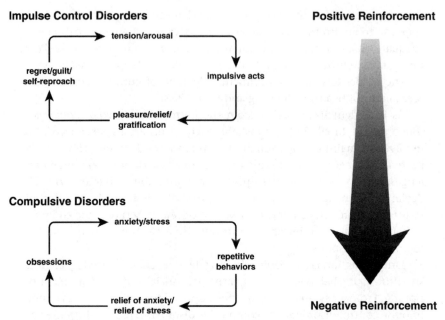

Figure 5.2: Diagram showing the stages of impulse control disorder and compulsive disorder cycles related to the sources of reinforcement. In impulse control disorders, an increasing tension and arousal occurs before the impulsive act, with pleasure, gratification, or relief during the act. Following the act there may or may not be regret or guilt. In compulsive disorders, there are recurrent and persistent thoughts (obsessions) that cause marked anxiety and stress followed by repetitive behaviors (compulsions) that are aimed at preventing or reducing distress (American Psychiatric Association, 1994). Positive reinforcement (pleasure/gratification) is more closely associated with impulse control disorders. Negative reinforcement (relief of anxiety or relief of stress) is more closely associated with compulsive disorders. [Taken with permission from Koob, 2004. Reprinted from *The Nebraska Symposium on Motivation 50* by permission of the University of Nebraska Press. © 2004 by the Board of Regents of the University of Nebraska.]

neurocircuits and specific neuropharmacological activity. A brain reward circuit for the acute positive reinforcing effects of drugs of abuse is well delineated and has as its focal point the basal forebrain structure known as the extended amygdala and its midbrain, pallidal and limbic connections (Koob et al., 1998; Fig. 5.3). However, as addiction proceeds and opponent processes develop the brain stress circuits are engaged, and the brain circuits associated with compulsive behavior (an obsessive compulsive disorder) are engaged contributing to the development of negative reinforcement. Superimpose on these shifts in circuitry a dramatic neurochemical dysregulation where one has decreases in the neurotransmitters that contribute to the positive reinforcing effects of drugs of abuse and recruitment of the brain stress systems (or de-recruitment of the brain anti-stress systems) and one

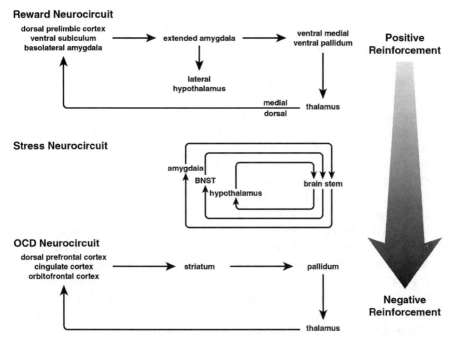

Figure 5.3: Three brain circuits hypothesized to be recruited at different stages of the addiction cycle as addiction moves from positive reinforcement to negative reinforcement. The top circuit refers to the brain reward system with a focus on the extended amygdala/lateral hypothalamic loop and extended amygdala/ventral pallidum loop. The middle circuit refers to the brain stress circuits in feed-forward loops, and the bottom circuit refers to the obsessive compulsive loop of the dorsal striatum/pallidum and thalamus.

has a powerful mechanism for the negative reinforcement associated with the "dark side" of addiction.

Recent neuroanatomical data and new functional observations have provided support for the hypothesis that the neuroanatomical substrates for many of the motivational effects of drugs of abuse may involve a common neural circuitry that forms a separate entity within the basal forebrain termed the "extended amygdala" (Alheid and Heimer, 1988). Originally described by Johnston (1923), the term "extended amygdala" represents a macrostructure that is composed of several basal forebrain structures: the bed nucleus of the stria terminalis, the central medial amygdala, a transition zone in the posterior part of the medial nucleus accumbens (e.g., posterior shell; Heimer and Alheid, 1991) and the area termed the sublenticular substantia innominata. There are similarities in morphology, immunohistochemistry and connectivity in these structures (Alheid and Heimer, 1988), and they receive afferent connections from limbic cortices, hippocampus, basolateral amygdala, midbrain, and lateral hypothalamus. The efferent connections from this complex include the posterior medial (sublenticular) ventral pallidum, medial ventral tegmental area, various brain stem projections, and perhaps most intriguing from a functional point of view, a considerable projection to the lateral hypothalamus (Heimer et al., 1991).

Specific elements of the extended amygdala have been implicated in the acute positive reinforcing effects of the major drugs of abuse, including cocaine, opiates, and alcohol (Koob et al., 1998). Even more compelling for the present conceptualization is evidence for the extended amygdala in the transition from the positive reinforcing effects of drugs of abuse to the negative reinforcing effects associated with the development of dependence. Parts of the extended amygdala are involved in the aversive stimulus effects of opiate withdrawal. Blockade of CRF receptors in the central nucleus of the amygdala blocks conditioned place aversion produced by opiate withdrawal (Heinrichs et al., 1995). Changes in sensitivity to opiate antagonists in opiate-dependent rats have been observed in the nucleus accumbens, bed nucleus of the stria terminalis and central nucleus of the amygdala (Stinus et al., 1990; Koob et al., 1989), and opiate-dependent rats self-administering heroin are particularly sensitive to opiate antagonism in the bed nucleus of the stria terminalis (Walker et al., 2000). Significant evidence exists for a role of CRF in the aversive stimulus and motivational effects of cocaine and ethanol withdrawal, and stress-induced reinstatement of drug self-administration in animals extinguished from self-administration (Richter and Weiss, 1999; Valdez et al., 2002; Erb et al., 2001; Sarynai et al., 2001), and these effects are hypothesized to be localized to overactivity of CRF systems also in the extended amygdala. This neuronal circuit, and specific neuropharmacological elements, is well situated to form a heuristic model for exploring the mechanisms associated with the transition from drug use to compulsive use associated with addiction (see Table 5.2). In addition, the construct of the extended amygdala ultimately may link the recent developments in the neurobiology of drug reward with existing knowledge of the substrates for emotional behavior (Davis, 1997), essentially bridging what have been largely independent research pursuits.

The basis for our allostatic view of brain reward dysfunction during the development of addiction derives originally from a neurobehavioral perspective where it was hypothesized that in brain motivational systems the initial acute effect of an emotional stimulus or a drug is opposed or counteracted by homeostatic changes in brain systems. An affect control system was hypothesized to suppress or reduce all departures from hedonic neutrality via a single negative feedback, or opponent, loop that opposes the stimulus-aroused affective state (Solomon and Corbit, 1974; Siegel, 1975; Poulos and Cappell, 1991). In this opponent process theory, affective

Table 5.2: Neurotransmitters implicated in the motivational effects of withdrawal from drugs of abuse

Neurotransmitter	Functional activity	Motivational effects
Dopamine	↓	"dysphoria"
Opioid peptides	↓	pain, "dysphoria"
Dynorphin	↑	"dysphoria"
Serotonin	↓	pain, depression, "dysphoria"
GABA	↓	anxiety, panic attacks
CRF	↑	stress
Neuropeptide Y	↓	stress
Norepinephrine	↑	stress

states – pleasant or aversive – were hypothesized to be automatically opposed by centrally mediated mechanisms that reduce the intensity of these affective states, and tolerance and dependence are inextricably linked (Solomon and Corbit, 1974). When a subject takes a drug such as an intravenous opiate, the first few self-administrations of the drug produce a pattern of motivational changes similar to that observed in Figure 5.4A. Euphoria results from the onset of the drug effect, which forms the *a-process*, and this euphoria subsequently declines in intensity. An aversive craving state then emerges and reflects the *b-process*, or opponent process (Fig. 5.4B).

An allostatic view of the brain motivational systems associated with drug addiction has been proposed to explain the persistent changes in motivation that are associated with vulnerability to relapse in addiction (Koob and Le Moal, 2001), and this model may generalize to other

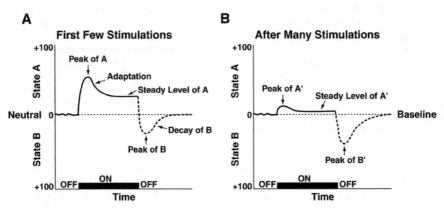

Figure 5.4: (A) The standard pattern of affective dynamics produced by a relatively novel unconditioned stimulus. (B) The standard pattern of affective dynamics produced by a familiar, frequently repeated unconditioned stimulus. [Taken with permission from Solomon 1980.]

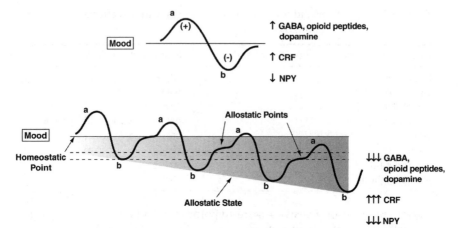

Figure 5.5: Diagram illustrating an extension of Solomon and Corbit's (1974) opponent-process model of motivation to outline the conceptual framework of the allostatic hypothesis. Both panels represent the affective response to the presentation of a drug. (Top) This diagram represents the initial experience of a drug with no prior drug history. The *a-process* represents a positive hedonic or positive mood state, and the *b-process* represents the negative hedonic or negative mood state. The affective stimulus (state) has been argued to be a sum of both an *a-process* and a *b-process*. An individual whom experiences a positive hedonic mood state from a drug of abuse with sufficient time between re-administering the drug is hypothesized to retain the *a-process*. In other words, an appropriate counteradaptive opponent-process (*b-process*) that balances the activational process (*a-process*) does not lead to an allostatic state. (Bottom) The changes in the affective stimulus (state) in an individual with repeated frequent drug use that may represent a transition to an allostatic state in the brain reward systems and, by extrapolation, a transition to addiction. Note that the apparent *b-process* never returns to the original homeostatic level before drug-taking is reinitiated, thus creating a greater and greater allostatic state in the brain reward system. In other words, the counteradaptive opponent-process (*b-process*) does not balance the activational process (*a-process*) but in fact shows a residual hysteresis. While these changes are exaggerated and condensed over time in the present conceptualization, the hypothesis here is that even during post-detoxification, a period of "protracted abstinence," the reward system is still bearing allostatic changes. In the nondependent state, reward experiences are normal, and the brain stress systems are not greatly engaged. During the transition to the state known as addiction, the brain reward system is in a major underactivated state while the brain stress system is highly activated. Small arrows refer to increased or decreased functional activity of the neurotransmitters. *DA* = dopamine; *CRF* = corticotropin-releasing factor; *GABA* = γ-aminobutyric acid. The following definitions apply: *allostasis,* the process of achieving stability through change; *allostatic state,* a state of chronic deviation of the regulatory system from its normal (homeostatic) operating level; *allostatic load,* the cost to the brain and body of the deviation, accumulating over time, and reflecting in many cases pathological states and accumulation of damage. [Modified with permission from Koob and Le Moal, 2001.]

psychopathology associated with dysregulated motivational systems. Such an allostatic view of the motivation for drug addiction also provides a physiological basis for opponent process theory via the domains of neurocircuitry and neurobiology (Koob and Le Moal, 2001). In this framework, the counteradaptive processes that form opponent-process are hypothesized to be part of the normal homeostatic limitation of reward function. The counteradaptive process (*b-process*) fails to return within the normal homeostatic range and is hypothesized to form an allostatic state. This allostatic state is further hypothesized to be reflected in a chronic deviation of reward setpoint that is fueled not only by dysregulation of reward circuits per se but also by recruitment of brain and hormonal stress responses (Fig. 5.5).

IMPLICATIONS OF THE ALLOSTATIC VIEW FOR MOTIVATION AND PSYCHOPATHOLOGY

The allostatic view discussed in the previous *Neuropsychopharmacology* article (Koob and Le Moal, 2001) has implications not only for drug addiction but also for motivational processes in general. Motivation can be defined as stimulation that arouses activity of a particular kind (Hebb, 1949), and presumably some kind of reward or reinforcement system is key to any motivational conceptual framework. Such a natural reward system, however, may have a biological limitation in that excessive engagement of a brain reward system may leave an organism vulnerable to environmental challenge, be it a sudden change in weather or the appearance of a predator. Under such a formulation, brain reward is limited, and the studies with drugs of abuse suggest that this limitation involves not only dysregulation of brain transmitters important for directing appetitively motivated behavior but also recruitment of brain stress systems that limit appetitively motivated behavior.

A biological perspective of the brain's reward system in the context of non–drug states suggests that it may be a limited resource (Koob and Le Moal, 1997). Early, Carl Jung conceptualized a hedonic system with limited energy in which the psyche was regarded as a relatively closed system. Jung described a closed system near that of "entropy." In an entropy situation, the system is closed and no energy from the outside can be fed into it (Jung, 1948). The term "libido" was used to describe a limited general life instinct or psychic energy. One can expend the psychic energy hedonic resource rapidly in a binge of compulsive behavior, but at the risk of entering into the spiraling dysregulation of the addiction cycle, triggering allostatic mechanisms and ultimately allostatic load. A more regulated "hedonic Calvinistic" approach in which the brain's reward system is allowed the time and resources to return to a near homeostatic setpoint would prevent the development of allostatic load and the subsequent spiraling distress associated with an addiction-like cycle (Koob and Le Moal, 1997).

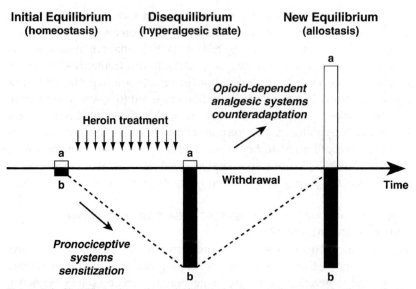

Figure 5.6: Simplified schematic model illustrating the putative neuroadaptive continuum model linking sensitization, apparent tolerance, and the withdrawal symptom of hyperalgesia. Before the first exposure to heroin, an initial equilibrium (homeostasis) is associated with a low-level balance between the opioid-dependent analgesic system **(a)** and N-methyl-D-aspartate (NMDA)–dependent pronociceptive systems **(b)** as indicated by the naloxone ineffectiveness in precipitating hyperalgesia. The dotted line represents the sum of the system activity of **a** and **b**. Functioning levels of the latter are represented by the height of the column. Repeated heroin administrations induce a gradual decrease of the nociceptive threshold (pronociceptive system sensitization), leading to a hyperalgesic state. This progressively shifts the unchanged analgesic response, giving the impression of less analgesia (apparent tolerance). After heroin treatment is stopped (withdrawal), the return to predrug nociceptive threshold is not mediated by a deactivation of pronociceptive systems but is supported by an endogenous opioid system counteradaptation. The new equilibrium (allostasis) is associated with a high-level balance between opioid-dependent analgesic systems and NMDA-dependent pronociceptive systems leading to long-term pain vulnerability. Reprinted with permission from Celerier et al., 2001.

How tolerance and sensitization processes – two apparently opposite phenomena – can concomitantly modify one given biological process has recently received empirical support (Celerier et al., 2001; Laulin et al., 2002; Rivat et al., 2002). The fact that opiates produce not only analgesia but also long-lasting hyperalgesia suggests that tolerance to the analgesic effect of an opiate could in part be the result of an actual sensitization of pronociceptive systems. First, both magnitude and duration of heroin-induced delayed hyperalgesia increase with intermittent opiate (heroin) administration, leading to an apparent decrease in the analgesic effectiveness of a given heroin dose. Second, a small dose of heroin, ineffective for triggering delayed hyperalgesia in nontreated rats, induced an enhancement in pain sensitivity for several days after a series of heroin administrations.

These data are in agreement with the sensitization hypothesis (that is, sensitization is the secondary effect). Third, the effectiveness of naloxone to precipitate hyperalgesia in rats that had recovered their predrug nociceptive value after single or repeated heroin administrations indicates that heroin-deprived rats were in a new biological state associated with a high-level balance between opioid-dependent analgesia systems and pronociceptive systems. Fourth, the N-methyl-D-aspartate receptor antagonist MK801 prevented both long-lasting enhancement in pain sensitivity and naloxone-precipitated hyperalgesia. Tolerance, sensitization, and one of the opiate withdrawal symptoms (hyperalgesia) are hypothesized to result from a single neuroadaptive process (i.e., a new allostatic equilibrium; see Fig. 5.6).

Allostasis as a concept may apply to any number of pathological states that are challenged by external and internal events. Bernard Carroll cogently argued that aging and major affective disorders reflect lifetime stress and that the brain's response to stress has all the features of an allostatic mechanism (Carroll, 2002). There is increased activity of the hypothalamic pituitary adrenal axis, impaired glucocorticoid feedback and regulation, circadian abnormalities, and decreased neurogenesis in the dentate gyrus with other brain system projection losses (Duman et al., 1997). Such effects of chronic stress are also seen with aging. In addition, it is argued that depression is an established outcome of stress and in turn an ongoing depressive episode can constitute chronic stress. One can argue, therefore, that depression also fits an allostatic model (Carroll, 2002). The neuroendocrinology of human depression closely resembles that of chronic stress in the laboratory, including increased hypothalamic-pituitary-adrenal axis activity, reduced glucocorticoid feedback, and dysregulated diurnal rhythms or cortisol (Checkley, 1996). Closing the physiological loop, there is a strong relationship between the number of depressive symptoms and premature mortality, and depression is associated with many cardiovascular risk factors (Carroll, 2002). Similar connections have been made between allostatic changes in brain stress systems and posttraumatic stress disorder and anxiety disorders (Schulkin et al., 1994; Lindy and Wilson, 2001). In addition, one can reasonably see how both developmental and genetic domains can modify allostatic load that may determine vulnerability to pathology (McEwen, 2000). The challenge for future research will be to explore how these brain stress systems integrate with brain motivational systems and to broaden allostasis as a concept from its current focus on the activation of the hypothalamic-pituitary-adrenal axis.

REFERENCES

Alheid, G. F., Heimer, L. (1988). New perspectives in basal forebrain organization of special relevance for neuropsychiatric disorders: The striatopallidal, amygdaloid, and corticopetal components of substantia innominata. *Neuroscience* 27:1–39.

American Psychiatric Association (1994). *Diagnostic and Statistical Manual of Mental Disorders,* 4th edition. Washington DC: American Psychiatric Press.

Cannon, W. B. (1929). Organization for physiological homeostasis. *Physiol Rev* 9:399–431.

Cannon, W. B. (1932). *The Wisdom of the Body.* New York: Norton.

Cannon, W. B. (1935). Stresses and strains of homeostasis. *Am J Med Sci* 189:1–14.

Carroll, B. J. (2002). Ageing, stress and the brain. In: *Endocrine Facet of Ageing,* vol. 242 of *Novartis Foundation Symposium* (ed. D. J. Chadwick, J. A. Goode), pp. 26–45. Chichester: John Wiley.

Celerier, E., Laulin, J.-P., Corcuff, J.-B., Le Moal, M., Simonnet, G. (2001). Progressive enhancement of delayed hyperalgesia induced by repeated heroin administration: A sensitization process. *J Neurosci* 21:4074–80.

Celerier, E., Rivat, C., Jun, Y., Laulin, J.-P., Larcher, A., Reynier, P., Simonnet, G. (2000). Long-lasting hyperalgesia induced by fentanyl in rats. *Anesthesiology* 92: 308–9.

Checkley, S. (1996). The neuroendocrinology of depression and chronic stress. *Br Med Bull* 52:597–617.

Davis, M. (1997). Neurobiology of fear responses: The role of the amygdala. *J Neuropsychiatry Clin Neurosci* 9:382–402.

Duman, R. S., Heninger, G. R., Nestler, E. J. (1997). A molecular and cellular theory of depression. *Arch Gen Psychiatry* 54:597–606.

Erb, S., Salmaso, N., Rodaros, D., Stewart, J. (2001). A role for the CRF-containing pathway from central nucleus of the amygdala to bed nucleus of the stria terminalis in the stress-induced reinstatement of cocaine seeking in rats. *Psychopharmacology* 158:360–5.

Hebb, D. O. (1949). *Organization of Behavior: A Neuropsychological Theory.* New York: John Wiley & Sons.

Heimer, L., Alheid, G. (1991). Piecing together the puzzle of basal forebrain anatomy. In: *The Basal Forebrain: Anatomy to Function,* vol. 295 of *Advances in Experimental Medicine and Biology* (eds. T. C. Napier, P. W. Kalivas, I. Hanin), pp. 1–42. New York: Plenum.

Heimer, L., Zahm, D.S., Churchill, L., Kalivas, P.W., Wohltmann, C. (1991). Specificity in the projection patterns of accumbal core and shell in the rat. *Neuroscience* 41:89–125.

Heinrichs, S. C., Menzaghi, F., Schulteis, G., Koob, G. F., Stinus, L. (1995). Suppression of corticotropin-releasing factor in the amygdala attenuates aversive consequences of morphine withdrawal. *Behav Pharmacol* 6:74–80.

Johnston, J. B. (1923). Further contributions to the study of the evolution of the forebrain. *J Comp Neurol* 35:337–481.

Jung, C. G. (1948). *Die Beziehungen der Psychotherapie zur Seelsorge* [The Structure and Dynamics of the Psyche]. Zurich: Rascher.

Koob, G. F. (1999). Corticotropin-releasing factor, norepinephrine and stress. *Biol Psychiatry* 46:1167–80.

Koob, G. F. (2004). Allostatic view of motivation: implications for psychopathology. In: *Motivational Factors in the Etiology of Drug Abuse,* vol. 50 of *Nebraska Symposium on Motivation* (eds. R. Bevins and M. T. Bardo , Lincoln NE) in press. University of Nebraska Press.

Koob, G. F., Le Moal, M. (1997). Drug abuse: Hedonic homeostatic dysregulation. *Science* 278:52–8.

Koob, G. F., Le Moal, M. (2001). Drug addiction, dysregulation of reward, and allostasis. *Neuropsychopharmacology* 24:97–129.

Koob, G. F., Sanna, P. P., Bloom, F. E. (1998). Neuroscience of addiction. *Neuron* 21:467–76.

Koob, G. F., Wall, T. L., Bloom, F. E. (1989). Nucleus accumbens as a substrate for the aversive stimulus effects of opiate withdrawal. *Psychopharmacology* 98:530–4.

Laulin, J.-P., Maurette, P., Corcuff, J.-B., Rivat, C., Chauvin, M., Simonnet, G. (2002). The role of ketamine in preventing fentanyl-induced hyperalgesia and subsequent acute morphine tolerance. *Anesth Analg* 94:1263–9.

Lindy, J. D., Wilson, J. P. (2001). An allostatic approach to the psychodynamic understanding of PTSD. In: *Treating Psychological Trauma and PTSD* (ed. J. P. Wilson, M. J. Friedman, J. D. Lindy), pp. 125–138. New York: Guilford Press.

McEwen, B. S. (1998). Protective and damaging effects of stress mediators. *New Engl J Med* 338:171–9.

McEwen, B. S. (2000). Allostasis and allostatic load: Implications for neuropsychopharmacology. *Neuropsychopharmacology* 22:108–24.

Poulos, C. X., Cappell, H. (1991). Homeostatic theory of drug tolerance: A general model of physiological adaptation. *Psychol Rev* 98:390–408.

Richter, R. M., Weiss, F. (1999). In vivo CRF release in rat amygdala is increased during cocaine withdrawal in self-administering rats. *Synapse* 32:254–61.

Rivat, C., Laulin, J.-P., Corcuff, J.-B., Celerier, E., Pain, L., Simonnet, G. (2002). Fentanyl enhancement of carrageenan-induced long-lasting hyperalgesia in rats. *Anesthesiology* 96:381–91.

Sarnyai, Z., Shaham, Y., Heinrichs, S. C. (2001). The role of corticotropin-releasing factor in drug addiction. *Pharmacol Rev* 53:209–43.

Schulkin, J., McEwen, B. S., Gold, P. W. (1994). Allostasis, amygdala, and anticipatory angst. *Neurosci Biobehav Rev* 18:385–96.

Siegel, S. (1975). Evidence from rats that morphine tolerance is a learned response. *J Comp Physiol Psychol* 89:498–506.

Solomon, R. L., Corbit, J. D. (1974). An opponent-process theory of motivation: 1. Temporal dynamics of affect. *Psychol Rev* 81:119–45.

Sterling, P., Eyer, J. (1988). Allostasis: A new paradigm to explain arousal pathology. In: *Handbook of Life Stress, Cognition and Health* (ed. S. Fisher, J. Reason), pp. 629–47.

Stinus, L., Le Moal, M., Koob, G. F. (1990). Nucleus accumbens and amygdala are possible substrates for the aversive stimulus effects of opiate withdrawal. *Neuroscience* 37:767–73.

Valdez, G. R., Roberts, A. J., Chan, K., Davis, H., Brennan, M., Zorrilla, E. P., Koob, G. F. (2002). Increased ethanol self-administration and anxiety-like behavior during acute withdrawal and protracted abstinence: regulation by corticotropin-releasing factor. *Alcohol Clin Exp Res* 26:1494–1501.

Walker, J. R., Ahmed, S. H., Gracy, K. N., Koob, G. F. (2000). Microinjections of an opiate receptor antagonist into the bed nucleus of the stria terminalis suppress heroin self-administration in dependent rats. *Brain Res* 854:85–92.

6 Adaptive Fear, Allostasis, and the Pathology of Anxiety and Depression

Jeffrey B. Rosen and Jay Schulkin

The traumatic events of September 11, 2001, have dramatically changed our lives forever. Americans and people around the world were shocked by the unprecedented and horrific attacks on civilians. The anthrax scare further shook the security and complacency of Americans. We must all live with heightened vigilance and increased fear of more unpredictable attacks.

With the attacks and their aftermath, many Americans experienced levels of emotion, particularly fear and anxiety, to which they are unaccustomed. Many experienced acute posttraumatic symptoms (Schuster et al., 2001; Galea et al., 2002; Schlenger et al., 2002; Silver, et al., 2002). Although some people do not recover well from the trauma and develop chronic anxious or depressive conditions, in the vast majority of cases, these feelings recede, and people go on with their normal lives. These people will not return to their previous state, however – the external world has changed, and their internal psychological states have changed, too.

What are the psychological and biological mechanisms for the rapid change in emotion, and then the return to a normal although changed life, or, in some cases, to a new life of pathological states of anxiety or depression?

This chapter explores the processes by which mammals maintain viability following physiological change due to adversity. These processes are captured by the concept of allostasis – a means of achieving psychological and biological states appropriate to changed circumstances. We address this concept within the context of the neurobiology of the normal, adaptive emotion of fear and pathological states of anxiety and depression, which can result from sustained or overexpressed normal feed-forward mechanisms associated with fear. We begin with definitions of allostasis, adaptive fear, and pathological anxiety. Several learning and conditioning mechanisms

This research was supported in part by NSF #IBN 0129809.

164

that are involved in adaptive fear are discussed next. The brain systems and circuits involved in fear, particularly those of the amygdala and associated structures, are described in a context of adaptive processes. We also describe feed-forward mechanisms regulated by glucocorticoids to explain how normal, adaptive biological processes of fear can transform into pathological states of anxiety and depression. These steroids are not molecules of fear, but molecules that regulate and sustain fear, acting as hypersensitizing agents that can overload adaptive processes.

ALLOSTASIS IN ADAPTIVE FEAR AND ALLOSTATIC OVERLOAD IN PATHOLOGICAL ANXIETY

The term allostasis (Sterling and Eyer, 1981, 1988) was introduced to refer to the process of insuring viability in the face of a challenge and change – conditions that cannot be explained by homeostasis. Allostasis means maintaining viability through change of state (bodily variation). It is broader than the concept of homeostasis in that homeostasis accounts for physiological stability around a consistent setpoint, such as body temperature, whereas allostasis allows for setpoints to fluctuate as a result of experience. In addition, allostasis was originally conceptualized to involve whole brain and body, rather than simple local feedback circuits, and is a far more complex form of regulation than homeostasis (Sterling and Eyer, 1988). Peripheral processes (hormones, autonomic activity) contribute inhibitory and stimulatory input by which they can diminish and suppress or enhance and prolong neural activation (Fig. 6.1). The concept of allostasis allows for the continuous development and fluctuation of new physiological and behavioral setpoints. At the heart of the concept of allostasis is the depiction of change to maintain (or achieve) a state appropriate to the current and anticipated circumstances. Both feed-forward and inhibitory mechanisms contribute to modification and fluctuation of central states to respond to changing circumstances.

Importantly, allostasis highlights anticipatory processes and one's ability to adapt to or cope with impending future events (Schulkin et al., 1994). States of fear, when limited in duration, are quintessential allostatic states, as discussed later. During a fear-evoking situation, both central and peripheral processes are highly active in response to threat and in anticipation of further harm. If one survives the danger, the fear motivates changes in behavior in anticipation of encountering a similar danger in the future. A new setpoint or threshold for activation and elicitation of defensive responses is attained.

Allostatic anticipatory processes that maintain organismic viability are adaptive, but if the system becomes overexpressed, it can lead to allostatic

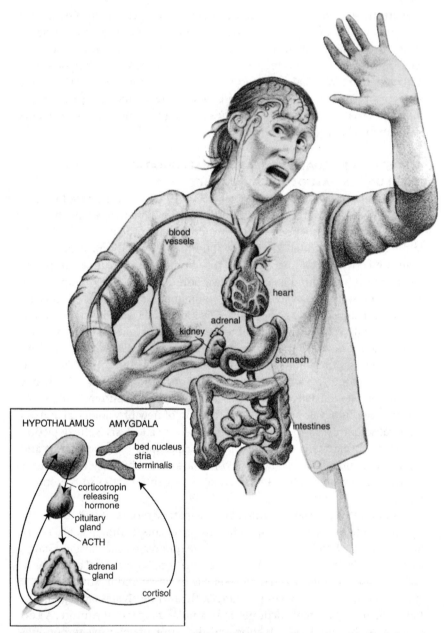

Figure 6.1: Artist's drawing of a person in a state of fear. The brain and several visceral organs are drawn to emphasize the integration of brain and body processes during fear. The inset shows the HPA axis (the hypothalamus, pituitary, and adrenal glands), along with the amygdala and bed nucleus of the stria terminalis working together. (ACTH = adrinocorticotropin hormone.) Artist: John Yansen.

overload and then pathology. In other words, allostatic overload via several mechanisms of overactive or inefficient responding can cause systems to breakdown (McEwen, 1998). This has been described for a number of physiological disorders including obesity, type 2 diabetes, and cardiovascular disease (Seeman et al., 1999). For anxiety and depression, overactivity or prolonged responding of feed-forward processes (e.g., glucocortiocoid secretion, neurotransmitter and neuropeptide release) that function normally during adaptive fear, change thresholds for future encounters by sensitizing fear systems so that they respond with exaggeration or at inappropriate times. These overloading processes are discussed following a description of normal, adaptive fear, the psychological processes of fear and pathological anxiety, and the underlying neural substrates of fear.

THE EMOTION OF FEAR

Fear states are emotional and motivational. They organize activity and behavior to promote problem solving and adaptation to danger. Fear is a prototypical exemplar of a central state – a state of the brain. Although systemic physiological changes influence the state of fear (James, [1890]1950), peripheral changes are not sufficient for the expression of emotional behaviors such as fear (Cannon, [1915] 1929; Bard, 1939). It is the physiological change in the brain that is linked to the state of fear. We are afraid when we perceive danger. Nevertheless, bodily events influence and reinforce this state (James, [1890]1950; Damasio, 2000). Changes in heart rate, blood pressure, respiration, facial muscles, and catecholamines, both peripheral and central, all influence the state of fear (see LeDoux, 1996; Rosen and Schulkin, 1998).

Learning, attention, and assessment of relevant information are also processes incorporated in fear (MacIntosh, 1975; Dickinson, 1980), and are important in predicting future outcomes (Rescorla and Wanger, 1972). Learning to fear a dangerous object or situation can occur rapidly, frequently in a single exposure. One probable reason for this is that animals do not need to learn new behaviors but learn the significance of stimuli and then respond in stereotypic, prepared ways that promote survival (Bolles, 1970). Thus, the central state of fear is linked to action tendencies (readiness to respond in stereotypic ways; Arnold, 1960), to attention (Lang, 1995), to appraisal and learning about threatening environmental stimuli (Lazarus, 1991; Rosen and Schulkin, 1998; Lane and Nadel, 2000).

Fear, as the perception of danger, is an adaptive response and fundamental in problem solving and survival. In fact, fear is an emotion that likely evolved as part of problem solving (Darwin, [1972] 1965). Emotions like fear prepare the animal to respond to danger by heightening vigilant attention (Gallagher and Holland, 1994; Lang, 1995) and motivating behavior (Bindra, 1978; Mowrer, 1947) that includes defensive behaviors (Bolles and Fanselow, 1980). The state of fear is one in which there is a readiness to perceive events as dangerous or alarming (LeDoux, 1987; Rosen et al., 1996). The intensity of fear can vary from mild, as in arousal during encounters with ambiguous events (lack of knowledge of whether stimuli are potentially threatening or not), to terror when life is almost certainly going to end in violence. At most times during everyday life, fear is induced to facilitate responding to changing circumstances that cause momentary threat and expedite anticipation of these circumstances in the future.

Fear states are thus entwined in learning about what is dangerous, what is safe, and what is neither, as well as acquiring informational value of stimuli that has predictive properties for the animal to respond in appropriate ways (Rescorla and Wagner, 1972; Dickinson, 1980). Fear functions to alert the animal to danger, preparing the animal for defensive action (Blanchard and Blanchard, 1972; Bolles and Fanselow, 1980). The motivated animal seeks relief from this aversive state, and, with the elimination of fear, there is the sense of relief (Mowrer, 1947). Although we have discussed fear as a unitary emotion, one must keep in mind that there is more than one kind of fear (Hebb, 1946; Kagan and Schulkin, 1995).

PATHOLOGICAL ANXIETY AND DEPRESSION

We have previously argued that pathological anxiety disorders and depression develop from normal adaptive fear (Rosen and Schulkin, 1998). Normal fear and its adaptive processes of anticipation enhance the ability to achieve and maintain appropriate responses to impending danger. Fear states increase in threatening circumstances and can remain elevated for long periods of time, but also should diminish when the danger has passed or the environment becomes safe. Nevertheless, these allostatic states can be overloaded and exceed their abilities to adapt. With pathological anxiety these allostatic mechanisms that normally respond and adjust to danger become hyperexcitable. This includes not only fear-related autonomic and behavioral responses that are activated during pathological anxiety, but, importantly, the perceptual fear response of greater vigilance. Appraisal mechanisms become overactive, leading to increased perception of fear that then leads to anxious thought and maladaptive behavior (e.g., Rosen and Schulkin, 1998; Davis and Whalen, 2001). In psychological terms, both anxious and depressive states have a common core of heightened negative affect (Mineka et al., 1998), likely a product of overactivity of neural systems instantiating fear (Rosen and Schulkin, 1998).

Animals evolved to respond to external stimuli. Responses can be flexible, but in many cases, responses have developed evolutionarily in a prepotent or fixed manner (Tinbergen, [1951] 1969; Lorenz, 1981). This seems particularly true of fear responses (Bolles, 1970). If the perceptual-response system is primed and more sensitive or excitable, there is then a greater tendency for action. Arnold's (1960) notion of emotions is that an increased tendency exists to respond to stimuli that elicit particular responses. Increases in the readiness to respond would produce greater or exaggerated responses to stimulation and would allow for these responses to be elicited with lower intensity stimulation. The various cognitive biases (e.g., interpretive, attention or memorial, (see Eysenck, 1992; Hertel, 2002; Matthews and MacLeod, 1994) and increased startle responses (Allen et al., 1999; Grillon et al., 1991, 1994; Morgan et al., 1995) demonstrated by patients with anxiety and depressive disorders in response to threatening and positive stimuli indicate that neural fear systems are hyperexcitable in anxiety and depressive disorders (however, see Baas et al., 2002). Neurologically, this can be conceptualized as hyperexcitability of brain structures that evaluate exteroceptive, interoceptive, and proprioceptive stimuli as dangerous. Thus, external as well as internal autonomic and muscular events are evaluated more readily as signaling danger in a hyperexcitable danger evaluation system. In allostatic terms, the adaptive processes of anticipation and the ability to achieve a state appropriate to the circumstances are incapable of functioning properly. Allostatic overload renders the processes unable to achieve and maintain new, adaptive setpoints. With pathological anxiety, a sustained and inflexible hyperexcitable fear system will evaluate and respond inappropriately to moderately fearful or even nonfearful circumstances and events.

We argue later that sustained, chronic activity of hormonal systems via feed-forward and kindling-like mechanisms plays a key role in producing allostatic overload, hyperexcitability in the neural circuits of fear, and pathological anxiety states. In particular, overactivation or chronic, repeated activation of feed-forward hormonal processes such as glucocorticoids and catecholamines amplify information processing and behavioral responses in neural fear systems to the point where they are chronically hyperexcitable. This would manifest itself in the various anxiety and depression disorders. However, first we discuss the neural, biochemical, and cellular bases of fear that instantiate adaptation and allostasis (Fig. 6.2).

ADAPTIVE FEAR AND THE BRAIN
Learning about potentially harmful stimuli is essential for subsequent detection and defense. Importantly, additional feed-forward processes that can

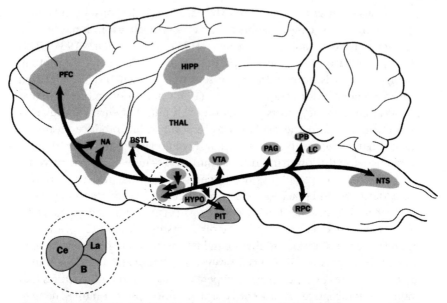

Figure 6.2: Schematic drawing of a neuroamatomic circuit of fear. Major areas and pathways are described, but not all are included. The amygdala (circle with the lateral [La], basal [B], and central [Ce] nuclei labeled) plays a central role in the circuit. As explained in the text, the lateral nucleus receives most sensory input via the thalamus (THAL) and cortex. The basal nucleus receives input from the lateral nucleus, hippocampus (HIPP), and cortex. The basal nucleus also send efferents to the lateral portion of the bed nucleus of the stria terimalis (BSTL), nucleus accumbens (NA), and prefrontal cortex (PFC). The central nucleus receives input form the lateral and basal nuclei and has extensive output to diencephelon, midbrain and brainstem. This includes the hypothalamus (HYPO), ventral tegmental area (VTA), periaqueductal gray (PAG), lateral parabrachial nucleus (PB), locus ceureleus (LC), reticularis pontis caudalis (RPC), and the nucleus of the solitary tract (NTS). As can be seen, many of these areas send reciprocal connects to the amygdala. Major inputs discussed in the text are the prefrontal cortex, nucleus of the solitary tract, and the locus ceureleus.

heighten anticipation and vigilance are necessary for sustaining these functions during times of high or continuous uncertainty or danger, discrepancy of events, alertness to novelty, and attention to the contours of the environment. Therefore, allostasis would include biochemical and molecular changes in fear circuits instantiating anticipation, and other feed-forward factors that can modulate the circuitry to sustain anticipation and vigilance over long periods. These processes are discussed within the context of the neuroanatomic circuitry for fear, primarily those involving the prefrontal cortex, the amygdala, and their connections. It is noted that these structures are involved in many functions, fear being one of them. Because of our long-term interest and research in amygdala circuitry, the amygdala is more fully discussed than the prefrontal cortex.

Prefrontal Cortex

A fundamental part of fear circuitry is the prefrontal cortex, primarily the orbitofrontal and ventromedial prefrontal cortices and cingulate cortex (e.g., Quirk et al., 2000; Davidson, 2002). In humans, the prefrontal cortex encompasses nearly a third of the brain. Medial and orbital aspects underlie a wide range of central states, including the representation of bodily impact (visceral representation at the level of the neocortex; Critchley et al., 2001) on decision making (Bechara et al., 2000; Critchley et al., 2002b; Damasio, 1994; Fuster, 2001) and executive control of behavior (Critchley et al., 2002a; Fuster, 2000). For example, in experiments using a gambling task, lesions of the ventromedial prefrontal cortex disrupt the production of anticipatory autonomic responses and informational processing of visceral information that contribute to correct decisions in the task (Bechara et al., 1997, 1999). Impulsivity, reflective of a lack of executive control of behavior, is also a problem in patients with orbitofrontal lesions (Damasio, 1994).

The medial prefrontal cortex also plays a role in inhibition of fear responses and extinction. Although some studies have found that lesions of the medial prefrontal cortex block short- and long-term extinction of fear-induced freezing in rats (Morgan and LeDoux, 1995; Quirk et al., 2000), other lesion studies have shown that medial prefrontal cortex lesions reduce full autonomic and behavioral expression of fear responses (Frysztak and Neafsey, 1991, 1994).

The ventromedial (infralimbic and paralimbic) and dorsomedial (cingulate) aspects of prefrontal cortex in rats may also play opposing roles in autonomic and hormonal regulation. The ventromedial prefrontal cortex tends to generate and accentuate autonomic and glucoccorticoid responses, whereas the dorsomedial prefrontal area dampens and negatively regulates these physiological responses (Sullivan and Gratton, 2002).

In rats, the orbital prefrontal cortex has been shown to display firing of neurons in coordination with firing of cells in the basolateral complex of the amygdala during discrimination learning and divergent firing with reverse learning (Schoenbaum et al., 1998, 2000). It is hypothesized that both the amygdala and orbitofrontal cells are encoding incentive value of stimuli. The orbitofrontal cortex is also functionally important for a combination of functions, including executive control of performance, incentive value of stimuli, and working memory (Schoenbaum et al., 2000; Schoenbaum and Setlow, 2001).

Right prefrontal cortex activation is linked to more negative emotions and withdrawal, whereas the left side of the prefrontal region is linked to more positive emotions (Davidson et al., 1995, 2000). For example, lesions of the left prefrontal cortical region are more closely associated with states of depression; the converse holds with damage to the contralateral region

(Davidson, 2002; Davidson et al., 2002). In functional brain imaging studies, the elicitation of positive emotions is closely linked to left prefrontal cortical activation (central states in which we prefer to stay), and negative emotions are closely linked to right prefrontal activation (states that we wish to leave) (Davidson, 2002; Davidson et al., 2002; Schmidt et al., 1999). In a number of contexts, activation of the left frontal cortex is tied to positive representations, experiences, and contexts; the converse holds true for negative representations (Davidson, 1998; Sutton and Davidson, 2000). Moreover, this cortical activation is associated with affective states; patients who are depressed have greater relative activation of right prefrontal cortex than those who are not (Davidson et al., 1999).

Amygdala

Although the amygdala has been known to be involved in the emotion of fear since the seminal studies of Kluver and Bucy (1939) showed a taming effect of amygdala lesions in monkeys, research in the last two decades has produced great advances in determining the neuroanatomy of fear circuits. Not only has the amygdala been found to be critical for many types of fear, but fear circuits that connect the amygdala to many other brain regions have been described that suggest that these circuits have evolved to function as neurobehavioral systems for particular kinds of cognitive and behavioral strategies.

Regions of the amygdala receive information from both cortical and subcortical regions (Krettek and Price, 1978; Amaral et al., 1992; Pitkanen, 2000). Specifically, the basolateral complex comprising the lateral, basal, and accessory basal nuclei are richly innervated by neocortical and subcortical uni- and polymodal sensory regions (e.g., Krettek and Price, 1978; Turner and Herkenham, 1991; Rosenkranz and Grace, 2002a; Stefanacci and Amaral, 2002), which then relay information to the central nucleus of the amygdala (Pitkanen et al., 1997). The central nucleus projects to numerous nuclei in midbrain and brainstem to orchestrate the rapid and primary behavioral, autonomic, and endocrine responses to threat and danger (Davis, 1997; Rosen and Schulkin, 1998; LeDoux, 2000). The central nucleus also receives visceral information from brainstem sites that include the solitary and parabrachial nuclei (Ricardo and Koh, 1978) and reciprocally project to these brainstem regions (e.g., Schwaber et al., 1982). The basal nucleus also directly projects to the nucleus accumbens, which led Nauta and Domesick (1984) to suggest an anatomic route by which motivation and motor control action are linked in the organization of active behavior (see also Yim and Mogenson, 1982; Swanson and Petrovich, 1998; Gray, 1999; Amorapanth et al., 2000; Swanson, 2000). Thus, the central nucleus of the amygdala via its projections to lower brain orchestrates reactive behavioral (freezing), autonomic and endocrine responses to fear, whereas efferents of the basal nucleus of the amygdala participate in active avoidance behaviors to fear (Amorapanth et al., 2000), likely through nucleus accumbens, striatum, and thalamus. In addition, the medial nucleus of the amygdala was shown recently to regulate endocrine responses via direct projections to the hypothalamus (Dayas et al., 1999).

A recent study also examined the effects of lesions of numerous visual cortical and subcortical areas on fear conditioning to a light stimulus (Shi and Davis, 2001). Combined lesions of lateral geniculate nucleus and the lateral posterior nucleus of the thalamus block visual fear-potentiated startle. The lateral geniculate projects to the primary visual cortex, which then projects to the temporal and perirhinal cortices. The lateral posterior thalamus projects directly to the temporal and perirhinal cortices. In conjunction with previous studies showing blockade of fear-potentiated startle following perirhinal lesions (Rosen et al., 1992; Campeau and Davis, 1995) sensory pathways for visual fear conditioning have been proposed to include the perirhinal and temporal cortices all impinging on the lateral nucleus of the amygdala (Shi and Davis, 2001). These thalamo-cortico-amygdaloid visual pathways for emotional conditioning parallel those described for auditory fear conditioning (LeDoux, 2000) suggesting similar organization of these two sensory-emotional circuits.

ACTIVATION OF THE AMYGDALA IN HUMANS

There is also a good deal of evidence in humans that the amygdala is linked to fear (e.g., Aggleton, 1992, 2000; LeDoux, 1996; Morris et al., 1996; Calder et al., 2001). Humans with lesions of the amygdala have impaired fear-related behavior and autonomic responses to conditioned stimuli (e.g., Nahm et al., 1993; LaBar et al., 1995; Angrilli et al., 1996). Several studies have found that lesions of the amygdala interfere with the recognition of fearful facial expression (Allman and Brothers, 1994; Adolphs et al., 1995). Furthermore, the humans with lesions of the left amygdala have impaired responding to fearful stimuli that are verbal in nature, whereas those with right amygdala lesions have deficits with visually presented fear stimuli (Funayama et al., 2001). Also, positron emission tomography (PET) imaging studies have shown greater activation of the amygdala during fear and anxiety-provoking stimuli (Ketter et al., 1996). Such PET studies have revealed that the amygdala is activated when presented with fearful versus happy faces (Morris et al., 1996; Dolan et al., 2000; Wright et al., 2001). With the use of functional magnetic resonance imaging (MRI), it has further been shown that the amygdala is activated and then habituates when shown fearful in contrast to neutral or happy faces (Breiter et al., 1996); but the amydala is also responsive to a variety of facial responses (Lane et al., 1997; Dolan et al., 2000; Wright et al., 2001). Numerous studies have also demonstrated that patients with anxiety disorder have excessive activation in the amygdala when presented with stimuli that provoke anxiety attacks (for review, see Davis and Whalen, 2001).

The amygdala is important not only for strong states of fear, but also for states of uncertainty. Whalen (1999) has presented a model of amygdala's primary function serving to regulate vigilance during times of ambiguity. Thus, the amygdala would be highly active not only during fear states, but also during novel experiences, changes in values of primary rewarding or punishing stimuli, and changes in predictive values of conditioned stimuli, (Baxter and Murray, 2002). These alterations in external environmental conditions involve anticipation of change and promote allostatic responses to accommodate the new uncertainties. The amygdala appears to be central to allostasis and psychological processes involved in everyday events and the more life-threatening circumstances that evoke fear. Knowledge of hormonal regulation of fear in the amygdala, and cellular and molecular events in amygdala neurons would increase our understanding of the basic machinery of fear and uncertainty. Some of these events are discussed later in the chapter.

Roles of the Amygdala and Prefrontal Cortex
in an Allostatic Framework

Neural mechanisms enable an animal to anticipate and learn ways to survive and maintain viability (allostasis) in the face of change, danger, and circumstances that induce fear. The amygdala and prefrontal cortex play distinct but overlapping roles in allostatic processes of fear. In general, the amygdala is involved in evaluation, detection, and anticipation of threatening events. It also orchestrates the emotional, behavioral, autonomic, and neuroendocrine response to danger. Although the prefrontal cortex is also important for these functions, particularly for anticipation of challenge and regulating neuroendocrine and autonomic physiology, it also acts to constrain emotional responses and is critical for reversing the emotional, and therefore behavioral, responses to previously learned stimuli (Milad and Quirk, 2002).

The prefrontal cortex is important for flexibility in responding and changing stimuli relationships that are important for allostasis. This has been shown in rats, monkeys, and humans (Schoenbaum and Setlow, 2001). Experimental paradigms in rats and monkeys that continuously change the predictive values of sensory stimuli (odors, lights, tones) for receiving reward or punishment for responding are severely disrupted by medial and orbital prefrontal cortex lesions. Unlike amygdala lesions, prefrontal cortex lesions do not disrupt the initial learning of predictive value of a stimulus but do interfere with the ability of animals to stop responding to the first stimulus relationship learned and learn a different relationship. In humans, a similar inability to change strategies is also evident. On a gambling task in which several decks of cards pay big wins but eventually produce overall losses, and other decks have small payoffs but overall produce winnings, people with ventromedial prefrontal cortex lesions choose the big payoff–eventual loss deck but cannot switch strategies and play the small payoff–eventual win, even though they know which decks they should be playing (Bechara et al., 1997). Autonomic, and possibly endocrine, systems appear to be important because people with ventromedial lesions playing the card game do not display normal skin conductance responses to making wrong decisions that may guide behavior to the proper choices (Bechara et al., 1997, 1999, 2000). Imaging techniques demonstrate similar viscerally related function of the prefrontal cortex in normal humans (Critchley et al., 2000, 2001, 2002a, 2002b).

The amygdala plays a central role in allostatic regulation of fear. In general, the amygdala is involved in evaluation, detection, and anticipation of threatening events (Davis and Whalen, 2001; LeDoux, 2000; Rosen and Schulkin, 1998). Although animals with lesions of the amygdala respond normally to painful stimuli, they do not learn to anticipate the events. The basolateral complex of the amygdala with its rich afferents from the thalamus and cortical regions is neuroataomically situated to connect information about neutral stimuli with those that produce pain or are harmful. From this privileged place the amygdala can participate in learning associations from crude sensory information from the thalamus or highly elaborated sensory information from the cortex that are crucial for anticipation of these events in the future. In this way information from neutral stimuli can be paired with painful or threatening events in the amygdala to form conditioned associations between these events. Following this associative process, an animal can then respond appropriately and defensively when the conditioned stimulus (previously a neutral stimulus) is detected in anticipation of the threatening event. Two of the most consistent experimental findings are the inability of amygdala-lesioned animals to learn new associations between neutral and painful stimuli and the loss of previously conditioned fear following amygdala lesions. These findings strongly argue for an anticipatory role of the amygdala, particularly the basolateral complex, in learning new fear-related associations and maintaining already learned fear-related associations.

In addition to projections from the central nucleus of the amygdala to midbrain and brainstem targets important for mounting quick behavioral, autonomic, and endocrine responses to danger, the amygdala projections to the cortex are also extensive. In rat, the sources are the lateral, basal, and accessory basal nuclei and are fairly restricted to the multisensory temporal lobe structures (perirhinal, pyriform, and entorhinal cortices) and prefrontal cortex (Pitkanen, 2000), although primary visual cortex in the primate brain also receives input from the amygdala (Amaral et al., 1992). These cortical structures also contribute the heaviest cortical input to the amygdala suggesting that many of the connections between the amygdala and cortex are reciprocal. This is particularly true of the amygdala and prefrontal cortex both anatomically (Amaral et al., 1992; Pitkanen, 2000) and functionally (for review, see Davidson et al., 2000). Thus, not only can sensory information influence amygdala function, but the amygdala can affect information processing in various association and sensory cortical areas, possibly contributing to the emotional coloring of information in these regions (LeDoux, 1996).

GLUCOCORTICOIDS AND THE REGULATION OF FEAR

While learning plays a key role in determining what stimuli signal potentially harmful events, fear responses are typically sustained beyond the dissipation or removal of fear-inducing stimuli. Prolonged fear responses can be sustained by a number of mechanisms. In most of the vertebrates that have been studied, fear is sustained by neuroendocrine factors (Jones et al., 1988; Sapolsky, 1992; Jones and Satterlee, 1996). The secretion of glucocorticoids helps to sustain a number of behavioral responses including fear-related behaviors (Richter, 1949; see review by Korte, 2001). Without glucocorticoids, as Richter (1949) noted in the mid-1900s, animals die under conditions of extreme duress (see also Selye, 1956). Adrenalectomized animals are unable to tolerate fear, duress, or chronic stress and suffer fatally. Glucocorticoids thus prepare the animal to cope with emergency and taxing environmental contexts (Cannon, [1915]1929; Richter, 1949).

Glucocorticoids (cortisol in humans and monkeys, corticosterone in rats) are also essential in the development of fear (Takahashi, 1994). They are secreted under a number of experimental conditions in which fear, anxiety, novelty, and uncertainty are experimental manipulations (Mason et al., 1957; Mason, 1968; Brier, 1989). In contexts in which there is loss of control, or the perception of a loss of control (worry is associated with the loss of control), glucocorticoids are secreted. This holds across a number of species, including humans (e.g., Brier, 1989). Perceived control reduces the levels of glucocorticoids that circulate. In rats, for example, predicting the onset of an aversive signal reduces the level of circulating glucocorticoids (Kant et al., 1992). Within the clinical sphere, one of the most consistent findings in fearful, depressed patients is elevated levels of cortisol and an enlarged adrenal cortex (Carroll, 1976; Nemeroff et al., 1992). These findings are congruent with those of Richter (1949), who observed an enlarged adrenal gland in stressed, fearful wild rats when compared with unstressed laboratory analogues.

From a biological view the chronic activation of glucocorticoid hormones is costly. The subordinate male macaque has elevated cortisol levels but lower levels of testosterone than the dominant one (Sapolsky, 1992, 2000). The lower level of testosterone decreases its reproductive fitness in the short term. The cost of chronic subordination is perhaps more fearfulness and uncertainty of attack as well as a further decrease in the likelihood of successful reproduction. This phenomenon of high corticosterone and low testosterone has been demonstrated in a number of species (e.g., Lance and Elsey, 1986).

Sustained fear is also metabolically costly. Although glucocorticoids are essential in the development of neuronal tissue and in adapting to duress (Gould et al., 1991), if the elevation of glucocorticoids is sustained over time, tissue (e.g., brain, muscle, and bone) will begin to deteriorate (Sapolsky, 1992; McEwen and Stellar, 1993). Chronic glucocorticoid activation, for example, increases the likelihood of neurotoxicity and neural endangerment through the loss of glucocorticoid receptors (Sapolsky, 1992).

Glucocorticoids are not the molecules of fear. They do, however, play a fundamental role in energy balance by regulating glucose utilization (hence their name). They are important is sustaining high plasma levels of glucose, a breakdown of fat and protein for neoglucogenesis when energy is needed for extended periods of time. Glucocorticoids are secreted in young children who are energetic (Tout et al., 1998), they play a role in attachment behaviors in humans and other animals (Carter et al., 1997; DeVries, Taymans, and Carter, 1997), and they facilitate a number of behavioral events (e.g., Denton, 1982; Schulkin, 1991; Sumners et al., 1991; Dallman and Bhatnager, 2000) by their actions in the brain and the induction of neuropeptides and neurotransmitters (Herbert and Schulkin, 2002).

Although glucocorticoids are certainly not the molecules of fear and anxiety, they are associated with fear, anxiety, and trauma – highly metabolically demanding events. With this caveat, extremely shy, socially withdrawn children may be vulnerable to anxiety disorders and perhaps depression throughout their lives (Hirshfeld et al., 1992). They should be vulnerable to allostatic load, for example vulnerability to allergic symptoms (Kagan et al., 1991) or vascular disease (Bell et al., 1993), perhaps because of the chronic worry that they experience in social contexts or in unfamiliar environments. Interestingly, high cortisol levels have been linked to both fearfulness in childhood and depression in adulthood (Brown et al., 1996).

An analogous phenomenon to that of shyness and fearful behavior in children has been observed in a subset of young, fearful rhesus monkeys that have high levels of cortisol. This subset also freezes for longer periods of time than other rhesus monkeys (Champoux et al., 1989). In adult rhesus monkeys, high levels of cortisol, in addition to high levels of corticotropin-releasing hormone (CRH) from the cerebrospinal fluid, is associated with behavioral inhibition (Habib et al., 2000; Kalin et al., 2000). In addition, when faced with an unknown intruder in an adjacent cage, macaques increase their CRH expression (Habib et al., 2000). Increases (or sensitization) of CRH in the brain occurs after stress, abuse, and maternal deprivation in macaques (Habib et al., 2001).

A subset of these macaques not only have higher levels of CRH and cortisol than normal animals, but also demonstrate greater fearful temperament and greater activation of the right hemisphere. High activation of right hemisphere has been linked to withdrawal and negative perception of events (Davidson, 1998; Kalin et al., 2000; Habib et al., 2001). Differences in temperamental expression to a number of unconditioned fear-related stimuli may reflect frontal neocortical activation (Kalin et al., 2001) because ibotenic acid lesions (cell body destroyed and fibers left intact) of the macaque amygdala left a number of unconditioned behavioral traitlike responses intact (Kalin et al., 2001), in addition to the normal asymmetry associated with traitlike dispositions.

To understand the impact of fear on the body and brain, a new conceptual framework that involves the concepts allostasis, allostatic state, and allostatic overload was developed. These can serve for understanding states of fear, chronic fear, anxiety, and depression and the eventual breakdown of biological tissue and function. Mechanisms, including glucocorticoids, play a role in learning and memory of fear, as well as sustaining fear for long periods when needed, whereas prolonged, chronic overexpression of these mechanisms represents an allostatic overload that can lead to sensitization and pathological states.

Although we focus the rest of the chapter on glucocorticoids in allostatis and allostatic overload of fear, corticosteroids are not the only peripheral signals acting in the brain that play a role in fear. Epinephrine, norepinephrine, and the autonomic nervous system are also crucial systems (Fig. 6.3). Epinephrine and norepinephrine are essential for learning

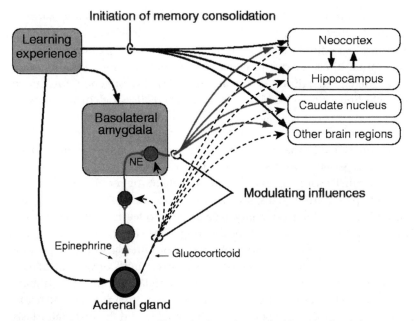

Figure 6.3: Schematic diagram illustrating how glucocorticoids and epinephrine can modulate memory consolidation processes via the amygdala. A learning experience will activate brain processes in numerous brain regions. It will also activate glucocorticoid secretion that can influence peripheral arousal mechanisms, such as epinephrine, and numerous brain regions. The basolateral complex of the amygdala plays a central role by being modulated by glucocorticoids and in turn modulating memory consolidation processes in other brain regions. Norepinephrine modulation of amygdala function is illustrated, but other neurotransmitter systems, such as dopamine, serotonin, and acetylcholine, also influence the amygdala. Adapted from McGaugh (2000).

and memory for aversive events primarily by direct and indirect action in the amygdala (McGaugh and Roozendaal, 2002).

In addition to peripheral inputs, neurotransmitter systems within brain also modulate actions in forebrain regions important for fear. For example, dopamine input to neurons in the lateral nucleus of the amygdala originating from cells in the midbrain is an important electrophysiological change in lateral nucleus neurons with fear conditioning (Rosenkranz and Grace, 2002b). Norepinephrine projecting from the brainstem also modulates fear conditioning by action in the basolateral complex of the amygdala (McGaugh and Roozendaal, 2002). In addition, cholinergic systems originating in the basal forebrain project to amygdala, cortex, and other forebrain areas to influence them and the psychological process of attention (Gallagher and Holland, 1994). Furthermore, amygdala and prefrontal cortex also project to the cholinergic basal forebrain to increase activity and output of this cholinergic system to further influence amygdala, cortex, and forebrain structures. This would increase attentional and vigilance processes increasing the ability to detect and appraise potentially dangerous stimuli. Allostatic overload may also change the long-term functioning of the cholinergic system by alternative slicing of gene transcripts of aceytlcholine esterase (Meshorer et al., 2002). Following chronic stress or repeated corticosterone treatment, there was a long-lasting shift in production from a functional synaptic form of acetylcholine esterase to a soluble form sequestered in the cytoplasm of neurons. This shift in form rendered the cells hypersensitive. A second example is a recent study demonstrating that individuals with copies of the short allele of the serotonin transporter promoter polymorphism display increased fear- and anxiety-related behaviors and also exhibit greater amygdala activity to fearful stimuli compared with individuals homozygous for the long allele (Hariri et al., 2002). This polymorphism is associated with reduced serotonin transporter expression and function, possibly allowing for increase excitability of the amygdala to fearful stimuli. These examples stress the interrelationship between many brain systems to set a tone for whole organism responses to stress and fearful stimuli. Some of these responses may be variable and fluctuate (allostasis) or may become inflexible (allostatic overload) due to acquired or genetic predispositions (or, most likely, an environmental–genetic interaction).

GLUCOCORTICOIDS AND THE ALLOSTATIC REGULAION OF FEAR: THE BASOLATERAL COMPLEX OF THE AMYGDALA

Glucocorticoids are important in allostatic regulation (McEwen, 1998). This is particularly evident in fear and aversive conditioning within the amygdala where glucocorticoids appear to be necessary and part of the normal processes of fear and aversive learning. Adrenalectomized rats have deficits in associative fear conditioning (Suboski et al., 1970; Sakaguchi et al., 1984; Pugh et al., 1997). Roozendaal, McGaugh and others have extensively studied the role of glucocorticoids with peripheral and central administration of glucocorticoids, inhibitors, agonists, and antagonists (for review, see Roozendaal, 2000; McGaugh and Roozendaal, 2002). In brief, glucocorticoids in the basolateral complex of the amygdala appear to be necessary for aversive and fear conditioning. For example, injection of the glucocorticoid receptor antagonist RU-486 into the basolateral complex of the amygdala blocks the consolidation of aversive conditioning. Other experiments have shown that glucocorticoid injections into the amygdala can facilitate aversive conditioning. Experiments like these, which use posttraining injection procedures, demonstrate that glucocorticoids are necessary for consolidation of the memory of aversive conditioning and may facilitate the memory

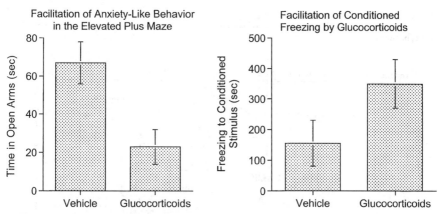

Figure 6.4: Glucocorticoids increase anxiety like behavior and fear conditioning. The left graph demonstrates that following chronic administration of corticosterone into the amygdala, rats spent significantly less time in the open arms of an elevated plus maze (increased anxiety-like behavior; Shepard et al., 2000). In the right graph, chronic corticosterone was given systemically (Corodimas et al., 1994). On the fifth day of administration, rats were given tone-shock pairings. In a retention test, fear-related freezing (an index of conditioned fear) was greater for rats that received chronic corticosterone.

process. Interestingly, chronic high levels of glucocorticoids given before conditioning disrupt long-term memory of fear conditioning but leave immediate short-term memory intact. Thus, although aversive or fear conditioning is facilitated by glucocorticoids, they can interfere with retrieval of long-term memory in both rats and humans (Luine et al., 1994; de Quervain et al., 1998, 2000), suggesting that glucocorticoids are necessary for aversive learning but may produce temporary allostatic overload that interfere with previously learned fear. Extinction of learned fear may also be disrupted by excessive amounts of glucocorticoids (Corodimas et al., 1994), thus interfering with normal processes regulating appropriate and adaptive levels of fear when stimuli no longer predict aversive events (Fig. 6.4). Finally, amygdala infusion of corticosterone also increases milder forms of anxiety as measured with rats in the elevated plus maze (Shepard et al., 2000).

CELLULAR AND MOLECULAR PROCESSES IN FEAR: THE BASOLATERAL COMPLEX OF THE AMYGDALA

Glucocorticoids act to affect neural function at the extracellular membrane and nuclear loci (Baulieu, 2000). While the molecular consequences of glucocorticoids in fear are not yet well understood, neural changes important for fear are beginning to be uncovered at the cellular and molecular levels. These include neurotransmitters, receptors, signal transduction molecules, gene transcription, and protein synthesis. This section of the chapter

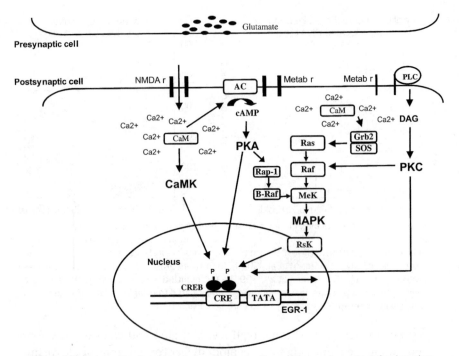

Figure 6.5: Schematic drawing of signal transduction pathways important for learning and memory of fear. Four signal transduction pathways are illustrated all leading to activation of cyclic adenosine monophosphate (cAMP) responsive element binding protein (CREB)-mediated transcription of the immediate-early gene *egr-1*. Glutamate N-methyl-D-aspartate (NMDA) receptors are known to be important for learning and memory of fear by increasing the influx of calcium (Ca^{2+}). Ca^{2+} acts as a second messenger to activate calcium/calmodulin-dependent kinase (CaMK) and protein kinase C (PKC), modulate cAMP synthesis, protein kinase A activity, and the mitogen-activated protein kinase (MAPK) pathway. Each of these signal transduction pathways can phosphorylate CREB, which regulates gene transcription. The figure shows CREB binding to the promoter region of *egr-1* to regulate its transcription.

describes a cascade of some of the molecular events in amygdala neurons initiated by fear conditioning that appears responsible for long-term changes in memory (Fig. 6.5).

As with other types of learning and models of learning (i.e., long-term potentiation, spaital learning, recall of previously learned tasks), glutamate receptors are involved in the first steps of long-term memory of fear conditioning (Blair et al., 2001). N-methyl-D-aspartate (NMDA)-type glutamate receptors are unique in that they require both glutamate and depolarization of the cell to open their channels to allow Ca^{2+} influx that can initiate numerous intracellular processes. NMDA receptors are therefore coincident detectors and may be prime candidates for detection of simultaneous input from a conditioned and unconditioned stimulus producing learning (Martin et al., 2000). Initial studies done in Michael Davis' lab demonstrated that NMDA antagonists injected into the amygdala before a series of light footshock pairings blocked the acquisition of fear to the light as measured by fear-potentiated startle (Miserendino et al., 1990; Campeau et al., 1992). Other labs have replicated these findings with other types of fear conditioning but also suggest that blocking NMDA receptors in the amygdala not only interferes with learning but also inhibits neuronal transmission in the amygdala (Maren et al., 1996; Fendt, 2001; Lee, Choi, Brown, and Kim, 2001). However, an amygdala injection of a specific NMDA antagonist to the 2B subunit of the NMDA receptor blocked long-term memory of fear conditioning without interfering with expression of fear (Rodrigues et al., 2001), supporting a role for amygdala NMDA receptors in learning of fear. Studies administering NMDA antagonists in the lateral ventricles find normal immediate short-term memory of fear conditioning but disrupted long-term memory (Malkani and Rosen, 2001). This effect is thought to occur in the amygdala.

Some of the intracellular events important for long-term memory of fear are also known. One signal transduction pathway is through protein kinase A (PKA). Increases in cyclic adenosine monophosphate (cAMP) via Ca^{2+} influx through NMDA receptors or voltage-gated calcium channels or by stimulation of modulatory receptors that activate adenylate cyclase to form cAMP. PKA is then activated and translocates into the nucleus to phosphorylate cAMP response element binding protein (CREB). Phosphorylated CREB regulates gene transcription, and gene transcription and protein synthesis are thought to be critical steps in long-term memory processes. Injection of PKA inhibitors into the lateral ventricles or amygdala block long-term memory of fear conditioning (Goosens et al., 2000; Malkani and Rosen, submitted; Schafe et al., 1999). Interestingly, the inhibitors have no effect on short-term memory of the conditioning (Malkani and Rosen, submitted; Schafe et al., 1999), suggesting that transduction pathways for short-term and long-term memory are different (Abel and Lattal, 2001).

Other signal transduction pathways in amygdala neurons also contribute to learning and memory of fear. Inhibitors of protein kinase C, mitogen-activating protein kinase (MAPK), calcium/calmodulin kinases (CaMK), and protein kinase B have all been shown to disrupt fear conditioning (Schafe et al., 1999; Goosens et al., 2000; Lin et al., 2001; Malkani and Rosen, submitted). Mutant mice either without Raf (activates MAPK), MAPK, or that overexpress CaMKII have deficits in fear conditioning (Mayford et al., 1996; Brambilla et al., 1997; Selcher et al., 2001).

These signal transduction systems clearly play a role in allostasis and possibly allostatic overload and the enhancement of fear conditioning. CREB has been studied most extensively in this regard. CREB is a constitutive transcription factor that when phosphorylated acts to regulate the transcription of new genes. Mutant mice lacking a CREB gene have deficits in fear conditioning and other types of learning. Interestingly, these deficits are not seen in short-term memory, but in long-term retention (Bourtchouladze et al., 1994; Gass et al., 1998; however, see Graves et al., 2002). CREB knockout mice tested up to 4 hours after initial fear conditioning display normal levels of retention, but tested 24 hour after conditioning show severe memory loss (Bourtchouladze et al., 1994). These data, along with findings in other species (Emptage and Carew, 1993; Vianna et al., 2000; Dubnau and Tully, 2001), strongly suggest that processes involved in short-term and long-term memory are different and function in parallel instead of in a serial fashion (Abel and Lattal, 2001; Vianna et al., 2000). Long-term memory is particularly important for allostasis (e.g., the anticipation and vigilance for future encounters with harmful stimuli). CREB may also play a role in allostatic overload and the development of pathological anxiety. A recent study demonstrated that levels of footshock that normally do not produce fear learning can induce robust fear in rats that overproduce CREB in the amygdala (Josselyn et al., 2001). The overabundance in CREB in the amygdala may increase transcriptional mechanisms that enhance or sensitize amygdala's processing of fearful events.

Inducible genes also play a role in long-term memory of fear conditioning. Immediate-early genes (IEGs) are a class of genes that typically have low

Figure 6.6: Demonstrates that *egr-1* expression in the amygdala is increased following fear conditioning. Top: The graph shows contextual fear conditioning occurring only in rats that received a footshock after being in the test chamber for 3 minutes (delayed-shock group). Fear-related freezing was found in a 3-minute postshock period and in a retention test 24 hours after conditioning. Another group that received a footshock immediately upon being placed in the chamber (immediate-shock group). Although receiving a shock, these rats did not display fear-related freezing. It has been suggested that an animal needs to create a representation of an environment for it to become a conditioned stimulus following footshock (Faneslow, 1986, 1990). Allowing rats time in the chamber before footshock (delayed shock) gives them time to create a representation of the context and thus to be associated with the footshock. A third group placed in the chamber but not receiving footshock did not display fear conditioning (context group). Middle: *egr-1* mRNA in the lateral nucleus of the amygdala is significantly increased in the delayed-shock group compared with a handled-only group and the context and immediate-shock groups. As befitting an immediate-early gene, the *egr-1* increase peaked at 30 minutes following footshock and was short lived. Bottom: Image analysis demonstrates that *egr-1* mRNA increase was localized to the dorsolateral portion of the lateral nucleus of the amygdala (LaDL). Changes were not found in other areas examined including the cortex and hippocampus. Data adapted from Malkani and Rosen (2000b).

Contextual Fear Conditioning in the Delayed-Shock Group

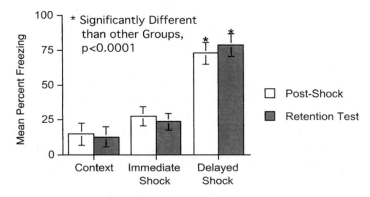

* Significantly Different than other Groups, p<0.0001

□ Post-Shock
■ Retention Test

EGR-1 mRNA Expression is Increased in the Lateral Nucleus of the Amygdala Following Contextual Fear Conditioning

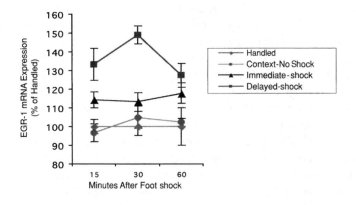

- Handled
- Context-No Shock
- Immediate-shock
- Delayed-shock

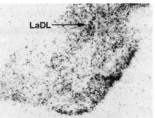

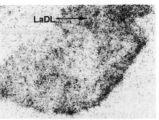

Handled-no shock

Context-no shock

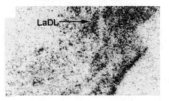

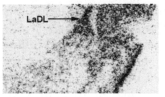

basal levels but are rapidly expressed after cellular activation. Unlike constitutive factors, these genes are transcribed and translated de novo. Many of them act as transcription factors that regulate the expression of genes that play diverse functions in cells such as receptors, enzymes, peptides, growth factors, and cytoskeletal proteins. Thus, inducible transcription factors may act as transcriptional switches to activate a host of cellular events that changes the responsivity and structure of neurons following activation that contribute to long-term memory. For illustration, changes in one of these factors, early growth response gene 1 (*egr-1*), during fear conditioning is described briefly (Fig. 6.6).

egr-1 in the amygdala is particularly responsive to fear conditioning. It may regulate the expression of synapsin I and II (Thiel et al., 1994; Petersohn et al., 1995; but not the synaptobrevin II gene, Petersohn and Thiel, 1996), which have been found to be regulated by *egr-1* in vitro. Synapsin I and II may play a role in various types of potentiation phenomena and learning and memory of fear (Beckman and Wilce, 1997). *egr-1* is therefore an interesting IEG that may regulate several proteins affecting neurotransmitter release, neuronal excitability, and synaptic plasticity.

Expression of *egr-1* in the amygdala with fear conditioning is also linked to early processes in the cellular cascades of fear learning described earlier. Both an NMDA antagonist and inhibitors of PKA and PKC injected intracerebroventircularly (ICV) blocked fear conditioning and the associated increase in expression of *egr-1* mRNA in the lateral nucleus of the amygdala (Fig. 6.7; Malkani and Rosen, 2001, submitted) but not in other brain regions such as the hippocampus or cortex. Thus a signal transduction pathways have been shown to regulate expression of a target gene involved in fear conditioning and brings us a step closer to delineating cellular pathways for learning and memory of fear from an extracellular signal via second messenger transduction to transcription.

GLUCOCORTICOID INVOLVEMENT IN CELLULAR AND MOLECULAR PROCESSES IN THE BASOLATERAL COMPLEX OF THE AMYGDALA

How glucocorticoids interact with cellular and molecular processes of fear conditioning such as those described earlier is not well known and is just beginning to be studied. Glucocorticoid and other steroid receptors are part of a major class of DNA binding factors that regulate gene transcription (Brivanlou and Darnell, 2002). Glucocorticoids are lipophillic, pass through the blood-brain barrier, and bind to intracellular high- and low-affinity corticosteroid receptors to form homodimers, which then regulate gene expression by binding directly to DNA. These corticosteroid-receptor complexes regulate transcription of numerous genes in most organs of the body and

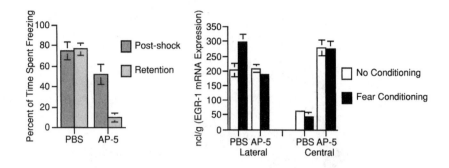

PKA Inhibitor, Rp-cAMPs, Administered ICV Blocks EGR-1 Expression in the Lateral Nucleus of the Amygdala Induced by Contextual Fear Conditioning

PKC Inhibitor, Chelerythrine Chloride, Administered ICV Blocks EGR-1 Expression in the Lateral Nucleus of the Amygdala Induced by Contextual Fear Conditioning

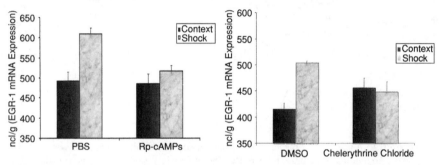

Figure 6.7: Demonstration that glutamate N-methyl-D-aspartate (NMDA) receptors and protein kinases are involved in fear conditioning and the related increase in *egr-1* expression in the amygdala. Top left: NMDA antagonist, AP-5, blocks long-term memory (retention test 24 hours after conditoning) but not short-term memory of fear conditioning (postshock). Top right: AP-5 inhibits fear-conditioning-induced *egr-1* expression in the lateral nucleus of the amygdala. AP-5 also induces *egr-1* expression in the central nucleus. Bottom: Blockade of *egr-1* expression in the lateral nucleus of the amygdala by protein kinase A (PKA; left) and C (PKC; right) inhibitors. Long-term memory of fear was also disrupted by both drugs. Data adapted from Malkani and Rosen (2001; submitted).

brain, including several inducible transcriptional factors in the hippocampus (Vreugdenhil et al., 2001). Korte (2001) has reviewed research on the ability of corticosteroids to strengthen or weaken neural signaling involved in numerous fear- and anxiety-inducing paradigms. In general, Korte concluded that low circulating levels of corticosteroids acting via high-affinity type I receptors regulate acute freezing behavior, whereas high levels via low-affinity type II (glucocortiocoid) receptors enhance learning and memory mechanisms during stressful experiences. A study demonstrated that corticosterone in the basolateral complex of the amygdala affects aversive

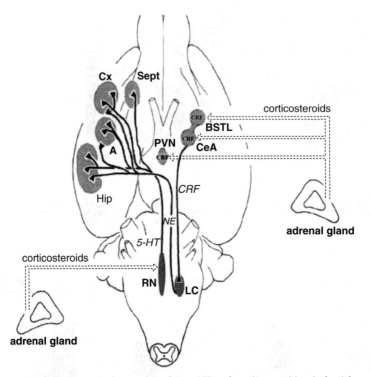

Figure 6.8: A schematic drawing of the ability of corticosteroids, via feed-forward mechanisms, to modulate the brain to increase fear and anxiety. Corticosteroids can stimulate corticotropin-releasing hormone (CRH) release and synthesis in the paraventricular nucleus of the hypothalamus (PVN), central nucleus of the amygdala (CeA), and bed nucleus of the stria terminalis (BSTL). Glucocorticoid modulation of the raphe nucleus (RN) can regulate serotonertic neurotransmission during fear and anxious states in numerous forebrain regions including the amygdala (A), hippocampus (Hip), septum (Sep), and the cortex (Cx). Adapted from Korte (2001).

learning via the CAMP signal transduction pathway (Roozendaal et al., 2001). Another possible inducible transcription factor involved in fear conditioning that may be regulated by glucocorticoids is *ngfi-b* (nerve growth factor induced gene-B). *ngfi-b* is an immediate early gene belonging to an orphan subclass of the nuclear receptor superfamily (Maruyama et al., 1998) that has 50–60% amino acid homology in its DNA-binding domain with glucocorticoid, mineralocorticoid, estrogen, vitamin D, and retinoic acid receptors and therefore may be regulated by these steroids. Similar to *egr-1*, we have found that *ngfi-b* is rapidly and transiently induced in the lateral nucleus of the amygdala specifically with fear conditioning, but unlike *egr-1*, is also induced in the hippocampus and cortex (Malkani and Rosen, 2000a).

Whether glucocorticoids regulate the fear-induced expression of *egr-1* and *ngfi-b* awaits further experimentation.

Feed-forward regulation of fear by glucocorticoids also promotes potentiation of fear (Fig. 6.8; Korte, 2001). Glucocorticoids via type I corticosteroid receptors may regulate unconditioned or previously learned fear in short, acute dangerous situations. However, when environmental circumstances require adaptation to new fear-inducing stimuli, allostatic processes regulated by glucocorticoid receptor activation may be invoked. This mechanism may be particularly important for development of dysfunction in anticipatory processes of fear (Rosen and Schulkin, 1998; Korte, 2001). This is discussed in the last sections of this chapter.

CENTRAL NUCLEUS OF THE AMYGDALA AND FEAR

In addition to the basolateral nucleus of the amygdala, the central nucleus of the amygdala also plays a unique role in conditioned fear. As discussed earlier, whereas much of the plasticity induced by classical fear conditioning may reside in the basolateral complex (LeDoux, 2000; Davis and Whalen, 2001), the nuclei of the complex (the lateral, basal, and accessory basal nuclei) all project to the central nucleus of the amygdala (Pitkanen, 2000). The central nucleus can orchestrate behavioral responses related to fear via its direct connections to numerous midbrain and brainstem regions and circuits instantiating various fear-related behaviors (Davis et al., 1987, 1991; LeDoux et al., 1988; LeDoux, 2000). Lesions or stimulation of the central nucleus are known to respectively block or enhance behaviors associated with fear, such as freezing, fear-potentiated startle, heart rate, and blood pressure (Kapp et al., 1979; Hitchcock and Davis, 1986; LeDoux et al., 1988; Rosen and Davis, 1988). Stimulation of the central nucleus of the amygdala, for example, activates the neural circuitry within the brainstem underlying the startle response via direct and indirect amygdala pathways and amplifies this fear-related reaction (Rosen and Davis, 1990; Rosen et al., 1991; Yeomans and Pollard, 1993). Stimulation of the central nucleus of the amygdala via projections to acetylcholine containing cells in the basal forebrain, midbrain dopamine cells, and brainstem norepinephrine and serotonin neurons may heighten attention toward events that are perceived as fearful by projecting back to the amygdala and other cortical areas (Kapp et al., 1992; Gallagher and Holland, 1994). For example, one of these projections from the central nucleus of the amygdala is to the cholinergic basal forebrain located in the nucleus basalis of Meyeart (Alheid et al., 1995). Cholinergic cells in the nucleus basalis send extensive projections back to the basal nucleus of the amygdala and all areas of the cortex including those involved in attentional processes. Thus, the basal forebrain cholinergic system, by activation via the central nucleus, can produce rapid attention and vigilance during states of fear and uncertain and aversive events (Kapp et al., 1992; Gallagher and Holland, 1994). In addition, autonomic and arousal-invoking information initially activated by central nucleus projects back to brain to modulate neural mechanisms that increase the likelihood that an event will be perceived as fearful (Bechara et al., 1997, 2000; Critchley et al., 2000, 2002b). Central nucleus activation has also been linked to anticipatory angst (Schulkin et al., 1994). Neurons within the amygdala are receptive to fearful signals (Armony et al., 1995) and are influenced by prefrontal cortex (Davidson et al., 2000). Thus, the central nucleus and its associated neural circuitry are engaged in the evaluation of fearful signals and orchestrate varying states and setpoints for behavioral and autonomic responses.

The central nucleus and the related extended amygdala and bed nucleus of the stria terminalis are also distinguished from the basolateral complex as being part of an autonomic brain system (Swanson and Petrovich, 1998). The central nucleus and the bed nucleus of the stria terminalis are not only the major amygdalar efferent sources of input to midbrain and brainstem targets controlling autonomic responses to fear but are also the main recipients of autonomic information into the amygdala and extended amygdala, particularly from the nucleus of the solitary tract and parabrachial nucleus. Several studies have indicated that the central nucleus responds to autonomic or systemic stimulation by increasing gene expression. For example, many drugs that have profound autonomic effects, such as the anithypertension drug nitroprusside, induce immediate-early gene expression in the central nucleus of the amygdala (Dun et al., 1995; Potts et al., 1997; Xie et al., 2000).

Although the central nucleus of the amygdala and the bed nucleus of the stria terminalis have similar inputs and outputs, there appears to be differences in function. Several experiments have suggested that the central nucleus is important for expression of conditioned fear, whereas the bed nucleus of the stria terminalis may be important for expression of unconditioned fear (Walker and Davis, 1997). This is discussed more extensively later.

CORTICOTROPIN-RELEASING HORMONE, AMYGDALA, AND FEAR

Being highly integrated with the autonomic nervous system, the central nucleus has one of the highest concentrations of neuropeptides in the brain. Numerous peptides that are synthesized in neurons of the central nucleus of the amygdala, including enkephalin, neuropeptide Y, and many others, also play a role in fear, but corticotropin-releasing hormone (CRH) and its role in fear has generated the most interest among these peptides. We therefore limit our discussion to CRH.

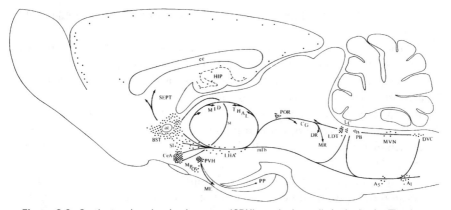

Figure 6.9: Corticotropin-releasing hormone (CRH)-producing cells is the brain. The dots represent cell bodies that produce CRH. Arrows show projection of the cells. A very high density of cells is found in the paraventricular nucleus of the hypothalamus (PVN) and central nucleus of the amygdala (CeA). Other areas of high density are the bed nucleus of the stria terminalis (BST), locus ceureleus (LC), and parabrachial nucleus (PB). Adapted from Swanson et al. (1983).

CRH is well known to be both a peptide that regulates pituitary and adrenal function and an extrahypothalamic peptide hormone linked to a number of behaviors, including behavioral expressions of fear (Koob et al., 1993; Kalin and Takahashi, 1994).

CRH is a 41 amino acid peptide hormone initially isolated from the PVN of the hypothalamus that facilitates adrenocorticotropin hormone secretion from the anterior pituitary (Vale et al., 1981). In addition, CRH is linked to immune, sleep, and appetitive functions (Owens and Nemeroff, 1991; Koob et al., 1993).

CRH cell bodies are widely distributed in the brain (Fig. 6.9; Palkovits et al., 1983; Swanson et al., 1983; Palkovits et al., 1985; Gray, 1990). The majority of CRH neurons within the PVN are clustered in the parvicellular division. Other regions with predominant CRH-containing neurons are the lateral bed nucleus of the stria terminalis and the central division of the central nucleus of the amygdala. To a smaller degree, there are CRH cells in the lateral hypothalamus, prefrontal cortex, and cingulate cortex. In brainstem regions, CRH cells are clustered near the locus coeruleus (Barringtons nucleus; Valentino et al., 1993, 1994), parabrachial region, and regions of the solitary nucleus.

In addition to CRH, several other of its family members have now been discovered – urocortin I, II, and III. Although less research has been conducted with the urocortins to date, there are data suggesting that they play a role in animal models of fear, anxiety, and depression (e.g., Sajdyk,

Schober, Gehlert, and Shekhar, 1999; Reul and Holsboer, 2002; Spina et al., 2002).

The CRH family has two receptors, CRH_1 and CRH_2, localized in rodent and primate brain (Lovenberg et al., 1995; Dautzenberg et al., 2002). Activation of both the CRH_1 and CRH_2 receptors is linked to a G protein and activates adenylate cyclase cascade and an increase in intracellular cAMP and calcium levels (Perrin and Vale, 1999). CRH appears to bind selectively to CRH_1 receptors. Urocortin binds to both receptors with high afinity, whereas urocortin II and III are selective for the CRH_2 receptor (Reyes et al., 2001).

The distribution of CRH_1 receptor sites includes regions of the hippocampus, septum, amygdala (medial and lateral region), neocortex, the ventral thalamic and medial hypothalamic sites, and sparse receptors are located in the PVN and the pituitary gland. The distribution is widespread in cerebellum in addition to brainstem sites such as major sensory nerves and the solitary nucleus (Potter et al., 1994). The distribution of CRH_2 receptors is more limited than that of CRH_1 receptors and is found primarily in subcortical regions including the amygdala, septum, bed nucleus of the stria terminalis, and PVN and the ventral medial nucleus of the hypothalamus (Primus et al., 1997). Both types of receptors have been linked to allostatic regulation (Coste et al., 2001).

Central CRH activation has been consistently and reliably linked to the induction of fear in animal studies (Kalin and Takahashi, 1990; Koob et al., 1993). Central infusions of CRH induce or potentiate a number of fear-related behavioral responses (Takahashi et al., 1989), and infusion of CRH antagonists both within and outside the amygdala reduce fear-related responses (Swiergiel et al., 1992; Koob et al., 1993). A recent study also reported that injection of a CRH antagonist into the basolateral complex of the amygdala immediately following footshock diminished retention of aversive conditioning in an inhibitory avoidance task (Roozendaal et al., 2002). It was shown that expression of CRH in the amygdala increased 30 minutes after footshock. The results indicated that, similar to glucocorticoids and norepinephrine, CRH in the amygdala modulates learning and memory for aversive events.

Startle responses are enhanced by CRH infusions (Swerdlow et al., 1989). Importantly, lesions of the central nucleus of the amygdala, and not the PVN, disrupt CRH-potentiated conditioned fear responses (Liang et al., 1992). That is, only lesions of the amygdala and not the hypothalamus disrupt the behavioral response, suggesting that CRH-induced or -facilitated fear behavior is generated through extrahypothalamic brain regions independently of CRH's role in the HPA axis. Moreover, peripheral blockade of corticotopin/glucocorticoids does not disrupt central CRH-related fear responses (Pich et al., 1993), further reinforcing this notion.

GLUCOCORTICOID FACILITATION OF CRH IN THE BED NUCLEUS OF THE STRIA TERMINALIS: IMPORTANCE FOR UNCONDITIONED FEAR AND ANXIETY

Whereas we have emphasized the CRH and glucocorticoids functioning in the amygdala for fear conditioning, there is a growing body of literature suggesting that the bed nucleus of the stria terminalis may be important for unconditioned fear (Davis et al., 1997). Lesions of the bed nucleus of the stria terminalis do not interfere with conditioned fear-related responses, unlike lesions of amygdala regions, which interfere with fear-potentiated startle or freezing (Hitchcock and Davis, 1991; Lee and Davis, 1997). Neither does stimulation of this region facilitate fear-potentiated startle responses, whereas stimulation of the central nucleus of the amygdala does (Davis, 1997; Rosen and Davis, unpublished observations). However, inactivation of the bed nucleus of the stria terminalis can interfere with unconditioned startle response (Walker and Davis, 1997) and with long-term CRH effects on behavior (Walker and Davis, 1997). The bed nucleus of the stria terminalis also interferes with unconditioned freezing of rats to a fox odor (Fendt et al., 2003), whereas amygdala lesions do not (Wallace and Rosen, 2001; Fendt et al., 2003).

This dissociation has led Davis (1998) to suggest that fear and anxiety are distinct constructs with different neuroanatomic circuities. They assume the classic distinction between fear as a learned, stimulus-specific process and anxiety as a nonspecific, generalized phenomenon. Thus the bed nucleus of the stria terminalis would be linked unconditioned adaptive anxiety (Lee and Davis, 1997), to general anxiety associated with drug abuse (Erb et al., 2001; Koob and Le Moal, 2001) and to symptoms associated with pathological generalized anxiety disorder (Stout et al., 2001).

Importantly, CRH may have different effects in the amygdala and bed nucleus of the stria terminalis. Infusions of CRH directly into the bed nucleus of the stria terminalis facilitate fear- and anxiety-related behavioral responses, whereas antagonists of CRH do the converse (Lee and Davis, 1997). CRH or CRH antagonists injected into the central nucleus of the amygdala do not effect fear behavior such as startle. Nevertheless, lesions of the amygdala block the enhancement of startle by CRH infusions into the bed nucleus of the stria terminalis (Liang et al., 1992). This would suggest that CRH input into the bed nucleus originates in the amygdala, possibly from CRH cells in the central nucleus. How the circuitry of the amygdala and bed nucleus of stria terminalis (Dong et al., 2001) interact to influence conditioned and unconditioned fear-anxiety differentially will be interesting to follow in the future.

GLUCOCORTICOIDS DIFFERENTIALLY REGULATE CRH IN THE HYPOTHALAMUS

Glucocorticoid hormones also have one well-known function: to restrain the HPA axis by negative feedback mechanisms (Munck et al., 1984). This negative feedback is a fundamental way in which the HPA axis is restrained during stress and activity (Munck et al., 1984; Sapolsky et al., 2000) and is understood in the context of negative feedback regulation, or homeostasis (e.g., Goldstein, 1995). One should also note that recent evidence suggests negative restraint of CRH may not be confined solely to the PVN. For example, it may also constrain CRH activity in the locus coeruleus (Pavcovich and Valentino, 1997).

The restraint of HPA activation by glucocorticoids is rapid and profound (Dallman et al., 1987). It is also specific; mineralocorticoids do not produce these effects (Sawchenko, 1987; Watts and Sanchez-Watts, 1995). Moreover, glucocorticoids directly control neuronal excitability (Joels et al., 1994). Some of the glucocorticoid effects on the brain are rapid, suggesting that corticosterone has nongenomic membrane effects via GABAergic mechanisms (Orchinik et al., 2001), in addition to its well-characterized genomic effects.

The degree of the HPA activation is coordinated by both humoral and neural mechanisms. Efferent pathways from the hippocampus and amygdala regulate the expression of CRH in the PVN (Jacobson and Sapolsky, 1991; Herman and Cullinan, 1997). For example, lesions or stimulation of the central nucleus of the amygdala either decrease or increase HPA activation respectively (Beaulieu et al., 1987). Neurons within the lateral bed nucleus of the stria terminalis and within the PVN may activate or inhibit PVN function via GABAergic mechanisms (Herman and Cullinan, 1997; Herman et al., 2002).

Although the profound effect of inhibition is indisputable, there are neuronal populations within the PVN that project to the brainstem that are not decreased by glucocorticoids and some which are actually enhanced (Swanson and Simmons, 1989; Tanimura and Watts, 1998). That is, CRH neurons en route to the pituitary are restrained by glucocorticoids, but CRH en route to other regions of the brain appears not to be restrained (Swanson and Simmons, 1989; Watts and Sanchez-Watts, 1995; Watts, 1996; Palkovits et al., 1998).

CORTICOSTERONE FACILITATION OF CRH AND BEHAVIORAL ACTIVATION OF FEAR

High levels of systemic glucocorticoids are associated with sustained fear (or the perception of adverse events) in a number of species (Mason et al., 1957; Jones et al., 1988; Brier, 1989; Takahashi, 1994; Kalin et al., 1998; Buchanan et al., 1999; see also review by Korte, 2001). For example, rats received conditioning trials in which the unconditioned stimulus (footshock) was presented concurrently with the conditioned stimulus (auditory tone). For several days after conditioning, the rats were treated with corticosterone (Corodimas et al., 1994). This corticosterone treatment is known to increase CRH gene expression in the central nucleus of the amygdala and bed nucleus of the stria terminalis (Makino et al., 1994), and it also facilitated conditioned fear-induced freezing in rats (Fig. 6.4; Corodimas et al., 1994).

In a subsequent study, we looked at contextual fear conditioning in groups of rats that were chronically treated with corticosterone or given a vehicle treatment. CRH expression was differentially regulated in the central nucleus of the amygdala and the parvocellular region of the PVN (Thompson et al., 2004; Fig. 6.10). One week after the completion of the conditioning and the last corticosterone injection, the rats were tested for the retention of conditioned fear. The corticosterone-treated rats displayed more fear conditioning than the vehicle-treated rats. The data suggest that repeated high levels of corticosterone can facilitate the retention of contextual fear conditioning, perhaps by the induction of CRH gene expression in critical regions of the brain such as the amygdala.

CRH also facilitates startle responses. This response does not depend on the adrenal glands because centrally delivered CRH facilitates startle responses in the absence of the adrenal glands (Lee et al., 1994). In that study, Lee and colleagues demonstrated that chronic high plasma levels of corticosterone in adrenally intact rats facilitated CRH-induced startle responses. Perhaps what occurs normally is that the glucocorticoids, by increasing CRH gene expression, increase the likelihood that something will be perceived as a threat, which results in a startle response. Thus, a dose of CRH, given intraventricularly, did not produce a startle response, but when the adrenally intact rats were maintained at high levels of corticosterone for several days prior to the CRH injection, the same dose did produce a startle response.

CRH centrally infused at high doses into the lateral ventricle facilitates seizures linked to amygdala function (Weiss et al., 1986). We found that a dose of CRH, which by itself does not induce kindling, does elicit seizures in a background of high glucocorticoid levels (Fig. 6.11). That is, instead of reducing seizures, as was predicted by corticosterone's restraint on the

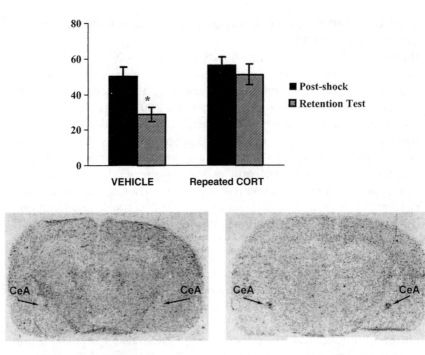

Figure 6.10: Demonstration that repeated corticosterone administration facilitates long-term memory of conditioned fear and increase expression of corticotropin-releasing hormone (CRH) in the amygdala. Rats were treated with corticosterone for 5 days and then given a single contextual fear conditioning trial. During a retention test 24 hours after conditioning, corticosterone-treated rats displayed a greater amount of time freezing than vehicle-treated rats. The brain sections show that the 5 days of corticosterone treatment increased CRH mRNA expression in the central nucleus of the amygdala (CeA). (Adapted from Thompson et al., 2004.)

HPA axis, it actually potentiated the seizures in adrenally intact rats (Rosen et al., 1994). One way to understand these findings is that by increasing CRH expression in the brain, the glucocorticoids lower the threshold for the induced seizure. It is now known that long-term effects on CRH gene expression results from adverse experiences (Bruijnzeel et al., 2001) creating perhaps a long-term allostatic state.

Shepard et al. (2000) have furthermore demonstrated that implants of corticosterone directly into the amygdala resulted in an increase in CRH expression in the central nucleus of the amygdala and a reduction in open field exploratory behavior (Fig. 6.4). Rats are typically hesitant at first to explore new environments, and this was exacerbated with the induction of CRH expression in the central nucleus when corticosterone was directly

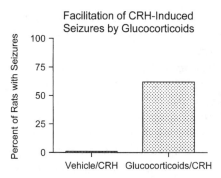

Facilitation of CRH-Induced
Seizures by Glucocorticoids

Figure 6.11: Potentiation of corticotro pin-releasing hormone (CRH)-induced seizures by dexamethasone. A 10-μg dose (ICV) of CRH did not produce seizures on its own, but with systemic 1 mg (subcutaneous) dexamethasone, CRH induced significantly more seizures. (Adapted from: Rosen et al., 1994.)

delivered into the amygdala. In addition, the corticosterone implants increased levels of CRH expression in the central nucleus, without effecting vasopressin levels, in the parvocellular region of the paraventricular nucleus of the hypothalamus. Moreover, gastric pathology, one consequence of allostatic overload, was apparent as a result of the corticosterone (Black, 1988). In further tests, pretreatment with the type-1 receptor CRH antagonist abolished these effects (Gabry et al., 2002; also see Aubry et al., 1997; Smith et al., 1998; Arborelius et al., 2000) for the role of the CRH type-1 receptor, and Bale (2000), the role for the type II receptor.

DEVELOPMENT OF ANXIETY AND DEPRESSION: SENSITIZATION AND THE PSYCHOLOGICAL PROCESSES OF ALLOSTATIC OVERLOAD

The concept of allostasis – achieving stability or viability through change – is intimately tied to learning and memory processes. The associative processes, Pavlovian and instrumental conditioning, are fundamental mechanisms and paradigms for understanding allostasis of fear and possibly pathological anxiety. Pavlovian conditioning and cellular and molecular processes of fear were discussed earlier in the chapter. Nonassociative habituation or desensitization procedures are used extensively in behavioral and cognitive therapies for anxiety disorders (e.g., Zinbarg et al., 1992). Sensitization is particularly pertinent for the development of allostatic overload and pathological anxiety (Rosen and Schulkin, 1998). Important for anxiety, sensitization to threatening or harmful stimuli may produce increased or inappropriate fear responses to innate or associatively learned fear stimuli. Although simple associative conditioning may explain learning of some fears, it is unable to explain the development of pathological anxiety from adaptive fear (e.g., Mineka and Zinbarg, 1996; Pitman et al., 1993; Bouton et al., 2001; Rosen and Schulkin, 1998). Mineka and Zinbarg (1996) and Bouton and colleagues (2001) have made strong cases that simple conditioning mechanisms must be viewed in a dynamic context to explain some of the etiology of anxiety disorders. Several factors that may sensitize fear responses, including one's history of exposure to uncontrollable and unpredictable stressors, the nature of the stressors and conditioned stimuli (how conditionable they are), and one's temperament (Kagan, 1994; Schmidt et al., 1997), may influence learning processes (Mineka and Zinbarg, 1996). These factors may sensitize neural fear circuits to associative conditioning processes and thus facilitate responses to threatening stimuli. For example, experience with uncontrollable or unpredictable stressors can increase subsequent fear conditioning (Peterson et al., 1993; Servatius and Shors, 1994). Male rats given repeated unpredictable tailshock subsequently learn classical conditioning of the eye-blink response to an eye shock much faster than rats not experiencing tailshocks prior to conditioning. These results appear to be contradictory to those of the well-known learned helplessness model, in which prior inescapable shock interferes with subsequent learning (Peterson et al., 1993). However, they are not incompatible because whether there is a facilitation or retardation of learning depends on what type of learning is being tested. Inescapable shock induces fear and thus facilitates subsequent Pavlovian conditioning to aversive events using preexisting responses, but hinders learning of instrumental behavioral responses that are not innate and need to be acquired. During times of stress and fear, automatic, stereotypic behaviors are relied on, and learning of more complicated, new behavior patterns are contraindicated.

Sensitization can also enhance previously learned associations. Fear responses to a conditioned tone paired to a weak footshock can be strengthened or inflated by subsequent random exposure to a more intense footshock (Rescorla, 1974). The greater the time interval between the conditioning and the exposure to the higher intensity shock, the greater the effect of the intense shock (Henderson, 1985). It is as if the memory of the fearful event becomes more intense over time.

Sensitization induced by aversive early life events can also have long-term consequences on adaptation and pathology in adulthood. For example, rat pups cared for by mothers that lick them at low rates or nurse them with an arched back instead of a curved back grew up to display higher levels of anxiety-like behavior and abnormal levels of glucocorticoids compared with pups cared for by more nurturing mothers (Liu et al., 1997). Infant monkeys growing up in conditions of uncertain food supplies were found to have high levels of stress hormones (glucocorticoids and CRH) and displayed more anxiety-like behavior in adulthood than monkeys growing up with either constantly abundant or scarce sources of food (Coplan et al., 1996, 2001). Maternal separation in early life may also contribute to neuronal cell loss in the hippocampus in aged animals (Meaney et al., 1988).

Sensitization, in addition to being a behavioral process, is also a neural process (Robinson and Berridge, 2000; Koob and Le Moal, 2001). Neural sensitization produced by repeated electrical stimulation of the amygdala that induces seizure discharge enhances fear of previously learned fear stimuli (Rosen et al., 1996). We have previously hypothesized (Rosen and Schulkin, 1998) that hyperexcitability in the amygdala is central to turning normal, adaptive fear responses into exaggerated fear that underlies pathological anxiety. Theoretically, a process similar to sensitization called kindling, in which repeated overactivation of the amygdala lowers the threshold in the amygdala for stimulation from environmental stimuli or events, produces this hyperexcitability. A number of experimental kindling paradigms have produced an hyperexcitable amygdala and found subsequent exaggerated fear responses. Because fear is one of the most often observed responses with temporal lobe epilepsy generated from amygdala seizure activity, we and others have demonstrated that experimentally induced kindled seizure activity in the amygdala facilitates both learned and innate fear responses. In our study (Rosen et al., 1996), rats were trained to be fearful of a light, and then over the next 2 days the amygdala or hippocampus were electrically stimulated (kindling stimulation) (Fig. 6.12). The following day, rats were tested for fear using the fear-potentiated startle paradigm. Rats kindled in the amygdala displayed greater fear responses than hippocampal kindled rats or nonkindled control rats. This demonstrates that hyperexcitability induced by localized seizures in the amygdala produced exaggerated fear. Others, particularly Adamec and colleagues, have shown similar effects with amygdala kindling (Adamec and Young, 2000; Adamec, 1990; Kalynchuk, 2000) or hyperexcitability by antagonizing benzodiazepine receptors in the amygdala (Adamec, 1993).

Shenkar has also developed experimental methods to induce hyperexcitability in the amygdala (Sanders et al., 1995; Sajdyk and Shekhar, 2000). Repeated sub-seizure-inducing injections of excitatory agents (e.g., benzodiazepine inverse agonists, noradrenergic agonists and antagonists) injected directly into the amygdala produce long-lasting paniclike responses in rats. Thus, either removal of inhibition or increased excitation in the amygdala produces hyperexcitability in the amygdala that models what is seen in anxiety disorder patients – increased vigilance to threatening stimuli, facilitated attention or learning to dangerous circumstances, and exaggerated fear responses.

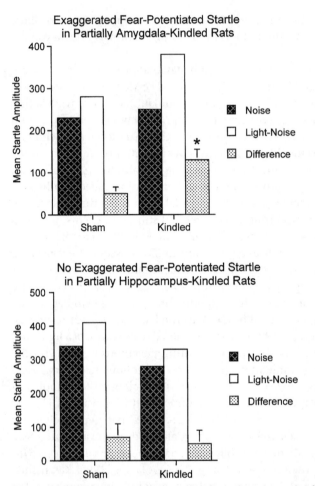

Figure 6.12: Exaggerated fear-potentiated startle as a result of partial amygdala kindling. **Top:** Rats were conditioned to be fearful of a light (light-footshock pairing). Rats received two amygdala kindling stimulation or sham stimulation over the next 2 days. Fear-potentiated startle tested 24 hours after partial amygdala kindling was significantly increased compared with sham-kindled rats ($p < .03$). There were no differences between the groups when baseline startle was elicited in the absence of the fear-conditioned stimulus (noise alone). Differences between the groups were found only when fear was induced (i.e., acoustic startle stimulus was presented while the fear conditioned stimulus was on (light-noise). **Bottom:** Exaggeration of fear-potentiated startle was not produced by partial kindling of the hippocampus. Although both groups displayed fear-potentiated startle, the hippocampus-kindled group was not different from the sham-kindled group. Adapted from Rosen et al. (1996).

These kindling models suggest that the normal neural mechanisms that instantiate the adaptive, allostatic state of fear can become overloaded via sensitization mechanisms, produced either behaviorally or neurally, to induce exaggerated fear and, in humans, anxiety and depressive disorders. The cellular and molecular processes in the amygdala discussed earlier are some of the likely key neurobiological players in the development of allostatic overload and the pathology of anxiety and depression.

NORMAL ALLOSTATIC REGULATION AND OVERLOAD: ROLE OF GLUCOCORTICOIDS IN THE AMYGDALA

In this and previous articles (Schulkin et al., 1994; Rosen and Schulkin, 1998), we suggest that glucocorticoids play a role in sustaining fear and additionally induce a chronic or hyperresponsive state in the amygdala that leads to excessive fear to threatening stimuli and circumstances. One mechanism for glucocorticoids to induce allostatic changes in the amygdala is via CRH. Support for this notion is found in a recent study demonstrating that the CRH response in the amygdala of sheep to a natural (dog) and unnatural (footshock) stressor is regulated by glucocorticoids (Cook, 2002). Following acute exposure to a dog for 6 minutes, both venous and amygdala levels of cortisol increased after 10–30 minutes. Amygdala CRH had a large increase during exposure to the dog and a second peak 10–30 minutes after corresponding to the increase in cortisol (Fig. 6.13). Similar dual peaks of CRH release were found with a footshock. Administration of a glucocorticoid receptor antagonist blocked the second CRH peak in the amygdala without affecting the first peak. These data indicate that the initial response of CRH in the amygdala to an acute stressor is independent of cortisol, but the second delayed peak is cortisol-dependent. In addition, and most interesting, the initial CRH response to a stressor following repeated inescapable exposure to the dog came under the control of cortisol. Sheep were given seven days of repeated exposure to the dog, either with the ability to escape or not to escape from the dog. On the eighth day, the sheep were given a footshock. whereas venous and amygdala cortisol levels in response to the footshock were identical in escape and nonescape groups, both peaks of CRH release in the amygdala were higher in the repeated nonescape group compared to the escape group and became regulated by cortisol. We interpret these findings to indicate that during normal acute danger, CRH in the amygdala increases rapidly to participate in mounting fear responses. This is similar to effects of exogenously applied CRH and is not under the control of glucocorticoids. This is the normal allostatic function of CRH in the amygdala. However, with repeated stress glucocorticoids sensitize the amygdala CRH cells so that they release exaggerated amounts of CRH. The

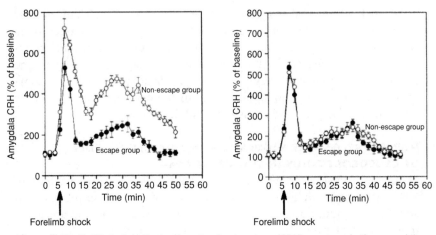

Figure 6.13: Facilitated corticotropin-releasing hormone (CRH) response in the amygdala of sheep to a stressor (footshock) following inescapable exposure to a dog is blocked by a glucocorticoid receptor antagonist. The graph on the left shows that CRH (collected by microdialysis) in the amygdala of sheep exposed to a footshock is greater following inescapable experience with a dog. The graph on the right shows that mifepristone, a glucocorticoid receptor antagonist, blocks the effects of inescapable exposure to a dog. Adapted from Cook (2002).

psychological stressor of inescapable, repeated danger produces allostatic overload in the CRH amygdala system. Taken together with other experimental data demonstrating that high levels of glucocorticoids increase CRH mRNA expression in the central nucleus of the amygdala (Swanson and Simmons, 1989; Makino et al., 1994; Watts and Sanchez-Watts, 1995; Thompson and Schulkin, 2004) and that the basolateral complex of the amygdala is necessary for inescapable-shock-induced facilitation of aversive learning (Shors and Mathew, 1998), the amygdala appears to be a site of convergence for psychological and endocrine information for both allostasis and allostatic overload.

We have previously argued that high levels of glucocorticoids produce a hyperexcitable amygdala that leads normal fear to develop into pathological anxiety and depression (Rosen and Schulkin, 1998). Although CRH is important for normal adaptive fear, increased production and release of glucocorticoids during times of trauma, heightened fear, and stress can facilitate release and synthesis of CRH in the central nucleus of the amygdala and is a likely mechanism for allostatic overload and pathology. The mechanisms for allostatic overload can be cellular and molecular without gross morphological changes, but high levels of glucocortiocoids are know to produce morphological changes in brain, typically decreases in hippocampal

Hippocampus   Amygdala

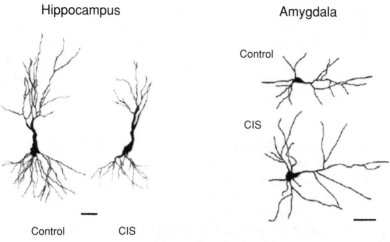

Figure 6.14: Hipocampus and amygdala pyramidal neurons respond differently to chronic inescapable shock (CIS). The dendritic branches of hippocampal CA1 pyramidal neuron shrink or atrophy in response to CIS. In contrast, pyramidal neurons in the basolateral complex of the amygdala hypertrophy with the same CIS. Adapted from Vyas et al. (2002).

and prefrontal neurons' dendritic trees (Woolley et al., 1990; Wellman, 2001). A recent report indirectly linked increased glucocorticoid production to changes in neuronal morphology in the basolateral complex of the amygdala following repeated stress (Vyas et al., 2002). Repeated immobilization stress increased the dendritic length and arborization of pyramidal and stellate cells of basolateral neurons (Fig. 6.14). This finding is even more remarkable because hippocampal pyramidal neurons in the same rats had decreased dendritic length and branching. Also interesting is that the immobilized rats had adrenal hypertrophy, suggesting that they had increased synthesis and release of glucocorticoids. Whereas the shrinkage of hippocampal cells in response to high levels of glucocorticoids and repeated stress has been shown before (e.g., Woolley et al., 1990), the increase in basolateral dendritic growth is exciting because it demonstrates that allostatic overload can also lead to hypertrophy, and not always atrophy. Having more dendrites in the amygdala allows for more synaptic connections between neurons of the amygdala and for more synapses for input from thalamus, cortex, and brainstem to influence amygdala neural activity. Speculatively, the growth in dendritic branching would facilitate processing of information in the amygdala and possibly increase vigilance and the propensity to interpret stimuli as threatening. Because the amygdala is critical for fear conditioning, increased dendritic branching induced by stress may facilitate fear learning. Indeed, repeated immobilization stress does facilitate fear conditioning and is independent of hippocampal atrophy (Conrad et al., 1999). Shors and Mathew (1998) have shown that NMDA receptors in the basolateral complex are necessary for facilitation of aversive conditioning by repeated stress. It will be interesting to see if increased branching of basolateral neurons following repeated stress (Vyas et al., 2002) is dependent on NMDA receptors and the molecular cascades described earlier.

ALLOSTATIC OVERLOAD IN HUMANS: ROLE OF GLUCOCORTICOIDS IN THE AMYGDALA

Do the amygdala and glucocorticoids play a role in allostatic overload in humans? Although the data are still sparse, early indications appear to be affirmative. First, there is a substantial number of findings of increased activity in the amygdala of depressed subjects (Drevets, 2001; Whalen et al., 2002), correlating with negative affect in other medication-free depressed subjects (Abercrombie et al., 1998) and patients suffering from a number of anxiety disorders (see Davis and Whalen, 2001). The prefrontal cortex activity is also correlated with anxiety and depressive disorders (Davidson et al., 1999). These effects are largely lateralized in both amygdala and prefrontal cortex (Davidson et al., 2000; Drevets et al., 2002). Second, robust findings in depressive patients, particularly in those with comorbid anxiety (Gold et al., 1988), are hypercortisolemia, hyperactivity in the HPA axis, and high levels of CRH in cerebrospinal fluid (Nemeroff, 1984; Arborelius et al., 1999; Holsboer, 2000). Melancholic depressives (those with hyperarousal, fear, and anhedonia) have been reported to have a positive correlation between abnormally high levels of cortisol and high but normal levels of cerebrospinal fluid CRH (Wong et al., 2000), indicating a lack of negative feedback control of CRH by cortisol. In addition to this apparent dysfunction of negative feedback in the HPA axis, hypercortisolemia may sustain overactivity in the amygdala via feed-forward processes. A recent study has found a significant positive correlation ($r = .69$) between activity in the amygdala measured by PET and plasma cortisol levels in both unipolar and bipolar depression (Fig. 6.15; Drevets et al., 2002). This correlation may directly reflect either the effect of amygdala activity on CRH secretion or cortisol actions in amygdala. It is intriguing to speculate that findings showing enlarged amygdala in patients with a first-episode depression (Frodl et al., 2002) may be due to increased chronic levels of glucocorticoids and blood flow in the amygdala.

A recent functional MRI study showed interesting differences between healthy and depressed subjects (Siegle et al., 2002). Whereas the amygdala

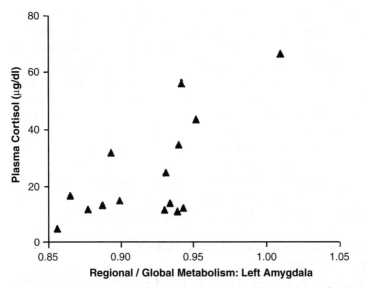

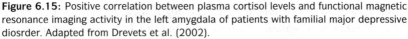

Figure 6.15: Positive correlation between plasma cortisol levels and functional magnetic resonance imaging activity in the left amygdala of patients with familial major depressive diosrder. Adapted from Drevets et al. (2002).

in both healthy and depressed subjects responded to aversive stimuli, the amygdala response of healthy subjects habituated quickly, whereas the amygdala of familial pure depressives remained active significantly longer. These data suggest that in familial pure depressive patients, the amygdala is in allostatic overload, hyperresponsive or sensitized, and does not habituate to fear-provoking stimuli. Whether CRH and cortisol are involved in the sensitized responses awaits further study.

Data on anxiety also indicate that the amygdala and cortisol are interactive in several anxiety disorders. Although the research has developed along two separate paths, activity in the amygdala in a number of anxiety disorders has been shown to be highly reactive to triggers that evoke anxious reactions (Davis and Whalen, 2001), and the HPA axis is hyperresponsive in anxiety disorders, particularly PTSD (Mason et al., 1988; Yehuda et al., 1991; Yehuda, 2002). PTSD patients also have high norepinephrine-to-cortisol ratios (Mason et al., 1988; Baker et al., 2001). In research on cortisol measures, posttraumatic stress disorder (PTSD) patients have basal hypocortisolemia but increased reactivity of the HPA axis to cortisol, suggesting that CRH and corticotropin-secreting cells are sensitized to cortisol in PTSD patients (Yehuda, 2002). Indeed, CRH has been found to be elevated in cerebral spinal fluid of PSTD patients (Bremner et al., 1997; Baker et al., 2001).

Patients with social phobia may also have increases in cortisol secre-tion during stress. Patients with generalized social phobia were shown to hypersecrete cortisol during a public performance of a mental arithmetic test (Condren et al., 2002). The amygdala research demonstrates a similar phenomenon. Patients with PTSD and social phobia have normal resting (nonprovoked) levels of amygdala activity, but the amygdala is highly re-sponsive to anxiety provocation (Rauch et al., 1996, 2000; Shin et al., 1997; Birbaumer et al., 1998; Schneider et al., 1999; Stein et al., 2002). Although most of these studies do not demonstrate an abnormal response of the amygdala per se, particularly because healthy humans also demonstrate in-creased amygdala activity to fearful or aversive stimuli (however, see Schnei-der et al., 1999), they do suggest that the amygdala has a lower threshold for responding to fearful stimuli in patients with anxiety disorder.

CONCLUSIONS: ALLOSTASIS IN ADAPTIVE AND PATHOLOGICAL FEAR

The dominant theme that led to the concept of allostasis was that chronic arousal of the brain drives increases in regulatory physiology to a point beyond what is acceptable for normal function. The concept of allostasis emphasizes multiple systems in both the adaptive phase and the decline into pathology. The diverse physiological systems for maintaining bodily viability in the presence of acute challenges are reflected in mobilization of cardiovascular function, activation of metabolic fuels, activation of immune defense, and engagement of central nervous systems function (McEwen, 1998). The chronic condition, however, can result in allostatic overload and cardiovascular, metabolic, immunological, and neuronal pathology.

Normal fear is an adaptation to danger; chronic anxiety and depression is the overexpression of the neural systems involved in adaptation to danger. Coping with anxious depression is metabolically expensive; expectations of adversity predominate. Moreover, this is a condition in which there can be both high systemic cortisol and elevated CRH in the cerebrospinal fluid (e.g., Nemeroff, 1992; Michelson et al., 1996; Drevets et al., 2002).

Anxious depressed patients also tend to have increased glucose metabolic rates in the amygdala (Drevets et al., 2002). The cortisol that reg-ulates CRH gene expression in the amygdala may underlie the fear and anx-iety of the anxiously depressed person (Schulkin et al., 1994). The chronic inability to turn off cortisol might result in the exaggerated amygdala re-sponse, however, and may therefore contribute to the depressed person's al-tered perception and experience of the world. Other consequences of chron-ically elevated levels of cortisol in the anxiously depressed patient include vulnerability to cardiovascular and bone pathology. For example, one of the pathological consequences of high cortisol is an increased likelihood of bone loss and, perhaps, increased levels of bone fractures (Michelson et al., 1996).

ALLOSTASIS – A PLAUSIBLE CONCEPT

The concept of allostasis forces one to broaden one's view of maintaining internal viability amid changing and uncertain circumstances, but it also highlights the regulatory costs, in terms of the initial protective mechanisms and the longer-term damaging consequences of allostatic overload. What occurs naturally is the link between normal variation in use and descent into overload.

The larger intellectual context is that of regulatory physiology, and the concepts of both homeostasis and allostasis are essentially related to physiological function. One could argue that the concept of homeostasis was construed too narrowly, that homeostasis can reflect both predicative and reactive mechanisms (Moore-Ede, 1986), both short-term and long-term regulation of the internal milieu. In fact, it does. The whole-body regulatory concepts of homeostasis and allostasis function in our lexicon as integrative terms for understanding physiological and behavioral systems. They reflect our need to understand how internal viability is maintained amid a changing environment (see also Mrosovsky, 1990).

Allostasis is tied to the central nervous system as it supervenes in the regulation of bodily viability. The concepts of homeostasis and allostasis force us to consider end-organ systems together as functioning wholes in maintaining bodily health. They serve to unify and bring into perspective a wide range of physiological and behavioral data. Allostasis functions as a heuristic for inquiry, somewhat as other broad biological categories do (e.g., adaptation). One can stretch and continue to expand the concept of homeostasis. We, however, think the concept of allostasis is a conceptual advance in understanding whole-body adaptation with an emphasis on cephalic regulation of local systems. Moreover, positive feed-forward regulation is not an aberration but endemic and pervasive in the hormonal regulation of behavior. Allostasis reflects more flexible behavioral and physiological options. Heuristically, allostasis integrates central and peripheral mechanisms into a single whole-body approach for understanding emotions in general and fear in particular. Interpreting the data within an allostatic frameworks should facilitate the enterprise.

REFERENCES

Abel, T., Lattal, K. M. (2001). Molecular mechanisms of memory acquisition, consolidation and retrieval. *Curr Opin Neurobiol* 11:180–7.

Abercrombie, H. C., Schaefer, S. M., Larson, C. L., Oakes, T. R., Lindgren, K. A., Holden, J. E., Perlman, S. B., Turski, P. A., Krahn, D. D., Benca, R. M., Davidson, R. J. (1998). Metabolic rate in the right amygdala predicts negative affect in depressed patients. *Neuroreport* 9:3301–7.

Adamec, R., Young, B. (2000). Neuroplasticity in specific limbic system circuits may mediate specific kindling induced changes in animal affect-implications for understanding anxiety associated with epilepsy. *Neurosci Biobehav Rev* 24:705–23.

Adamec, R. E. (1990). Does kindling model anything clinically relevant? *Biol Psychiatry* 27:249–79.

Adamec, R. E. (1993). Partial limbic kindling – brain, behavior, and the benzodiazepine receptor. *Physiol Behav* 54:531–45.

Adolphs, R., Tranel, D., Damasio, H., Damasio, A. R. (1995). Fear and the human amygdala. *J Neurosci* 15:5879–91.

Aggleton, J. P. (1992). *The Amygdala: Neurobiological Aspects of Emotion, Memory, and Mental Dysfunction.* New York: Wiley-Liss.

Aggleton, J. P. (2000). *The Amygdala: A Functional Analysis.* New York: Oxford University Press.

Alheid, G. F., de Olmos, J. S., Beltramino, C. A. (1995). Amygdala and extended amygdala. In: *The rat nervous system,* vol. 2 (ed. G. Paxinos), pp. 495–578. San Diego: Academic Press.

Allen, N. B., Trinder, J., Brennan, C. (1999). Affective startle modulation in clinical depression: preliminary findings. *Biol Psychiatry* 46:542–50.

Allman, J., Brothers, L. (1994). Neuropsychology. Faces, fear and the amygdala. *Nature,* 372:613–14.

Amaral, D. G., Price, J. L., Pitkanen, A., Carmichael, S. T. (1992). Anatomical organization of the primate amygdaloid complex. In: *The Amygdala: Neurobiological Aspects of Emotion, Memory, and Mental Dysfunction* (ed. J. P. Aggleton), pp. 1–66. New York: Wiley.

Amorapanth, P., LeDoux, J. E., Nader, K. (2000). Different lateral amygdala outputs mediate reactions and actions elicited by a fear-arousing stimulus. *Nat Neurosci* 3:74–9.

Angrilli, A., Mauri, A., Palomba, D., Flor, H., Birbaumer, N., Sartori, G., di Paola, F. (1996). Startle reflex and emotion modulation impairment after a right amygdala lesion. *Brain* 119:1991–2000.

Arborelius, L., Owens, M. J., Plotsky, P. M., Nemeroff, C. B. (1999). The role of corticotropin-releasing factor in depression and anxiety disorders. *J Endocrinol* 160:1–12.

Arborelius, L., Skelton, K. H., Thrivikraman, K. V., Plotsky, P. M., Schulz, D. W., Owens, M. J. (2000). Chronic administration of the selective corticotropin-releasing factor 1 receptor antagonist CP-154,526: behavioral, endocrine and neurochemical effects in the rat. *J Pharmacol Exp Ther* 294:588–97.

Armony, J. L., Servan-Schreiber, D., Cohen, J. D., LeDoux, J. E. (1995). An anatomically constrained neural network model of fear conditioning. *Behav Neurosci* 109:246–57.

Arnold, M. B. (1960). *Emotion and Personality,* vols. 1 and 2. New York: Columbia University Press.

Aubry, J. M., Turnbull, A. V., Pozzoli, G., Rivier, C., Vale, W. (1997). Endotoxin decreases corticotropin-releasing factor receptor 1 messenger ribonucleic acid levels in the rat pituitary. *Endocrinology* 138:1621–6.

Baas, J. M., Grillon, C., Bocker, K. B., Brack III, A. A., Morgan, C. A., Kenemans, J. L., Verbaten, M. N. (2002). Benzodiazepines have no effect on fear-potentiated startle in humans. *Psychopharmacology* 161:233–47.

Baker, D. G., Ekhator, N. N., Kasckow, J. W., Hill, K. K., Zoumakis, E., Dashevsky, B. A., Chrousos, G. P., Geracioti, T. D. J. (2001). Plasma and cerebrospinal fluid interleukin-6 concentrations in posttraumatic stress disorder. *Neuroimmunomodulation* 9:209–17.

Bale, T. L., Contarino, A., Smith, G. W., Chan, R., Gold, L. H., Sawchenko, P. E., Koob, G. F., Vale, W. W., Lee, K. F. (2000). Mice deficient for corticotropin-releasing hormone receptor-2 display anxiety-like behaviour and are hypersensitive to stress. *Nature Genetics* 24:410–14.

Bard, P. (1939). Central nervous mechanisms for emotional behavior patterns in animals. *Res Nerv Ment Dis* 29:190–218.

Baulieu, E. E. (2000). 'New' active steroids and an unforeseen mechanism of action. *Comptes Rendus Acad Sci Ser III* 323:513–18.

Baxter, M. G., Murray, E. A. (2002). The amygdala and reward. *Nat Rev Neurosci* 3:563–73.

Beaulieu, S., DiPaolo, T., Cote, J., Barden, N. (1987). Participation of the central amygdaloid nucleus in the response of adrenocorticotropin secretion to immobilization stress: Opposing roles of the noradrenergic and dopaminergic systems. *Neuroendocrinology* 45:37–46.

Bechara, A., Damasio, H., Damasio, A. R. (2000). Emotion, decision making and the orbitofrontal cortex. *Cereb Cortex* 10:295–307.

Bechara, A., Damasio, H., Damasio, A. R., Lee, G. P. (1999). Different contributions of the human amygdala and ventromedial prefrontal cortex to decision-making. *J Neurosci* 19:5473–81.

Bechara, A., Damasio, H., Tranel, D., Damasio, A. (1997). Deciding advantageously before knowing the advantageous strategy. *Science* 275:1293–5.

Beckman, A. M., Wilce, P. A. (1997). Egr transcription factors in the nervous system. *Neurochem Int* 31:477–510.

Bell, I. R., Martino, G. M., Meredith, K. E., Schwartz, G. E., Siani, M. M., Morrow, F. D. (1993). Vascular disease risk factors, urinary free cortisol, and health histories in older adults: Shyness and gender interactions. *Biol Psychol* 35:37–49.

Bindra, D. (1978). How adative behavior is produced: A perceptual-motivational alternative to response-reinforcement. *Behav Brain Sci* 1:41–91.

Birbaumer, N., Grodd, W., Diedrich, O., Klose, U., Erb, M., Lotze, M., Schneider, F., Weiss, U., Flor, H. (1998). fMRI reveals amygdala activation to human faces in social phobics. *Neuroreport* 9:1223–6.

Black, H. E. (1988). The effects of steroids upon the gastrointestinal tract. *Toxicol Pathol* 16:213–22.

Blair, H. T., Schafe, G. E., Bauer, E. P., Rodrigues, S. M., LeDoux, J. E. (2001). Synaptic plasticity in the lateral amygdala: A cellular hypothesis of fear conditioning. *Learn Mem* 8:229–42.

Blanchard, D. C., Blanchard, R. J. (1972). Innate and conditioned reactions to threat in rats with amygdaloid lesions. *J Comp Physiol Psychol* 81:281–90.

Bolles, R. C. (1970). Species-specific defensive reactions and avoidance learning. *Psychol Rev* 71:32–48.

Bolles, R. C., Fanselow, M. S. (1980). A percepetual-defensive-recuperative model of fear and pain. *Behav Brain Sci* 3:291–323.

Bourtchouladze, R., Frenguelli, B., Blendy, J., Cioffi, D., Schutz, G., Silva, A. J. (1994). Deficient long-term memory in mice with a targeted mutation of the cAMP-responsive element-binding protein. *Cell* 79:59–68.

Bouton, M. E., Mineka, S., Barlow, D. H. (2001). A modern learning theory perspective on the etiology of panic disorder. *Psychol Rev* 108:4–32.

Brambilla, R., Gnesutta, N., Minichiello, L., White, G., Roylance, A. J., Herron, C. E., Ramsey, M., Wolfer, D. P., Cestari, V., Rossi-Arnaud, C., Grant, S. G., Chapman, P. F., Lipp, H. P., Sturani, E., Klein, R. (1997). A role for the Ras signalling pathway in synaptic transmission and long-term memory. *Nature* 390:281–6.

Breiter, H. C., Etcoff, N. L., Whalen, P. J., Kennedy, W. A., Rauch, S. L., Buckner, R. L., Strauss, M. M., Hyman, S. E., Rosen, B. R. (1996). Response and habituation of the human amygdala during visual processing of facial expression. *Neuron* 17:875–87.

Bremner, J. D., Licinio, J., Darnell, A., Krystal, J. H., Owens, M. J., Southwick, S. M., Nemeroff, C. B., Charney, D. S. (1997). Elevated CSF corticotropin-releasing factor concentrations in posttraumatic stress disorder. *Am J Psychiatry* 154:624–9.

Brier, A. (1989). Experimental approaches to human stress research: Assessment of neurobiological mechanisms of stress in volunteers and psychiatric patients. *Biol Psychiatry* 26:438–62.

Brivanlou, A. H., Darnell, J. E. (2002). Signal transduction and the control of gene expression. *Science* 295:813–18.

Brown, L. L., Tomarken, A. J., Orth, D. N., Loosen, P. T., Kalin, N. H., Davidson, R. J. (1996). Individual differences in repressive-defensiveness predict basal salivary cortisol levels. *J Pers Social Psychol* 70:362–71.

Bruijnzeel, A. W., Stam, R., Compaan, J. C., Wiegant, V. M. (2001). Stress-induced sensitization of CRH-ir but not P-CREB-ir responsivity in the rat central nervous system. *Brain Res* 908:187–96.

Buchanan, T. W., al'Absi, M., Lovallo, W. R., (1999). Cortisol fluctuates with increases and decreases in negative affect. *Psychoneuroendocrinology* 24:227–41.

Calder, A. J., Lawrence, A. D., Young, A. W. (2001). Neuropsychology of fear and loathing. *Nat Rev Neurosci* 2:352–63.

Campeau, S., Davis, M. (1995). Involvement of subcortical and cortical afferents to the lateral nucleus of the amygdala in fear conditioning measured with fear-potentiated startle in rats trained concurrently with auditory and visual conditioned stimuli. *J Neurosci* 15:2312–27.

Campeau, S., Miserendino, M. J., Davis, M. (1992). Intra-amygdala infusion of the N-methly-D-Apartate receptor anatagonist AP5 blocks acquisition but not expression of fear-potentiated startle to an auditory conditioned stimulus. *Behav Neurosci* 106:469–574.

Cannon, W. B. ([1915] 1929). *Bodily Changes in Pain, Hunger, Fear, and Rage*. New York: Appleton-Century-Crofts.

Carroll, B. J. (1976). Urinal free cortisol excretion in depression. *Psychol Med* 6:43–50.

Carter, C. S., DeVries, A. C., Taymans, S. E., Roberts, R. L., Williams, J. R., Getz, L. L. (1997). Peptides, steroids, and pair bonding. *Ann N Y Acad Sci* 807:260–72.

Champoux, M. D., Coe, C. E., Shankberg, S. M., Kihn, C. M., Soumi, S. J. (1989). Hormonal effects of early rearing conditions in the infant rhesus monkey. *Am J Primatol* 19:111–18.

Condren, R. M., O'Neill, A., Ryan, M. C., Barrett, P., Thakore, J. H. (2002). HPA axis response to a psychological stressor in generalised social phobia. *Psychoneuroendocrinology* 27:693–703.

Conrad, C. D., LeDoux, J. E., Magarinos, A. M., McEwen, B. S. (1999). Repeated restraint stress facilitates fear conditioning independently of causing hippocampal CA3 dendritic atrophy. *Behav Neurosci* 113:902–13.

Cook, C. J. (2002). Glucocorticoid feedback increases the sensitivity of the limbic system to stress. *Physiol Behav* 75:455–64.

Coplan, J. D., Andrews, M. W., Rosenblum, L. A., Owens, M. J., Friedman, S., Gorman, J. M., Nemeroff, C. B. (1996). Persistent elevations of cerebrospinal fluid concentrations of corticotropin-releasing factor in adult nonhuman primates exposed to early-life stressors: Implications for the pathophysiology of mood and anxiety disorders. *Proc Nat Acad Sci* 93:1619–23.

Coplan, J. D., Smith, E. L., Altemus, M., Scharf, B. A., Owens, M. J., Nemeroff, C. B., Gorman, J. M., Rosenblum, L. A. (2001). Variable foraging demand rearing: sustained elevations in cisternal cerebrospinal fluid corticotropin-releasing factor concentrations in adult primates. *Biol Psychiatry* 50:200–04.

Corodimas, K. P., LeDoux, J. E., Gold, P. W., Schulkin, J. (1994). Corticosterone potentiation of learned fear. *Ann N Y Acad Sci* 746:392–3.

Coste, S. C., Murray, S. E., Stenzel-Poore, M. P. (2001). Animal models of CRH excess and CRH receptor deficiency display altered adaptations to stress. *Peptides* 22:733–41.

Critchley, H., Melmed, R., Featherstone, E., Mathias, C., Dolan, R. (2002a). Volitional control of autonomic arousal: A functional magnetic resonance study. *Neuroimage* 16:909.

Critchley, H. D., Elliott, R., Mathias, C. J., Dolan, R. J. (2000). Neural activity relating to generation and representation of galvanic skin conductance responses: A functional magnetic resonance imaging study. *J Neurosci* 20:3033–40.

Critchley, H. D., Mathias, C. J., Dolan, R. J. (2001). Neuroanatomical basis for first- and second-order representations of bodily states. *Nat Neurosci* 4:207–12.

Critchley, H. D., Mathias, C. J., Dolan, R. J. (2002b). Fear conditioning in humans: The influence of awareness and autonomic arousal on functional neuroanatomy. *Neuron* 33:653–63.

Dallman, M. F., Akana, S. F., Jacobson, L., Levin, N., Cascio, C. S., Shinsako, J. (1987). Characterization of corticosterone feedback regulation of ACTH secretion. *Ann N Y Acad Sci* 512:402–14.

Dallman, M. F., Bhatnager, S. (2000). Chronic stress: Role of the hypothalmo-pituitary-adrenal axis. In: *Handbook of Physiology* (ed. B. S. McEwen), pp. 179–210. New York: Oxford University Press.

Damasio, A. R. (1994). *Descartes' Error: Emotion, Reason, and the Human Brain*. New York: Grosset/Putnam.

Damasio, A. R. (2000). A second chance for emotion. In: *Cognitive Neuroscience of Emotion* (ed. R. D. Lane, L. Nadel). New York: Oxford Press.

Darwin, C. ([1972] 1965). *The Expression of the Emotions in Man and Animals*. Chicago: University of Chicago Press.

Dautzenberg, F. M., Higelin, J., Brauns, O., Butscha, B., Hauger, R. L. (2002). Five amino acids of the Xenopus laevis CRF (corticotropin-releasing factor) type 2 receptor mediate differential binding of CRF ligands in comparison with its human counterpart. *Mol Pharmacol* 61:1132–9.

Davidson, R. (1998). Anterior electrophysiological asymmetries, emotion, and depression: Conceptual and methodological conundrums. *Psychophysiology* 35:607–14.

Davidson, R. J. (2002). Anxiety and affective style: Role of prefrontal cortex and amygdala. *Biol Psychiatry* 51:68–80.

Davidson, R. J., Abercrombie, H., Nitschke, J. B., Putnam, K. (1999). Regional brain function, emotion and disorders of emotion. *Curr Opin Neurobiol* 9:228–34.

Davidson, R. J., Jackson, D. C., Kalin, N. H. (2000). Emotion, plasticity, context, and regulation: Perspectives from affective neuroscience. *Psychol Bull* 126:890–909.

Davidson, R. J., Pizzagalli, D., Nitschke, J. B., Putnam, K. (2002). Depression: Perspectives from affective neuroscience. *Ann Rev Psychol* 53:545–74.

Davis, M. (1997). Neurobiology of fear responses: The role of the amygdala. *J Neuropsychiatry Clin Neurosci* 9:382–402.

Davis, M. (1998). Are different parts of the extended amygdala involved in fear versus anxiety? *Biol Psychiatry* 44:1239–47.

Davis, M., Hitchcock, J. M., Rosen, J. B. (1987). Anxiety and the amygdala: Pharmacological and anatomical analysis of the fear-potentiated startle paradigm. In: *The Psychology of Learning and Motivation*, vol. 21 (ed. G. H. Bower), pp. 264–306. San Diego: Academic Press.

Davis, M., Hitchcock, J. M., Rosen, J. B. (1991). Neural mechanisms of fear conditioning measured with the acoustic startle reflex. In: *Neurobiology of Learning, Emotion and Affect* (ed. J. Madden) pp. 67–95. New York: Raven Press.

Davis, M., Walker, D. L., Lee, Y. (1997). Amygdala and bed nucleus of the stria terminalis: Differential roles in fear and anxiety measured with the acoustic startle reflex. In: *Biological and Psychological Perspectives on Memory and Memory Disorders* (ed. L. Squire, D. Schacter), pp. 305–31. Washington, DC: American Psychiatric Association Press.

Davis, M., Whalen, P. J. (2001). The amygdala: Vigilance and emotion. *Mol Psychiatry* 6:13–34.

Dayas, C. V., Buller, K. M., Day, T. A. (1999). Neuroendocrine responses to an emotional stressor: evidence for involvement of the medial but not the central amygdala. *Eur J Neurosci* 11:2312–22.

de Quervain, D. J., Roozendaal, B., McGaugh, J. L. (1998). Stress and glucocorticoids impair retrieval of long-term spatial memory. *Nature* 394:787–90.

de Quervain, D. J., Roozendaal, B., Nitsch, R. M., McGaugh, J. L., Hock, C. (2000). Acute cortisone administration impairs retrieval of long-term declarative memory in humans. *Nat Neurosci* 3:313–14.

Denton, D. A. (1982). *The Hunger for Salt*. New York: Springer-Verlag.

DeVries, A. C., Taymans, S. E., Carter, C. S. (1997). Social modulation of corticosteroid responses in male prairie voles. *Ann N Y Acad Sci* 807:494–7.

Dickinson, A. (1980). *Contemporary Animal Learning Theory*. Cambridge: Cambridge University Press.

Dolan, R. J., Lane, R., Chua, P., Fletcher, P. (2000). Dissociable temporal lobe activations during emotional episodic memory retrieval. *Neuroimage* 11:203–9.

Dong, H. W., Petrovich, G. D., Swanson, L. W. (2001). Topography of projections from amygdala to bed nuclei of the stria terminalis. *Brain Res Rev* 38:192–246.

Drevets, W. C. (2001). Neuroimaging studies of mood disorders. *Biol Psychiatry* 48:813–39.

Drevets, W. C., Price, J. L., Bardgett, M. E., Reich, T., Todd, R. D., Raichle, M. E., diagnostic, (2002). Glucose metabolism in the amygdala in depression: Relationship to diagnostic subtype and plasma cortisol levels. *Pharmacol Biochem Behav* 71:431–47.

Dubnau, J., Tully, T. (2001). Functional anatomy: From molecule to memory. *Curr Biol* 11:R240–3.

Dun, N. J., Dun, S. L., Shen, E., Tang, H., Huang, R., Chiu, T. H. (1995). c-fos expression as a marker of central cardiovascular neurons. *Biol Signals* 4:117–23.

Emptage, N. J., Carew, T. J. (1993). Long-term synaptic facilitation in the absence of short-term facilitation in Aplysia neurons. *Science* 262:253–6.

Erb, S., Salmaso, N., Rodaros, D., Stewart, J. (2001). A role for the CRF-containing pathway from central nucleus of the amygdala to bed nucleus of the stria terminalis in the stress-induced reinstatement of cocaine seeking in rats. *Psychopharmacology* 158:360–5.

Eysenck, M. (1992). *Anxiety: The cognitive perspective*. Hove, England: Lawrence Erlbaum.

Fanselow, M. S. (1986). Associative vs topographical accounts of the immediate shock freezing deficit in rats: Implication for the response selection rules governing species-specific defensive reactions. *Learn Motivation* 17:16–39.

Faneslow, M. S. (1990). Factors governing one trial contextual conditioning. *Anim Learning Behav* 18:264–70.

Fendt, M. (2001). Injections of the NMDA receptor antagonist aminophosphonopentanioc acid into the lateral nucleus of the amygdala block the expression of fear-potentiated startle and freezing. *J Neurosci* 21:4111–15.

Fendt, M., Endres, T., Apfelbach, R. (2003). Temporary inactivation of the bed nucleus of the stria terminalis but not of the amygdala blocks freezing induced by trimethylthiazoline, a component of fox feces. *J Neurosci* 23:23–8.

Frodl, T., Meisenzahl, E., Zetzsche, T., Bottlender, R., Born, C., Groll, C., Jager, M., Leinsinger, G., Hahn, K., Moller, H. J. (2002). Enlargement of the amygdala in patients with a first episode of major depression. *Biol Psychiatry* 51:708–14.

Frysztak, R. J., Neafsey, E. J. (1991). The effect of medial frontal cortex lesions on respiration, "freezing," and ultrasonic vocalizations during conditioned emotional responses in rats. *Cereb Cortex* 1:418–25.

Frysztak, R. J., Neafsey, E. J. (1994). The effect of medial frontal cortex lesions on cardiovascular conditioned emotional responses in the rat. *Brain Res* 643:181–93.

Funayama, E. S., Grillon, C., Davis, M., Phelps, E. A. (2001). A double dissociation in the affective modulation of startle in humans: Effects of unilateral temporal lobectomy. *J Cogn Neurosci* 13:721–9.

Fuster, J. M. (2000). Executive frontal functions. *Exp Brain Res* 133:66–70.

Fuster, J. M. (2001). The prefrontal cortex – an update: Time is of the essence. *Neuron* 30:319–33.

Gabry, K. E., Chrousos, G. P., Rice, K. C., Mostafa, R. M., Sternberg, E., Negrao, A. B., Webster, E. L., McCann, S. M., Gold, P. W. (2002). Marked suppression of gastric ulcerogenesis and intestinal responses to stress by a novel class of drugs. *Mol Psychiatry* 7:474–83.

Galea, S., Resnick, H., Ahern, J., Gold, J., Bucuvalas, M., Kilpatrick, D., Stuber, J., Vlahov, D. (2002). Posttraumatic stress disorder in Manhattan, New York City, after the September 11th terrorist attacks. *J Urban Health* 79:340–53.

Gallagher, M., Holland, P. C. (1994). The amygdala complex: Multiple roles in associative learning and attention. *Proc Nat Acad Sci* 91:11771–6.

Gass, P., Wolfer, D. P., Balschun, D., Rudolph, D., Frey, U., Lipp, H., Schutz, G. (1998). Deficits in memory tasks of mice with CREB mutations depend on gene dosage. *Learn Mem* 5:274–88.

Gold, P. W., Goodwin, F. K., Chrousos, G. P. (1988). Clinical and biochemical manifestation of depression: Relation to the neurobiology of stress (part 2 of 2 parts). *New Engl J Med* 319:348–53.

Goldstein, D. S. (1995). *Stress, Catecholamines, and Cardiovascular Disease*. New York: Oxford University Press.

Goosens, K. A., Holt, W., Maren, S. (2000). A role for amygaloid PKA and PKC in the acquisition of long-term conditional fear memories in rats. *Behav Brain Res* 114:145–52.

Gould, E., Woolley, C., McEwen, B. (1991). The hippocampal formation: Morphological changes induced by thyroid, gonadal and adrenal hormones. *Psychoneuroendocrinology* 16:67–84.

Graves, L., Dalvi, A., Lucki, I., Blendy, J. A., Abel, T. (2002). Behavioral analysis of the CREB aD mutation on a B6/129 F1 hybrid background. *Hippocampus* 12:18–26.

Gray, T. S. (1990). The organization and possible function of amygdaloid corticotropin-releasing hormone pathways. In: *Corticotropin-releasing hormone: Basic and clinical studies of a neuropeptide* (ed. E. B. De Souza, C. B. Nemeroff), New York: CRC.

Gray, T. S. (1999). Functional and anatomical relationships among the amygdala, basal forebrain, ventral striatum, and cortex. An integrative discussion. *Ann N Y Acad Sci* 877:439–44.

Grillon, C., Ameli, R., Goddard, A., Woods, S. W., Davis, M. (1994). Baseline and fear-potentiated startle in panic disorder patients. *Biol Psychiatry* 35:431–9.

Grillon, C., Ameli, R., Woods, S. W., Merkangas, K., Davis, M. (1991). Fear-potentiated startle in humans: Effects of anticipatory anxiety on the acoustic blink reflex. *Psychophysiology* 28:588–95.

Habib, K. E., Gold, P. W., Chrousos, G. P. (2001). Neuroendocrinology of stress. *Endocrinol Metab Clin North Am* 30:695–728.

Habib, K. E., Weld, K. P., Rice, K. C., Pushkas, J., Champoux, M., Listwak, S., Webster, E. L., Atkinson, A. J., Schulkin, J., Contoregg, C., Chrousos, G. P., McCann, S. M., Suomi, S. J., Higley, J. D., Gold, P. W. (2000). Oral administration of a corticotropin-releasing hormone receptor antagonist significantly attenuates behavioral, neuroendocrine, and autonomic responses to stress in primates. *Proc Nat Acad Sci* 97:6079–84.

Hariri, A. R., Mattay, V. S., Tessitore, A., Kolachana, B., Fera, F., Goldman, D., Egan, M. F., Weinberger, D. R. (2002). Serotonin transporter genetic variation and the response of the human amygdala. *Science* 297:400–3.

Hebb, D. O. (1946). Emotion in the man and animal: An analysis of the intuitive processes of recognition. *Psychol Rev* 53:88–106.

Henderson, R. (1985). Fearful memories: The motivational significance of forgetting. In: *Affect, Condtioning and Cognition: Essays in the Determinants of Behavior* (ed. F. R. Brush, Overmier), pp. 43–53. Hillsdale, NJ: Lawrence Erlbaum.

Herbert, J., Schulkin, J. S. (2002). Neurochemical coding of adaptive responses in the limbic system. In: *Hormones, Brain and Behavior* (ed. D. Pfaff). New York: Elsevier Press.

Herman, J. P., Cullinan, W. E. (1997). Neurocircuitry of stress: Central control of the hypothalamo-pituitary-adrenocortical axis. *Trends Neuroscience* 20:78–84.

Herman, J. P., Tasker, J. G., Ziegler, D. R., Cullinan, W. E. (2002). Local circuit regulation of paraventricular nucleus stress integration: Glutamate-GABA connections. *Pharmacol Biochem Behavior* 71:457–68.

Hertel, P. T. (2002). Cognitive biases in anxiety and depression: Introduction to the special issue. *Cogn Emotion* 16 321–30.

Hirshfeld, D. R., Rosenbaum, J. F., Biederman, J., Bolduc, E. A., Faraone, S. V., Snidman, N., Reznick, J. S., Kagan, J. (1992). Stable behavioral inhibition and its association with anxiety disorder. *J Am Acad Child Adolesc Psychiatry* 31:103–11.

Hitchcock, J. M., Davis, M. (1986). Lesions of the amygdala, but not the cerebellum or red nucleus, block conditioned fear as measured with the potentiated startle paradigm. *Behav Neurosci* 100:11–22.

Hitchcock, J. M., Davis, M. (1991). Efferent pathway of the amygdala involved in conditioned fear as measured with the fear-potentiated startle paradigm. *Behav Neurosci* 105:826–42.

Holsboer, F. (2000). The corticosteroid receptor hypothesis of depression. *Neuropsychopharmacology* 23:477–501.

Jacobson, L., Sapolsky, R. (1991). The role of the hippocampus in feedback regulation of the hypothalamic-pituitary-adrenocortical axis. *Endocr Rev* 12:118–34.

James, W. ([1890] 1950). *The Principles of Psychology,* vols. 1 and 2. New York: Dover Press.

Joels, M., Hesen, W., Karst, H., de Kloet, E. R. (1994). Steroids and electrical activity in the brain. *J Steroid Biochem Mol Biol* 49:391–8.

Jones, R. B., Beuring, G., Blokhuis, H. J. (1988). Tonic immobility and heterophil/lymphocyte responses of the domestic fowl to corticosterone infusion. *Physiol Behav* 42:249–53.

Jones, R. B., Satterlee, D. G. (1996). Threat-induced behavioural inhibition in Japanese quail genetically selected for contrasting adrenocortical response to mechanical restraint. *Br Poultry Sci* 37:465–70.

Josselyn, S. A., Shi, C., Carlezon, W. A., Jr., Neve, J. L., Nestler, E. J., Davis, M. (2001). Long-term memory is facilitated by cAMP response element-binding protein overexpression in the amygdala. *J Neurosci* 21:2404–12.

Kagan, J. (1994). *Galen's Prophecy: Temperament in Human Nature.* New York: Basic Books.

Kagan, J., Schulkin, J. (1995). On the concepts of fear. *Harv Rev Psychiatry* 3:231–4.

Kagan, J., Snidman, N., Julia-Sellers, M., Johnson, M. O. (1991). Temperament and allergic symptoms. *Psychosom Med* 53:332–40.

Kalin, N. H., Larson, C., Shelton, S. E., Davidson, R. J. (1998). Asymmetric frontal brain activity, cortisol, and behavior associated with fearful temperament in rhesus monkeys. *Behav Neurosci* 112:286–92.

Kalin, N. H., Shelton, S. E., Davidson, R. J. (2000). Cerebrospinal fluid corticotropin-releasing hormone levels are elevated in monkeys with patterns of brain activity associated with fearful temperament. *Biol Psychiatry* 47:579–85.

Kalin, N. H., Shelton, S. E., Davidson, R. J., Kelley, A. E. (2001). The primate amygdala mediates acute fear but not the behavioral and physiological components of anxious temperament. *J Neurosci* 21:2067–74.

Kalin, N. H., Takahashi, L. K. (1990). Fear-motivated behavior induced by prior shock experience is mediated by corticotropin-releasing hormone systems. *Brain Res* 509:80–4.

Kalin, N. H., Takahashi, L. K. (1994). Restraint stress increases corticotropin releasing hormone mRNA content of the amygdala and the paraventricular nucleus. *Brain Res* 656:182–6.

Kalynchuk, L. E. (2000). Long-term amygdala kindling in rats as a model for the study of interictal emotionality in temporal lobe epilepsy. *Neurosci Biobehav Rev* 24:691–704.

Kant, G. J., Bauman, R. A., Anderson, S. M., Mougey, E. H. (1992). Effects of controllable vs. uncontrollable chronic stress on stress-responsive plasma hormones. *Physiol Behav* 51:1285–8.

Kapp, B. S., Frysinger, R. C., Gallagher, M., Applegate, C. D. (1979). Amygdala central nucleus lesions: Effects on heart rate conditioning in the rabbit. *Physiol Behav* 23:1109–17.

Kapp, B. S., Whalen, P. J., Supple, W. F., Jr., Pascoe, J. P. (1992). Amygdaloid contributions to conditioned arousal and sensory information processing. In: *The Amygdala: Neurobiological Aspects of Emotion, Memory, and Mental Dysfunction* (ed. J. P. Aggleton), pp. 229–54. New York: Wiley.

Ketter, T. A., Andreason, P. J., George, M. S., Lee, C., Gill, D. S., Parekh, P. I., Willis, M. W., Herscovitch, P., Post, R. M. (1996). Anterior paralimbic mediation of procaine-induced emotional and psychosensory experiences. *Arch Gen Psychiatry* 53:59–69.

Kluver, H., Bucy, P. C. (1939). Preliminary analysis of functions of the temporal lobes in monkeys. *Arch Neurol Psychiatry* 42:979–1000.

Koob, G. F., Heinrichs, S. C., Pich, E. M., Menzaghi, F., Baldwin, H., Miczek, K., Britton, K. T. (1993). The role of corticotropin-releasing factor in behavioural responses to stress. *Ciba Foun Symp* 172:277–89.

Koob, G. F., Le Moal, M. (2001). Drug addiction, dysregulation of reward, and allostasis. *Neuropsychopharmacology* 24:91–129.

Korte, S. M. (2001). Corticosteroids in relation to fear, anxiety and psychopathology. *Neurosci Biobehav Rev* 25:117–42.

Krettek, J. E., Price, J. L. (1978). Amygdaloid projections to subcortical structures within the basal forebrain and brainstem in the rat and cat. *J Comp Neurol* 78:225–54.

LaBar, K. S., LeDoux, J. E., Spencer, D. D., Phelps, E. A. (1995). Impaired fear conditioning following unilateral temporal lobectomy in humans. *J Neurosci* 15:6846–55.

Lance, V. A., Elsey, R. M. (1986). Stress-induced suppression of testosterone secretion in male alligators. *J Exp Zoology* 239:241–6.

Lane, R. D., Nadel, L. (2000). *Cognitive Neurosci Emotion*. New York: Oxford University Press.

Lane, R. D., Reiman, E. M., Bradley, M. M., Lang, P. J., Ahern, G. L., Davidson, R. J., Schwartz, G. E. (1997). Neuroanatomical correlates of pleasant and unpleasant emotion. *Neuropsychologia* 35:1437–44.

Lang, P. J. (1995). The emotion probe: Studies of motivation and attention. *Am Psychol* 50:372–85.

Lazarus, R. S. (1991). Cognition and motivation in emotion. *Am Psychol* 46:352–67.

LeDoux, J. E., (1987). Emotion. In: *Handbook of Physiology*, vol. 5 (ed. F. Plum), pp. 419–59. Bethesda, MD: American Physiological Society.

LeDoux, J. E. (1996). *The Emotional Brain*. New York: Simon Schuster.

LeDoux, J. E. (2000). Emotion circuits in the brain. *Ann Rev Neurosci* 23:155–84.

LeDoux, J. E., Iwata, J., Cicchetti, P., Reis, D. J. (1988). Different projections of the central amygdaloid nucleus mediate emotional responses conditioned to acoustic stimuli. *J Neurosci* 8:17–29.

Lee, H. J., Choi, J.-S., Brown, T. H., Kim, J. J. (2001). Amygdalar NMDA receptors are critical for the expression of multiple conditioned fear responses. *J Neurosci* 21:4116–24.

Lee, Y., Davis, M. (1997). Role of the hippocampus, the bed nucleus of the stria terminalis, and the amygdala in the excitatory effect of corticotropin-releasing hormone on the acoustic startle reflex. *J Neurosci* 17:6434–46.

Lee, Y., Schulkin, J., Davis, M. (1994). Effect of corticosterone on the enhancement of the acoustic startle reflex by corticotropin releasing factor (CRF). *Brain Res* 666:93–9.

Liang, K. C., Melia, K. R., Campeau, S., Falls, W. A., Miserendino, M. J., Davis, M. (1992). Lesions of the central nucleus of the amygdala, but not the paraventricular nucleus of the hypothalamus, block the excitatory effects of corticotropin-releasing factor on the acoustic startle reflex. *J Neurosci* 12:2313–20.

Lin, C. H., Yeh, S. H., Lin, C. H., Lu, K. T., Leu, T. H., Chang, W. C., Gean, P. W. (2001). A role for the PI-3 kinase signaling pathway in fear conditioning and synaptic plasticity in the amygdala. *Neuron* 31:841–51.

Liu, D., Diorio, J., Tannenbaum, B., Caldji, C., Francis, D., Freedman, A., Sharma, S., Pearson, D., Plotsky, P. M., Meaney, M. J. (1997). Maternal care, hippocampal glucocorticoid receptors, and hypothalamic-pituitary-adrenal responses to stress. *Science* 277:1659–62.

Lorenz, K. Z. (1981). *The Foundations of Ethology*. New York: Springer-Verlag.

Lovenberg, T. W., Liaw, C. W., Grigoriadis, D. E., Clevenger, W., Chalmers, D. T., De Souza, E. B., Oltersdorf, T. (1995). Cloning and characterization of a functionally distinct corticotropin-releasing factor receptor subtype from rat brain. *Proc Nat Acad Science U S A* 92:836–40.

Luine, V. N., Spencer, R. L., McEwen, B. S. (1994). Effects of chronic corticosterone ingestion on spatial memory performance and hippocampal serotonergic function. *Brain Res* 639:65–70.

MacIntosh, N. C. J. (1975). A theory of attention: Variations of associations of stimulus and reinforcment. *Psychol Rev* 82:276–98.

Makino, S., Gold, P. W., Schulkin, J. (1994). Effects of corticosterone on CRH mRNA and content in the bed nucleus of the stria terminalis: Comparison with the effects in the central nucleus of the amygdala and the paraventricular nucleus of the hypothalamus. *Brain Res* 657:141–9.

Malkani, S., Rosen, J. B. (2000a). Induction of NGFI-B following contextual fear conditioning and its blockade by diazepam. *Mol Brain Res* 80:153–65.

Malkani, S., Rosen, J. B. (2000b). Specific induction of immediate early growth response gene 1 (EGR-1) in the lateral nucleus of the amygdala following contextual fear conditioning in rats. *Neuroscience* 97:693–702.

Malkani, S., Rosen, J. B. (2001). N-methyl-D-aspartate receptor antagonism blocks contextual fear conditioning and differentially regulates early growth response-mRNA expression in the amygdala: Implications for a functional amygdaloid circuit of fear. *Neuroscience* 102:853–61.

Malkani, S., Rosen, J. B. (submitted). Signal transduction systems regulating fear-conditioning-induced egr-1 (zif268) expression in the amygdala.

Maren, S., Aharonov, G., Stote, D. L., Fanselow, M. S. (1996). N-methyl-D-aspartate receptors in the basolateral amygdala are required for both acquisition and expression of conditional fear in rats. *Behav Neurosci* 110:1365–74.

Martin, S. J., Grimwood, P. D., Morris, R. G. (2000). Synaptic plasticity and memory: An evaluation of the hypothesis. *Ann Rev Neurosci* 23:649–711.

Maruyama, K., Tsukada, T., Ohkura, N., Bandoh, S., Hosono, T., Yamaguchi, K. (1998). The NGFI-B subfamily of the nuclear receptor superfamily [review]. *Int J Oncol* 12:1237–43.

Mason, J. W. (1968). A review of psychoendocrine research on the pituitary-adrenal cortical system. *Psychosom Med* 30(suppl.):576–607.

Mason, J. W., Brady, J. V., Sidman, M. (1957). Plasma 17-hydroxycorticosteroid levels and conditioned behavior in rhesus monkeys. *Endocrinology* 60:741–52.

Mason, J. W., Giller, E. L., Kosten, T. R., Harkness, L. (1988). Elevation of urinary norepinephrine/cortisol ratio in posttraumatic stress disorder. *J Nerv Ment Dis* 176:498–502.

Matthews, A., MacLeod, C. (1994). Cognitive approaches to emotion and emotional disorders. *Ann Rev Pscyhol* 45:25–50.

Mayford, M., Bach, M. E., Huang, Y. Y., Wang, L., Hawkins, R. D., Kandel, E. R. (1996). Control of memory formation through regulated expression of a CaMKII transgene. *Science* 274:1678–83.

McEwen, B. S. (1998). Protective and damaging effects of stress mediators. *New Engl J Med* 338:171–9.

McEwen, B. S., Stellar, E. (1993). Stress and the individual. Mechanisms leading to diseas. *Arch Int Med* 153:2093–101.

McGaugh, J. L. (2000). Memory – a century of consolidation. *Science* 287:248–51.

McGaugh, J. L., Roozendaal, B. (2002). Role of adrenal stress hormones in forming lasting memories in the brain. *Curr Opin Neurobiol* 12:205–10.

Meaney, M. J., Aitken, D. H., van Berkel, C., Bhatnagar, S., Sapolsky, R. M. (1988). Effect of neonatal handling on age-related impairments associated with the hippocampus. *Science* 239:766–8.

Meshorer, E., Erb, C., Gazit, R., Pavlovsky, L., Kaufer, D., Friedman, A., Glick, D., Ben-Arie, N., Soreq, H. (2002). Alternative splicing and neuritic mRNA translocation under long-term neuronal hypersensitivity. *Science* 295:508–12.

Michelson, D., Stratakis, C., Hill, L., Reynolds, J., Galliven, E., Chrousos, G., Gold, P. (1996). Bone mineral density in women with depression. *New Engl J Med* 335:1176–81.

Milad, M. R., Quirk, G. J. (2002). Neurons in medial prefrontal cortex signal memory for fear extinction. *Nature* 420:70–4.

Mineka, S., Watson, D., Clark, L. A. (1998). Comorbidity of anxiety and unipolar mood disorders. *Ann Rev Psychol* 49:377–412.

Mineka, S., Zinbarg, R. (1996). Conditioning and ethological models of anxiety disorders: Stress-in-dynamic-context anxiety models. In: *Perspectives on Anxiety, Panic and Fear*, volume 43 of the Nebraska Symposium on Motivation (ed. D. A. Hope), pp. 135–210. Lincoln: University of Nebraska Press.

Miserendino, M. J. D., Sananes, C. B., Melia, K. R., Davis, M. (1990). Blocking of acquisition but not expression of conditioned fear-potentiated startle by NMDA antagonists in the amygdala. *Nature* 345:716–18.

Moore-Ede, M. C. (1986). Physiology of the circadian timing system: Predictive versus reactive homeostasis. *Am J Physiol* 250:R737–52.

Morgan, C. A., Grillon, C., Southwick, S. M., Davis, M., Charney, D. S. (1995). Fear-potentiated startle in posttraumatic stress disorder. *Biol Psychiatry* 38:378–85.

Morgan, M. A., LeDoux, J. E. (1995). Differential contribution of dorsal and ventral medial prefrontal cortex to the acquisition and extinction of conditioned fear in rats. *Behav Neurosci* 109:681–8.

Morris, J. S., Frith, C. D., Perrett, D. I., Rowland, D., Young, A. W., Calder, A. J., Dolan, R. J. (1996). A differential neural response in the human amygdala to fearful and happy facial expressions. *Nature* 383:812–5.

Mowrer, O. H. (1947). On the dual nature of learning: Reinterpretation of conditioning and problem solving. *Har Educ Rev* 17:102–48.

Mrosovsky, N. (1990). *Rheostasis: The Physiology of Change*. Oxford: Oxford University Press.

Munck, A., Guyre, P. M., Holbrook, N. J. (1984). Physiological functions of glucocorticoids in stress and their relation to pharmacological actions. *Endocr Rev* 5:25–44.

Nahm, F. K., Tranel, D., Damasio, H., Damasio, A. R. (1993). Cross-modal associations and the human amygdala. *Neuropsychologia* 31:727–44.

Nauta, W. J., Domesick, V. B. (1984). Afferent and efferent relationships of the basal ganglia. *CIBA Found Symp* 107:3–29.

Nemeroff, C. B. (1984). Elevated concentrations of CSF corticotropin-releasing factor-like immunoreactivty in depressed patients. *Science* 26:1342–3.

Nemeroff, C. B. (1992). New vistas in neuropeptide research in neuropsychiatry: Focus on CRF. *Neuropsychopharmacology* 6:69–75.

Nemeroff, C. B., Krishnan, K. R., Reed, D., Leder, R., Beam, C., Dunnick, N. R. (1992). Adrenal gland enlargement in major depression: A computed tomographic study. *Arch Gen Psychiatry* 49:384–7.

Orchinik, M., Carroll, S. S., Li, Y. H., McEwen, B. S., Weiland, N. G. (2001). Hetero-geneity of hippocampal GABA(A) receptors: Regulation by corticosterone. *J Neurosci* 21:330–9.

Owens, M. J., Nemeroff, C. B. (1991). Physiology and pharmacology of corticotropin-releasing factor. *Pharmacol Rev* 43:425–73.

Palkovits, M., Brownstein, M. J., Vale, W. (1985). Distribution of corticotropin-releasing factor in rat brain. *Federation Proc* 44:215–9.

Palkovits, M., Brownstein, M. J., Vale, W. (1983). Corticotropin releasing factor (CRF) immunoreactivity in hypothalamic and extrahypothalamic nuclei of sheep brain. *Neuroendocrinology* 37:302–5.

Palkovits, M., Young, W. S., Kovacs, K., Toth, Z., Makara, G. B. (1998). Alterations in corticotropin-releasing hormone gene expression of central amygdaloid neu-rons following long-term paraventricular lesions and adrenalectomy. *Neuroscience* 85:135–47.

Pavcovich, L. A., Valentino, R. J. (1997). Regulation of a putative neurotransmitter effect of corticotropin-releasing factor: Effects of adrenalectomy. *J Neurosci* 17:401–8.

Perrin, M. H., Vale, W. W. (1999). Corticotropin releasing factor receptors and their ligand family. *Ann N Y Acad Sci* 885:312–28.

Petersohn, D., Schoch, S., Brinkmann, D. R., Thiel, G. (1995). The human synapsin II gene promotor. Possible role for the transcription factor zif268/egr-1, polyoma enhancer activator 3, and AP2. *J Biol Chem* 270:24361–9.

Petersohn, D., Thiel, G. (1996). Role of zinc-finger proteins Sp1 and zif268/egr-1 in transcriptional regulation of the human synaptobrevin II gene. *Eur J Biochem* 239:827–34.

Peterson, C., Maier, S. F., Seligman, M. E. P. (1993). *Learned Helplessness: A Theory for the Age of Personal Control.* New York: Oxford University Press.

Pich, E. M., Heinrichs, S. C., Rivier, C., Miczek, K. A., Fisher, D. A., Koob, G. F. (1993). Blockade of pituitary-adrenal axis activation induced by peripheral im-munoneutralization of corticotropin-releasing factor does not affect the behav-ioral response to social defeat stress in rats. *Psychoneuroendocrinology* 18:495–507.

Pitkanen, A. (2000). Connectivity of the rat amygdaloid complex. In: *The Amygdala: A Functional Analysis*, 2nd ed. (ed. J. P. Aggleton), pp. 31–116. New York: Oxford University Press.

Pitkanen, A., Savander, V., LeDoux, J. E. (1997). Organization of intra-amygdaloid circuitries in the rat: An emerging framework for understanding functions of the amygdala. *Trends Neurosci* 20:517–23.

Pitman, R. K., Orr, S. P., Shalev, A. Y. (1993). Once bitten, twice shy: Beyond the conditioning model of PTSD. *Biol Psychiatry* 33:145–6.

Potter, E., Sutton, S., Donaldson, C., Chen, R., Perrin, M., Lewis, K., Sawchenko, P. E., Vale, W. (1994). Distribution of corticotropin-releasing factor receptor mRNA expression in the rat brain and pituitary. *Proc Nat Acad Sci U S A* 91:8777–81.

Potts, P. D., Polson, J. W., Hirooka, Y., Dampney, R. A. (1997). Effects of sinoaortic den-ervation on Fos expression in the brain evoked by hypertension and hypotension in conscious rabbits. *Neuroscience* 77:503–20.

Primus, R. J., Yevich, E., Baltazar, C., Gallager, D. W. (1997). Autoradiographic localization of CRF1 and CRF2 binding sites in adult rat brain. *Neuropsychopharmacology* 17:308–16.

Pugh, C. R., Tremblay, D., Fleshner, M., Rudy, J. W. (1997). A selective role for corticosterone in contextual-fear conditioning. *Behav Neurosci* 111:503–11.

Quirk, G. J., Russo, G. K., Barron, J. L., Lebron, K. (2000). The role of ventromedial prefrontal cortex in the recovery of extinguished fear. *J Neurosci* 20:6225–31.

Rauch, S. L., van der Kolk, B. A., Fisler, R. E., Alpert, N. M., Orr, S. P., Savage, C. R., Fischman, A. J., Jenike, M. A., Pitman, R. K. (1996). A symptom provocation study of posttraumatic stress disorder using positron emission tomography and script-driven imagery. *Arch Gen Psychiatry* 53:380–7.

Rauch, S. L., Whalen, P. J., Shin, L. M., McInerney, S. C., Macklin, M. L., Lasko, N. B., Orr, S. P., Pitman, R. K. (2000). Exaggerated amygdala response to masked facial stimuli in posttraumatic stress disorder: A functional MRI study. *Biol Psychiatry* 47:769–76.

Rescorla, R. A. (1974). Effect of inflation of the unconditioned stimulus value following conditioning. *J Comp Physiol Psychol* 86:101–6.

Rescorla, R. A., Wagner, A. R. (1972). A theory of Pavlovian conditioning: Variations in the effectiveness of reinforcement and non-reinforcement. In: *Classical Conditioning: Current Research and Theory* (ed. W. J. Baker, W. Prokasy), pp. 64–99. New York: Appleton-Century-Crofts.

Reul, J. M., Holsboer, F. (2002). Corticotropin-releasing factor receptors 1 and 2 in anxiety and depression. *Curr Opin Pharmacol* 2:23–33.

Reyes, T. M., Lewis, K., Perrin, M. H., Kunitake, K. S., Vaughan, J., Arias, C. A., Hogenesch, J. B., Gulyas, J., Rivier, J., Vale, W. W., Sawchenko, P. E. (2001). Urocortin II: A member of the corticotropin-releasing factor (CRF) neuropeptide family that is selectively bound by type 2 CRF receptors. *Proc Nat Acad Sci U S A* 98:2843–8.

Ricardo, J. A., Koh, E. T. (1978). Anatomical evidence of direct projections from the nucleus of the solitary tract to the hypothalamus, amygdala, and other forebrain structures in the rat. *Brain Res* 153:1–26.

Richter, C. P. (1949). Domestication of the Norway rat and its implications for the problem of stress. *Proc Assoc Res Nerv Ment Dis* 29:19–30.

Robinson, T. E., Berridge, K. C. (2000). The psychology and neurobiology of addiction: An incentive-sensitization view. *Addiction* 95(suppl 2):S91–117.

Rodrigues, S. M., Schafe, G. E., LeDoux, J. E. (2001). Intra-amygdala blockade of the NR2B subunit of the NMDA receptor disrupts the acquisition but not the expression of fear conditioning. *J Neurosci* 21:6889–96.

Roozendaal, B. (2000). Glucocorticoids and the regulation of memory consolidation. *Psychoneuroendocrinology* 25;213–38.

Roozendaal, B., Brunson, K. L., Holloway, B. L., McGaugh, J. L., Baram, T. Z. (2002). Involvement of stress-released corticotropin-releasing hormone in the basolateral amygdala in regulating memory consolidation. *Proc Nat Acad Sci* 99:13908–13.

Roozendaal, B., Quirarte, G. L., McGaugh, J. L. (2001). Glucocorticoids interact with the basolateral amygdala beta-adrenoceptor – cAMP/cAMP/PKA system in influencing memory consolidation. *Eur J Neurosci* 15:553–60.

Rosen, J. B., Davis, M. (1988). Enhancement of acoustic startle by electrical stimulation of the amygdala. *Behav Neurosci* 102:195–202.

Rosen, J. B., Davis, M. (1990). Enhancement of electrically elicited startle by amygdaloid stimulation. *Physiol Behav* 48:343–9.

Rosen, J. B., Hamerman, E., Sitcoske, M., Glowa, J., Schulkin, J. (1996). Hyperexcitability: Exaggerated fear-potentiated startle produced by partial amygdala kindling. *Behav Neurosc* 110:43–50.

Rosen, J. B., Hitchcock, J. M., Miserendino, M. J. D., Falls, W. A., Campeau, S., Davis, M. (1992). Lesions of the perirhinal cortex, but not of the frontal, visual or insular cortex block fear-potentiated startle using a visual conditioned stimulus. *J Neurosci* 12:4624–33.

Rosen, J. B., Hitchcock, J. M., Sananes, C. B., Miserendino, M. J. D., Davis, M. (1991). A direct projection from the central nucleus of the amygdala to the acoustic startle pathway: Anterograde and retrograde tracing studies. *Behav Neurosci* 105:817–25.

Rosen, J. B., Pishevar, S. K., Weiss, S. R. B., Smith, M. A., Kling, M. A., Gold, P. W., Schulkin, J. (1994). Glucocorticoid potentiation of CRH-induced seizures. *Neurosci Lett* 174:113–16.

Rosen, J. B., Schulkin, J. (1998). From normal fear to pathological anxiety. *Psychol Rev* 105:325–50.

Rosenkranz, J. A., Grace, A. A. (2002a). Cellular mechanisms of infralimbic and prelimbic prefrontal cortical inhibition and dopaminergic modulation of basolateral amygdala neurons in vivo. *J Neurosci* 22:324–37.

Rosenkranz, J. A., Grace, A. A. (2002b). Dopamine-mediated modulation of odour-evoked amygdala potentials during Pavlovian conditioning. *Nature* 417:282–7.

Sajdyk, T. J., Schober, D. A., Gehlert, D. R., Shekhar, A. (1999). Role of corticotropin-releasing factor and urocortin within the basolateral amygdala of rats in anxiety and panic responses. *Behav Brain Res* 100:207–15.

Sajdyk, T. J., Shekhar, A. (2000). Sodium lactate elicits anxiety in rats after repeated GABA receptor blockade in the basolateral amygdala. *Eur J Pharmacol* 14:265–73.

Sakaguchi, A., LeDoux, J. E., Sved, A. F., Reis, D. J. (1984). Strain difference in fear between spontaneously hypertensive and normotensive rats is mediated by adrenal cortical hormones. *Neurosci Lett* 46:59–64.

Sanders, S. K., Morzorati, S. L., Shekhar, A. (1995). Priming of experimental anxiety by repeated subthreshold GABA blockade in the rat amygdala. *Brain Res* 699:250–9.

Sapolsky, R. M. (1992). Cortisol concentrations and the social significance of rank instability among wild baboons. *Psychoneuroendocrinology* 17:701–9.

Sapolsky, R. M. (2000). Glucocorticoids and hippocampal atrophy in neuropsychiatric disorders. *Arch Gen Psychiatry* 57:925–35.

Sapolsky, R. M., Romero, L. M., Munck, A. U. (2000). How do glucocorticoids influence stress responses? Integrating permissive, suppressive, stimulatory, and preparative actions. *Endocr Rev* 21:55–89.

Sawchenko, P. (1987). Adrenalectomy-induced enhancement of CRF and vasopressin immunoreactivity in parvocellular neurosecretory neurons: Anatomic, peptide, and steroid specificity. *J Neurosci* 7:1093–1106.

Schafe, G. E., Nadel, N. V., Sullivan, G. M., Harris, A., LeDoux, J. E. (1999). Memory consolidation for contextual and auditory fear conditioning is dependent on protein synthesis, PKA, MAP kinase. *Learn Mem* 6:97–110.

Schlenger, W. E., Caddell, J. M., Ebert, L., Jordan, B. K., Rourke, K. M., Wilson, D., Thalji, L., Dennis, J. M., Fairbank, J. A., Kulka, R. A. (2002). Psychological reactions to terrorist attacks: Findings from the National Study of Americans' Reactions to September 11. *JAMA* 288:581–8.

Schmidt, L. A., Fox, N. A., Goldberg, M. C., Smith, C. C., Schulkin, J. (1999). Effects of acute prednisone administration on memory, attention and emotion in healthy human adults. *Psychoneuroendocrinology* 24:461–83.

Schmidt, L. A., Fox, N. A., Rubin, K. H., Sternberg, E. M., Gold, P. W., Smith, C. C., Schulkin, J. (1997). Behavioral and neuroendocrine responses in shy children. *Dev Psychobiol* 30:127–40.

Schneider, F., Weiss, U., Kessler, C., Muller-Gartner, H. W., Posse, S., Salloum, J. B., Grodd, W., Himmelmann, F., Gaebel, W., Birbaumer, N. (1999). Subcortical correlates of differential classical conditioning of aversive emotional reactions in social phobia. *Biol Psychiatry* 45:863–71.

Schoenbaum, G., Chiba, A. A., Gallagher, M. (1998). Orbitofrontal cortex and basolateral amygdala encode expected outcomes during learning. *Nat Neurosci* 1:155–9.

Schoenbaum, G., Chiba, A. A., Gallagher, M. (2000). Changes in functional connectivity in orbitofrontal cortex and basolateral amygdala during learning and reversal training. *J Neurosci* 20:5179–89.

Schoenbaum, G., Setlow, B. (2001). Integrating orbitofrontal cortex into prefrontal theory: Common processing themes across species and subdivisions. *Learn Mem* 8:134–47.

Schulkin, J. (1991). *Sodium Hunger*. Cambridge: Cambridge University Press.

Schulkin, J., McEwen, B. S., Gold, P. W. (1994). Allostasis, amygdala, and anticipatory angst. *Neurosci Biobehav Rev* 18:385–96.

Schuster, M. A., Stein, B. D., Jaycox, L., Collins, R. L., Marshall, G. N., Elliott, M. N., Zhou, A. J., Kanouse, D. E., Morrison, J. L., Berry, S. H. (2001). A national survey of stress reactions after the September 11, 2001, terrorist attacks. *New Engl J Med* 345:1507–12.

Schwaber, J. S., Kapp, B. S., Higgins, G. A., Rapp, P. R. (1982). Amygdaloid and basal forebrain direct connections with the nucleus of the solitary tract and the dorsal motor nucleus. *J Neurosci* 2:1424–38.

Seeman, T. E., Singer, B. H., Rowe, J. W., Horwitz, R. I., McEwen, B. S. (1999). Price of adaptation – allostatic load and its health consequences. MacArthur studies of successful aging. *Arch Int Med* 157:2259–68.

Selcher, J. C., Nekrasova, T., Paylor, R., Landreth, G. E., Sweatt, J. D. (2001). Mice lacking the ERK1 isoform of MAP kinase are unimpaired in emotional learning. *Learn Mem* 8:11–19.

Selye, H. (1956). *The Stress of Life*. New York: McGraw-Hill.

Servatius, R. J., Shors, T. J. (1994). Exposure to inescapable stress persistently facilitates associative and nonassociative learning in rats. *Behav Neurosci* 108:1101–6.

Shepard, J. D., Barron, K. W., Myers, D. A. (2000). Corticosterone delivery to the amygdala increases corticotropin-releasing factor mRNA in the central amygdaloid nucleus and anxiety-like behavior. *Brain Res* 861:288–95.

Shi, C., Davis, M. (2001). Visual pathways involved in fear conditioning measured with fear-potentiated startle: Behavioral and anatomic studies. *J Neurosci* 21:9844–55.

Shin, L. M., Kosslyn, S. M., McNally, R. J., Alpert, N. M., Thompson, W. L., Rauch, S. L., Macklin, M. L., Pitman, R. K. (1997). Visual imagery and perception in posttraumatic stress disorder. A positron emission tomographic investigation. *Arch Gen Psychiatry* 54:233–41.

Shors, T. J., Mathew, P. R., et al. (1998). NMDA receptor antagonism in the lateral/basolateral but not central nucleus of the amygdala prevents the induction of facilitated learning in response to stress. *Learn Mem* 5:220–30.

Siegle, G. J., Steinhauer, S. R., Thase, M. E., Stenger, V. A., Carter, C. S. (2002). Can't shake that feeling: Event-related fMRI assessment of sustained amygdala activity in response to emotional information in depressed individuals. *Biol Psychiatry* 51:693–707.

Silver, R. C., Holman, E. A., McIntosh, D. N., Poulin, M., Gil-Rivas, V. (2002). Nationwide longitudinal study of psychological responses to September 11. *JAMA* 288:1235–44.

Smith, G. W., Aubry, J. M., Dellu, F., Contarino, A., Bilezkiianl, M., Gold, L. H., Chen, R. (1998). Corticotropin releasing factor receptor 1-deficient mice display decreased anxiety, impaired stress response, and aberrant neuroendocrine development. *Neuron* 20:1093–102.

Spina, G., Merlo-Pich, E., Akwa, Y., Balducci, C., Basso, M., Zorrilla, P., Britton, T., Rivier, J., Vale, W., Koob, F. (2002). Time-dependent induction of anxiogenic-like effects after central infusion of urocortin or corticotropin-releasing factor in the rat. *Psychopharmacology* 160:113–21.

Stefanacci, L., Amaral, D. G. (2002). Some observations on cortical inputs to the macaque monkey amygdala: An anterograde tracing study. *J Comp Neurol* 451:301–23.

Stein, M. B., Goldin, P. R., Sareen, J., Zorrilla, L. T., Brown, G. G. (2002). Increased amygdala activation to angry and contemptuous faces in generalized social phobia. *Arch Gen Psychiatry* 59:1027–34.

Sterling, P., Eyer, J. (1981). Biological basis of stress-related mortality. *Soc Sci Med* 15:3–42.

Sterling, P., Eyer, J. (1988). Allostasis: A new paradigm to explain arousal pathology. In: *Handbook of Life Stress, Cognition and Health* (ed. S. Fisher, J. Reason), pp. 629–49. New York: John Wiley and Sons.

Stout, R. L., Dolan, R., Dyck, I., Eisen, J., Keller, M. B. (2001). Course of social functioning after remission from panic disorder. *Compr Psychiatry* 42:441–7.

Suboski, M. D., Marquis, H. A., Black, M., P., P. (1970). Adrenal and amygdala function in the incubation of aversively conditioned responses. *Physiol Behav* 5:283–9.

Sullivan, R. M., Gratton, A. (2002). Prefrontal cortical regulation of hypothalamic-pituitary-adrenal function in the rat and implications for psychopathology: Side matters. *Psychoneuroendocrinology* 27:99–114.

Sumners, C., Gault, T. R., Fregly, M. J. (1991). Potentiation of angiotensin II-induced drinking by glucocorticoids is a specific glucocorticoid Type II receptor (GR)-mediated event. *Brain Res* 552:283–90.

Sutton, S. K., Davidson, R. J. (2000). Prefrontal brain electrical asymmetry predicts the evaluation of affective stimuli. *Neuropsychologia* 38:1723–33.

Swanson, L. W. (2000). Cerebral hemisphere regulation of motivated behavior (1). *Brain Res* 886:113–64.

Swanson, L. W., Petrovich, G. D. (1998). What is the amygdala? *Trends Neurosci* 21:323–31.

Swanson, L. W., Sawchenko, P. E., Rivier, J., Vale, W. W. (1983). Organization of ovine corticotropin-releasing factor immunoreactive cells and fibers in the rat brain: An immunohistochemical study. *Neuroendocrinology* 36:165–86.

Swanson, L. W., Simmons, D. M. (1989). Differential steroid hormone and neural influences on peptide mRNA levels in CRH cells of the paraventricular nucleus: A hybridization histochemical study in the rat. *J Comp Neurol* 285:413–35.

Swerdlow, N. R., Britton, K. T., Koob, G. F. (1989). Potentiation of acoustic startle by corticotropin-releasing factor (CRF) and by fear are both reversed by alpha-helical CRF (9–41). *Neuropsychopharmacology* 2:285–92.

Swiergiel, A. H., Takahashi, L. K., Rubin, W. W., Kalin, N. H. (1992). Antagonism of corticotropin-releasing factor receptors in the locus coeruleus attenuates shock-induced freezing in rats. *Brain Res* 587:263–8.

Takahashi, L. K. (1994). The organizing action of corticosterone on the development of behavioral inhibition. *Dev Brain Res* 81:121–7.

Takahashi, L. K., Kalin, N. H., Vanden Burgt, J. A., Sherman, J. E. (1989). Corticotropin-releasing factor modulates defensive-withdrawal and exploratory behavior in rats. *Behav Neurosci* 103:648–54.

Tanimura, S. M., Watts, A. G. (1998). Corticosterone can facilitate as well as inhibit corticotropin-releasing hormone gene expression in the rat hypothalamic paraventricular nucleus. *Endocrinology* 139:3830–6.

Thiel, G., Schoch, S., Petersohn, D. (1994). Regulation of synapsin I gene expression by the zinc finger transcription factor zif268/egr-1. *J Biol Chem* 269:15294–301.

Thompson, B. L., Erickson, K., Schulkin, J., Rosen, J. B. (2004). Chronic corticosterone enhances contextual fear conditioning and CRH mRNA expression in the amygdala. *Beh BR Res* 149:209–215.

Tinbergen, N. ([1951] 1969). *The Study of Instinct*. Oxford: Oxford University Press.

Tout, K., de Haan, M., Campbell, E. K., Gunnar, M. R. (1998). Social behavior correlates of cortisol activity in child care: gender differences and time-of-day effects. *Child Dev* 69:1247–62.

Turner, B. H., Herkenham, M. (1991). Thalamoamygdaloid projections in the rat: A test of the amygdala's role in sensory processing. *J Comp Neurol* 313:295–325.

Vale, W., Spiess, J., Rivier, C., Rivier, J. (1981). Characterization of a 41-residue ovine hypothalamic peptide that stimulates secretion of corticotropin and beta-endorphin. *Science* 213:1394–7.

Valentino, R. J., Foote, S. L., Page, M. E. (1993). The locus coeruleus as a site for integrating corticotropin-releasing factor and noradrenergic mediation of stress responses. *Ann N Y Acad Science* 697:173–88.

Valentino, R. J., Page, M. E., Luppi, P. H., Zhu, Y., Van Bockstaele, E., Aston-Jones, G. (1994). Evidence for widespread afferents to Barrington's nucleus, a brainstem region rich in corticotropin-releasing hormone neurons. *Neuroscience* 62:125–43.

Vianna, M. R., Izquierdo, L. A., Barros, D. M., Walz, R., Medina, J. H., Izquierdo, I. (2000). Short- and long-term memory: Differential involvement of neurotransmitter systems and signal transduction cascades. *An Acad Bras Scienc* 72:353–64.

Vreugdenhil, E., de Kloet, E. R., Schaaf, M., Datson, N. A. (2001). Genetic dissection of corticosterone receptor function in the rat hippocampus. *Eur Neuropsychopharmacol* 11:423–30.

Vyas, A., Mitra, R., Shankaranarayana Rao, B. S., Chattarji, S. (2002). Chronic stress induces contrasting patterns of dendritic remodeling in hippocampal and amygdaloid neurons. *J Neurosci* 22:6810–18.

Walker, D. L., Davis, M. (1997). Double dissociation between the involvement of the bed nucleus of the stria terminalis and the central nucleus of the amygdala in startle increases produced by conditioned versus unconditioned fear. *J Neurosci* 17:9375–83.

Wallace, K. J., Rosen, J. B. (2001). Neurotoxic lesions of the lateral nucleus of the amygdala decrease conditioned fear, but not unconditioned fear of a predator odor: Comparison to electrolytic lesions. *J Neurosci* 21:3619–27.

Watts, A. G. (1996). The impact of physiological stimuli on the expression of corticotropin-releasing hormone (CRH) and other neuropeptide genes. *Neuroendocrinology* 17:281–326.

Watts, A. G., Sanchez-Watts, G. (1995). Region-specific regulation of the neuropeptide mRNAs in rat limbic forebrain neurons by aldosterone and corticosterone. *J Physiol* 484:721–36.

Weiss, S. R., Post, R. M., Gold, P. W., Chrousos, G., Sullivan, T. L., Walker, D., Pert, A. (1986). CRF-induced seizures and behavior: Interaction with amygdala kindling. *Brain Res* 372:345–51.

Wellman, C. L. (2001). Dendritic reorganization in pyramidal neurons in medial prefrontal cortex after chronic corticosterone administration. *J Neurobiol* 49:245–53.

Whalen, P. (1999). Fear, vigilance and ambiguity: Initial neuroimaging studies of the human amygdala. *Curr Dir Psychol Sci* 7:177–88.

Whalen, P. J., Shin, L. M., Somerville, L. H., McLean, A. A., Kim, H. (2002). Functional neuroimaging studies of the amygdala in depression. *Sem Clin Neuropsychiatry* 7:234–42.

Wong, M. L., Kling, M. A., Munson, P. J., Listwak, S., Licinio, J., Prolo, P., Karp, B., McCutcheon, I. E., Geracioti, T. D. J., DeBellis, M. D., Ric, K. C., Goldstein, D. S., Veldhuis, J. D., Chrousos, G. P., Oldfield, E. H., McCann, S. M., Gold, P. W. (2000). Pronounced and sustained central hypernoradrenergic function in major depression with melancholic features: Relation to hypercortisolism and corticotropin-releasing hormone. *Proc Nat Acad Sci* 97:325–30.

Woolley, C. S., Gould, E., McEwen, B. S. (1990). Exposure to excess glucocorticoids alters dendritic morphology of adult hippocampal pyramidal neurons. *Brain Res* 531:225–31.

Wright, C. I., Fischer, H., Whalen, P. J., McInerney, S. C., Shin, L. M., Rauch, S. L. (2001). Differential prefrontal cortex and amygdala habituation to repeatedly presented emotional stimuli. *Neuroreport* 12:379–83.

Xie, Q., Li, L., Kawada, S. (2000). Hypotension after coronary arterial occlusion induces regional expression of c-Fos protein in the rat brain. *Ann Thorac Cardiovasc Surg* 6:319–25.

Yehuda, R. (2002). Current status of cortisol findings in post-traumatic stress disorder. *Psychiatr Clin North Am* 25:341–68.

Yehuda, R., Giller, E. L., Southwick, S. M., Lowy, M. T., Mason, J. W. (1991). Hypothalamic-pituitary-adrenal dysfunction in post-traumatic stress disorder. *Biol Psychiatry* 30:1031–48.

Yeomans, J. S., Pollard, B. A. (1993). Amygdala efferents mediating electrically evoked startle-like responses and fear potentiation of acoustic startle. *Behav Neurosci* 107:596–610.

Yim, C. Y., Mogenson, G. J. (1982). Response of nucleus accumbens neurons to amygdala stimulation and its modification by dopamine. *Brain Res* 239:401–15.

Zinbarg, R. E., Barlow, D. H., Brown, T. A., Hertz, R. M. (1992). Cognitive-behavioral approaches to the nature and treatment of anxiety disorders. *Ann Rev Psychol* 43:235–67.

7 A Chronobiological Perspective on Allostasis and Its Application to Shift Work

Ziad Boulos and Alan M. Rosenwasser

INTRODUCTION

Allostasis, or *stability through change,* refers to a model of physiological regulation proposed by Sterling and Eyer (1988) to account for epidemiological data linking contemporary patterns of morbidity and mortality to a range of psychosocial conditions characterized as involving disruptions of intimate social relationships. Thus, marital dissolution, bereavement, unemployment, migration, and stressful work conditions are all associated with increased age-specific mortality rates (Sterling and Eyer, 1981, 1988). Importantly, the higher mortality rates are from virtually all leading causes of death, including cardiovascular disease, diabetes, and cancer, as well as accidents and suicide. Recent studies have confirmed and extended these findings. For example, data reviewed by McEwen (2000a) show clear negative relationships between the prevalence of hypertension, cervical cancer, osteoarthritis, and chronic disease on one hand, and socioeconomic status on the other, as well as differences in the prevalence of mood and anxiety disorders and of substance abuse related to income and education gradients.

These social and psychological factors all result in chronic arousal, or stress. Stressful stimuli represent real or perceived threats to the integrity of the organism, and they elicit a set of coordinated neuroendocrine responses that serve to prepare the organism to meet the challenge and to defend homeostasis. The responses are initiated by the sympathetic nervous system and the hypothalamic-pituitary-adrenocortical (HPA) axis, which trigger widespread changes in cardiovascular, metabolic, and immune parameters. These adaptive changes are normally reversed after removal of the arousing stimulus, but under conditions of repeated or chronic stress, they can persist for extended durations and may even become irreversible. Thus, the central principle of allostasis is that "to maintain stability an organism must *vary* all the parameters of its internal milieu and match

228

them appropriately to environmental demands" (Sterling and Eyer, 1988, p. 636).

As an illustration of allostasis, Sterling and Eyer (1988) presented two continuous 24-hour records of arterial blood pressure, one from a normotensive and one from a hypertensive subject. Both records show a diurnal rhythm with lower levels during nighttime sleep, superimposed on which are several peaks and troughs associated with specific behavioral and environmental events, but the levels are generally higher in the hypertensive subject. Sterling and Eyer argued that both the chronic elevation that characterizes hypertension and the changes that accompany transitions in behavioral state reflect a resetting of blood pressure and that both are examples of allostasis. They also pointed out that such changes are not limited to blood pressure but occur in virtually all physiological parameters. Although these authors discussed at length both the health effects of chronic arousal and the underlying physiological mechanisms, the roles of diurnal rhythmicity and sleep in allostatic regulation were left unspecified.

Implicit in Sterling and Eyer's model is the notion that allostatic regulation entails a physiological price, paid for in a predisposition for cardiovascular, gastrointestinal, immune system, and other pathology. The price of allostasis was given explicit recognition by McEwen and Stellar (1993), who introduced the concept of allostatic load to refer to the wear and tear on organs and tissues resulting from overactivity of the physiological systems that are mobilized to defend homeostasis in the face of repeated or chronic stress (see also Schulkin et al., 1994; McEwen, 1998a, 1998b; McEwen and Seeman, 1999).

Another important property of allostatic regulation is that it benefits from experience and allows the organism to anticipate changes in need and to make appropriate adjustments in advance. This requires central coordination by the brain, which is able to override local negative feedback mechanisms and impose new regulatory setpoints (Sterling and Eyer, 1988). This aspect of allostasis was emphasized by Schulkin et al. (1994, 1998), who argued that the anticipation of negative events, shaped by an individual's past experience and perceptions, can engender chronic anxiety, fear, and hopelessness and lead to such pathologies as agitated depression, panic disorder, and posttraumatic stress.

Following Sterling and Eyer's lead, several authors have mentioned diurnal (or circadian) and sleep-wake–related variations in behavior and physiology, including variations in basal levels of adrenal hormones and other stress mediators, as examples of allostasis, but this aspect of the allostasis model has rarely been discussed or justified in any detail. In this chapter, we reexamine the concepts of allostasis and homeostasis from a chronobiological perspective. We conclude that circadian regulation cannot be considered a subtype of allostatic regulation without greatly diluting the concept of allostasis and detracting from its role in arousal pathology. Rather, we suggest that repeated or chronic *deviations* from normal temporal patterns of behavior and physiology, as orchestrated by a circadian timing system stably entrained to the day-night cycle, can entail a significant allostatic load, as indicated by the prevalence of gastrointestinal, cardiovascular, reproductive, and other pathology in shift workers.

HOMEOSTASIS, ALLOSTASIS, AND CIRCADIAN REGULATION

Homeostasis and Allostasis

The allostasis model was conceived by Sterling and Eyer (1988) as an alternative to homeostasis, which had been the dominant model of physiological regulation for the previous 100 years. Both allostasis and homeostasis address the central question of how organisms are able to maintain internal stability while meeting the challenges inherent to life in an uncertain and inconstant world. In the homeostasis model, this is achieved by holding the parameters of the body's internal environment constant, largely through local negative feedback mechanisms that oppose any changes in these parameters and force them back to fixed setpoints. In the original version of the allostasis model, however, stability is maintained by varying the parameters of the internal environment, which involves changing the setpoints around which these parameters are regulated in order to meet perceived and anticipated demand (Sterling and Eyer, 1988).

The chronic elevation of blood pressure observed in rodents and monkeys following prolonged exposure to social stressors provides a clear example of allostatic regulation (Sterling and Eyer, 1981). The initial response to the stressful stimulus is an acute rise in blood pressure, resulting primarily from an increase in cardiac output. However, if the stimulus is maintained for extended durations, the blood pressure elevation becomes chronic, maintained by an increase in vascular resistance rather than cardiac output. The higher resistance is largely due to a thickening of arteriolar smooth muscle and an increase in the vascular wall-to-lumen ratio, and these structural changes are accompanied by a resetting of arterial baroreceptors. Much the same changes are thought to occur in human hypertension (Julius, 1988).

A different point of view, first proposed by McEwen and Stellar (1993) and later expanded by McEwen (2000b), distinguishes between *allostatic systems,* such as the sympathetic nervous system and the HPA axis, which have no setpoints and can vary to meet changing demands, and *homeostatic systems,* which are maintained within a narrow range around fixed set-points. This view entails a very restrictive definition of homeostasis, which only applies to "a limited number of systems like pH, body temperature and oxygen tension, components of the internal milieu, that are truly essential for life" (McEwen, 2000b, p. 173).

To clarify their interpretation of allostasis, McEwen and Stellar (1993) suggested the analogy of a seesaw, with environmental challenges on one side and physiological responses on the other. Maintaining equilibrium (homeostasis) requires that repeated or chronic stress be balanced by com-mensurate changes in the activity of allostatic systems. If the same analogy is applied to the Sterling and Eyer model, the resetting of homeostatic set-points that characterizes allostasis would be represented by a shifting of the fulcrum of the seesaw toward the environmental end, allowing the same environmental load to be balanced by a smaller physiological response.

McEwen's distinction between homeostatic and allostatic systems is consistent with the idea, stated by Cannon more than 70 years ago (Can-non, 1929, 1930), that the autonomic nervous system, together with the adrenals and other endocrine glands, are activated to defend homeostasis. For Cannon, however, homeostatic regulation was not limited to pH and body temperature, but included a number of other variables such as blood electrolytes, glucose, fat, and protein. This wider view of homeostasis is reflected in the more recent scientific literature, as indicated by the use of such terms as osmostat, glucostat, lipostat, ponderostat, gonadostat, chemo-stat, and mechanostat. The proposed involvement of hedonic homeostasis and reward setpoints in drug addiction (Koob and Le Moal, 1997, 2001; Roberts et al., 2000) is consistent with this wider view. As discussed in more detail later, homeostatic regulation also applies to sleep, based in part on the occurrence of compensatory changes in sleep quality following sleep deprivation.

In the absence of more specific criteria, McEwen's proposed dichotomy between systems that are essential for life and ones that are not is bound to appear arbitrary. Similarly, although it is clear that physiological systems differ in the range of their activity, there is no obvious boundary between what constitutes a narrow range and what constitutes a wide range of ac-tivity. Furthermore, the ranges of some variables – body temperature, for example – show large species differences, and a bipartite classification of physiological systems as being either homeostatic or allostatic based on observations made on one species may not be applicable to others.

These points are illustrated in the following section on circadian regula-tion. Data reviewed in that section show that most if not all physiological systems, allostatic as well as homeostatic (by any definition), show circadian variations, that the daily ranges of these variations can differ widely between species, and that even in the case of homeostatic systems, the range of daily variations can be adapted to meet environmental challenges.

Circadian Regulation

Ubiquity and Adaptive Significance of Circadian Rhythms

Circadian rhythms are endogenously generated oscillations in physiology and behavior that persist (free run) in the absence of external time cues, at periods that generally deviate slightly from 24 hours. Such rhythms are synchronized (entrained) to the solar day by daily environmental cycles, or *zeitgebers* (time-givers), the most important of which is the daily light-dark cycle.

Circadian rhythms have been documented in all major eukaryotic taxa – animals, plants, and unicellular organisms – and, since the mid-1980s, are known to occur in some prokaryotes as well (Hastings et al., 1991; Johnson et al., 1996). Such ubiquity indicates that endogenous circadian timing represents a fundamental adaptation to the daily cycles in the environment (Pittendrigh, 1993). Initially, cyclic changes in the geophysical environment – light, temperature, and humidity – were probably the main agents of selection, but later the rhythmic activity patterns of prey, predators, and competitors would have provided added selection pressure for circadian organization.

One of the principal advantages conferred by endogenous circadian timing is the ability to anticipate and thus be prepared for cyclic environmental changes. For example, the rise in body temperature and corticosteroid secretion before the end of the night, while we are still asleep, prepares us for daytime activity (Wever, 1979). Similarly, rats adapted to restricted daily feeding schedules, with food presented at the same time each day, show pronounced anticipatory increases in behavioral activity, body temperature, corticosteroid levels, and in the activities of several intestinal and hepatic enzymes immediately preceding the scheduled feeding, preparing the animal to obtain and process food more efficiently (Boulos and Terman, 1980). Circadian regulation allows the temporal coordination of interdependent physiological, cellular, and biochemical events, as well as the temporal segregation of incompatible functions (Hastings et al., 1991).

Virtually all physiological systems show diurnal rhythms in at least some parameters; many of these rhythms have been shown to persist in constant conditions and are therefore truly circadian. This includes parameters that are universally acknowledged as being homeostatically regulated, such as body temperature, as well as parameters of what have been called allostatic systems, such as the HPA axis, the autonomic nervous system, and the immune system. Diurnal or circadian rhythms also characterize various aspects of brain function, including neuronal firing rates in most brain regions, neurotransmitter and neurohormone receptor activity, synthetic and catabolic enzyme activity, and gene expression.

Homeostatic and Circadian Regulation of Body Temperature

Thermoregulation is often considered a paradigmatic example of a homeostatic system; indeed, the fundamental distinction between homeothermic and poikilothermic organisms refers to the tendency of the former to maintain a relative constancy of body temperature, arbitrarily defined as $\pm 2°C$ (Bligh and Johnson, 1973), in the face of much wider fluctuations in environmental temperature. This constancy is achieved by means of autonomic and behavioral effectors, which activate heat loss or heat production mechanisms whenever body temperature exceeds or falls below a reference level, or setpoint (the term "setpoint" is used here in a descriptive sense, and does not refer to a particular control mechanism, cf. Satinoff, 1983). Often, the defended level is a range, rather than a specific value, with upper and lower thresholds for the activation of appropriate effectors, and with, in many cases, different thresholds for different effectors (Satinoff, 1978).

Regular daily fluctuations in body temperature have long been recognized, and the endogenous nature of the daily temperature rhythm has been known since the beginning of the previous century (Simpson and Galbraith, 1906). Body temperature rhythms are widely believed to reflect circadian variations in thermoregulatory setpoints. In humans, for example, immersion of the extremities in cold water in the morning, during the rising phase of the body temperature rhythm, leads to an earlier onset of shivering and longer rewarming times than during the falling phase in the evening (Hildebrandt, 1974). Body immersion in 30°C water is followed by a decrease in core body temperature in the evening, but leads to a small increase in the morning, indicating greater cold defense efficiency during the rising phase of the temperature rhythm (Cabanac et al., 1976). Conversely, sweating and peripheral vasodilation in response to heat loads have lower thresholds and longer latencies during the falling or low phase of the temperature rhythm than during the rising or high phase (Hildebrandt, 1974; Wenger et al., 1976; Stephenson et al., 1984; Stephenson and Kolka, 1985). These rhythmic changes in thermoregulatory responses are accompanied by changes in the subjective sensations evoked by cold and warm stimuli (Cabanac et al., 1976).

In the view of several authors, circadian changes in setpoint are not a contradiction of the principle of homeostasis. As noted by Halberg (1953),

the essential difference is that the defended level of a physiological parameter is a 24-hour curve, rather than a flat line, and it is deviations from this
cyclically varying setpoint that are kept in check by homeostatic mechanisms. In a similar vein, Moore-Ede (1986) has proposed that the concept
of homeostasis be extended to include circadian timing, which enables organisms to predict when environmental challenges are most likely to occur.
Moore-Ede distinguished between "reactive homeostasis," which describes
the negative feedback mechanisms activated in response to changes in the
internal milieu that have already occurred, and "predictive homeostasis,"
which refers to responses initiated in anticipation of predictably timed challenges and includes both circadian rhythms and the annual changes in behavior and metabolism observed in many species (Moore-Ede, 1986).

Daily Ranges of Body Temperature Rhythms

Data from a large number of mammalian and avian species, compiled by
Aschoff (1982), show that the daily range of the body temperature rhythm
increases with decreasing body weight. This relationship was obtained in
passerine and nonpasserine birds, and in primates and other mammals
weighing up to 10 kg. The results indicate that the temperature recorded
during the animals' active phase is similar across all weight classes within
each group and shows a relatively small difference between groups: $42.5 \pm
0.75°C$ in nonpasserine birds, $41.04 \pm 1.03°C$ in passerine birds, and $38.36 \pm
0.42°C$ in nonprimate mammals. However, body temperature recorded dur-
ing the inactive phase decreases with decreasing body weight. Aschoff concluded that there is an optimal body temperature, maintained during an
animal's active phase, that is similar across a wide variety of endotherms,
but that the relatively greater cost of thermoregulation in small animals,
resulting from a higher surface-to-mass ratio, is partially offset by a greater
reduction in body temperature during the inactive phase.

Some animals show considerable flexibility in the ranges of their diurnal
temperature rhythms when faced with unfavorable climatic conditions or
insufficient food and water. Thus, when deprived of food, some of the larger,
homeothermic bird species increase their daily ranges through a regulated
lowering of nighttime temperature proportional to body weight loss (Daan
et al., 1989; Graf et al., 1989). An interesting example has also been reported
in camels in the Algerian Sahara (Schmidt-Nielsen et al., 1957). In summer,
these animals show a diurnal rhythm of rectal temperature with a range
of about $2°C$, but when deprived of water, the range increases to $6°C$, with
higher daily maxima and lower minima. The authors of that study noted
that the increased heat storage during the day allows water conservation by

decreasing evaporative heat loss, and decreases heat gain by reducing the temperature gradient between the animal and its environment.

Masking Effects and the Role of Sleep

Although body temperature is subject to endogenously generated circadian oscillations, it is also influenced by various exogenous factors, which can affect the range and waveform of the daily temperature profile. These include environmental factors, such as ambient light and temperature, as well as behavioral factors, such as sleep, motor activity, and meal intake. The influence of sleep on daily rhythms is especially important in humans (and other primates), whose sleep is consolidated into one major episode lasting several hours, interrupted only by brief awakenings. In contrast, sleep in rodents and many other mammals consists of several bouts distributed across the 24-hour day, but with more frequent and longer bouts during one half of the circadian cycle than the other (e.g., Borbély and Neuhaus, 1978).

In the circadian rhythm literature, the direct effects of environmental and behavioral factors on rhythmic variables are known as external and internal masking, respectively. The term "masking" reflects the (often exclusive) focus of many chronobiologists on the endogenous component of daily rhythms, but masking effects are important in their own right and impart additional flexibility to daily temporal organization that would not be possible with strict pacemaker programming (Mrosovsky, 1999).

Assessment of the endogenous circadian contribution to daily rhythms requires that external and internal masking effects be either eliminated or distributed evenly across the circadian cycle. In humans, this can be achieved with constant routine protocols, during which subjects are kept awake in a resting position under constant environmental conditions for at least 24 hours, and their daily food and fluid intake is divided into small, equal portions and provided at regular intervals (usually hourly or bihourly) throughout the protocol (Mills et al., 1978; Czeisler et al., 1985).

Figure 7.1 shows daily profiles of several physiological and hormonal parameters recorded in groups of young adult men under normal entrained conditions, with sleep at night and moderate (indoor) activity during the day (left panels) and under constant routines (right panels). The bottom panels of the figure confirm the efficacy of the constant routine in eliminating daily variations in actigraphically monitored wrist activity.

The data in Figure 7.1 illustrate some of the effects of sleep and other exogenous factors on the daily range and waveform of endogenously generated circadian oscillations. For most variables, including body temperature, cortisol, urine volume, prolactin, and parathyroid hormone (PTH),

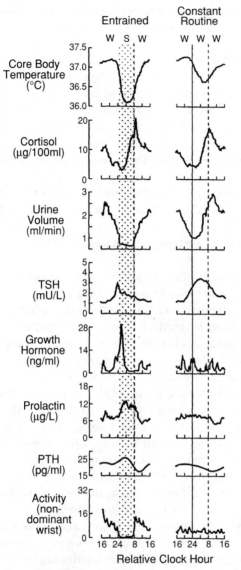

Figure 7.1: Temporal profiles of several physiological and behavioral variables recorded from young male subjects under normal entrained conditions with nighttime sleep (*stippling*) and daytime wakefulness (*left panels*), and from the same subjects under constant routines (*right panels*). TSH = thyroid stimulating hormone; PTH = parathyroid hormone. From Czeisler and Khalsa (2000).

exogenous influences reinforce the endogenous oscillation, widening the daily range of the rhythmic variable. This also implies that when sleep and activity are not scheduled at their normal times within the circadian cycle, as occurs during shift work, their effects oppose the endogenous oscillation, resulting in a narrowing of the daily range and a distortion of the daily waveform. In the case of thyroid-stimulating hormone (TSH), the endogenous, nocturnal increase observed during constant routines is partially suppressed by sleep, while the growth hormone profiles show a pronounced sleep-related increase in the early part of the night, which is associated with the relatively high levels of slow wave sleep and electroencephalographic (EEG) delta activity observed at the start of the sleep episode (Holl et al., 1991; Gronfier et al., 1997).

Constant routines have also been used to assess the contribution of endogenous circadian factors to the well-documented diurnal variations in heart rate and arterial blood pressure. Data from two studies showed clear circadian rhythmicity in heart rate during constant routine protocols, whereas the daily profiles of systolic, diastolic, and mean arterial blood pressure were essentially flat (Kerkhof et al., 1998; Van Dongen et al., 2001). These results suggest that the involvement of the circadian timing system in the regulation of blood pressure is entirely indirect and is mediated by its effects on the sleep-wake cycle.

Rheostasis

The chronic, stress-induced increase in blood pressure and the circadian oscillations in body temperature and other rhythmic parameters described in the preceding sections are only a few examples of physiological regulation around changing setpoints. Many others were described by Mrosovsky (1990), who argued that while such occurrences were an elaboration rather than a contradiction of homeostasis, they were sufficiently widespread as to merit a separate designation. Mrosovsky chose the term "rheostasis," which he defined as "a condition in which, at any one instant, homeostatic defenses are present but over a span of time, there is a change in the level that is defended" (p. 31). Mrosovsky also noted that changes in the level of a variable, even ones that take place with great regularity, do not necessarily qualify as rheostasis, unless the different levels of that variable are actively defended.

Mrosovsky distinguished between "programmed rheostasis" and "reactive rheostasis." Examples from both categories are shown in Table 7.1 (for a more complete listing, see table 8.1 in Mrosovsky, 1990). Programmed rheostasis includes cyclical changes in defended levels (circadian and seasonal variations and changes associated with estrous or menstrual cycles),

Table 7.1: Examples of programmed and reactive rheostasis (from Mrosovsky, 1990)

Programmed rheostasis	Reactive rheostasis
Circadian cycles	Infection: fever
Body temperature	Psychogenic fever
Glucocorticoid levels	Undernutrition:
Seasonal cycles	Gonadostat resetting
Weight changes	Body temperature and torpor
Gonadostat resetting	Glucocorticoid levels
Hibernation	Thyrostat resetting
Body temperature	High-fat diet: persistent obesity
Ph regulation (alphastat resetting)	Hypovolemia: osmostat resetting
Weight changes	Hypertension: blood pressure
Estrous cycles	High altitude: CO_2 chemostat
Weight changes	
Gonadostat resetting	
Menstrual cycles	
Body temperature	
Non-REM sleep:	
Body temperature	
CO_2 chemostat resetting	
Pregnancy	
Osmostat resetting	
CO_2 chemostat	
Puberty and aging	
Gonadostat resetting	

developmental changes occurring, for example, at puberty and during aging, and changes associated with sleep and with pregnancy and lactation. Reactive rheostasis, which includes stress-induced hypertension, occurs in response to stimuli that an organism may or may not encounter during its lifetime. A third category, termed "second-order rheostasis," includes the changes in the daily range of the circadian rhythm of body temperature described earlier.

One other type of regulatory mechanism related to programmed rheostasis was proposed by Bauman and Currie (1980) to underlie nutrient partitioning during pregnancy and lactation. This mechanism, termed "homeorhesis," was defined as "the orchestrated or coordinated changes in metabolism of body tissues necessary to support a physiological state" (Bauman and Currie, 1980, p. 1515). Bauman and Currie suggested that the regulation of nutrient partitioning during pregnancy involved homeorhetic controls arising from and assuring the growth of the fetus, whereas coordinated adaptations in rates of lipogenesis and lipolysis in adipose tissue represented homeorhetic controls necessary for milk synthesis during lactation.

Overlapping Concepts of Physiological Regulation

The various concepts reviewed in the preceding sections – allostasis, predictive homeostasis, homeorhesis, and rheostasis – all reflect the recognition that, while the parameters of the internal milieu are regulated, the levels around which they are regulated are frequently subject to change. Mrosovsky's concept of rheostasis is the most comprehensive because it encompasses all changes in regulated level and therefore includes predictive homeostasis, homeorhesis, and allostasis. Furthermore, the distinction he makes between programmed and reactive rheostasis is an important one, because it differentiates between obligatory changes that are "built into developmental or cyclical programs by evolutionary pressures" (Mrosovsky, 1990, p. 76) and changes that occur in response to certain stimuli and thus depend on the specific experiences of individual organisms.

Experience also plays a significant role in Sterling and Eyer's concept of allostasis; indeed, it is one of the features distinguishing allostatic from homeostatic regulation because only the former is affected by, and benefits from, experience (Sterling and Eyer, 1988). In that respect, allostasis and reactive rheostasis appear very similar, especially in view of the fact that most (although not all) of the examples of reactive rheostasis described by Mrosovsky are responses to prolonged exposure to such factors as psychological stress, high-fat diets, undernutrition, high altitude, and dehydration.

In contrast, circadian variations in regulated levels are examples of programmed rheostasis, and they occur independently of experience. For example, mice raised for six generations in constant illumination and in the absence of any cyclic changes in their external environment, continue to show normal circadian activity rhythms (Aschoff, 1960).

The differential role of experience in programmed and reactive changes in regulated levels has important implications for any definition of allostasis, which we return to in the final section of this chapter. First, however, we examine the physiological regulation of circadian timing and sleep, especially the part played by the brain because central coordination by the brain is another essential characteristic of allostasis. As will be apparent from the following sections, the neural mechanisms of chronobiological regulation are embedded within the hypothalamic and basal forebrain circuitry responsible for the defense of physiological setpoints as well as for their adaptive modification.

CIRCADIAN TIMING MECHANISMS

The Circadian Pacemaker

Soon after the initial report that hypothalamic damage can disrupt circadian rhythms of food and water intake (Richter, 1967), specific lesions of the hypothalamic suprachiasmatic nucleus (SCN) were shown to abolish the normal expression of both behavioral (Stephan and Zucker, 1972) and neuroendocrine (Moore and Eichler, 1972) rhythms in the rat. These early findings have been confirmed for numerous biobehavioral rhythms in a variety of mammalian species, and it has now been convincingly demonstrated that the SCN is the site of the primary circadian pacemaker in the mammalian brain (Klein et al., 1991; Buijs et al., 1996). More recently, studies using multielectrode neuronal recording in cultured SCN tissue slices (Welsh et al., 1995) and other techniques have provided compelling evidence that the fundamental circadian time generator is a cell-autonomous process, expressed within at least some, and possibly all, individual SCN neurons.

Typically, the SCN has been characterized anatomically as comprising distinct ventrolateral and dorsomedial subdivisions, but this scheme has been recently recast to comprise an SCN core and an SCN shell subnucleus (Moore, 1997; Moore and Silver, 1998). Whereas most SCN neurons contain the inhibitory amino acid neurotransmitter gamma-aminobutyric acid (GABA), core and shell subnuclei are distinguished at the cellular level by peptide phenotype. A great many neuropeptides have been localized in the SCN, but it is probably the concentrations of vasopressin-positive neurons in the shell and VIP (vasoactive intestinal peptide)-positive neurons in the core that best define these two divisions. The specific functions of chemically defined SCN cell populations are not fully known, but a reasonable heuristic is that the SCN core serves to collect and integrate pacemaker inputs (as described later, major SCN afferent systems tend to converge in the core), whereas the shell may be primarily responsible for generation and transmission of the circadian timing signal (Inouye and Shibata, 1994). This suggestion is consistent with findings that spontaneous circadian rhythmicity in peptide levels and gene expression are seen more reliably in the SCN shell than in the core (Inouye and Shibata, 1994; Sumova et al., 1998; Hamada et al., 2001).

In the last several years, analysis of the fundamental circadian oscillatory mechanism has been extended to the molecular genetic level. Following the initial identification of mammalian homologs of the Drosophila "clock gene," *Per*, evidence has accumulated rapidly indicating that a number of specific genes and proteins serve similar (although apparently not identical) functions in the core circadian mechanism of flies, mammals, and possibly all other animals (Dunlap, 1999; Young, 2000). Putative mammalian clock genes include three *Per* genes (*Per1*, *Per2*, *Per3*), *Clock*, *Bmal*, *CK1ϵ*, and two plant cryptochrome homologs (*Cry1* and *Cry2*), all of which are expressed within SCN neurons. These genes and their protein products interact to form autoregulatory transcription-translation feedback loops that define the core of the circadian oscillator (Shearman et al., 2000). For example, CLOCK-BMAL protein heterodimers exert positive drive on the transcription of *Per* and *Cry* genes, while PER and CRY proteins form both homo- and heterodimers that negatively regulate CLOCK and BMAL activity. This negative feedback loop results not only in the rhythmic transcription of specific clock genes, such as *Per*, but also in the rhythmic expression of a large number of clock-controlled genes (i.e., genes that are controlled by, but are not part of, the core oscillator loop), which serve as the basis for oscillator outputs to other cellular processes. Further downstream, dramatic circadian rhythm abnormalities have been observed in animals after mutation or genetic knockout of specific clock genes including *Per1* or *Per2* (or both), *Clock*, *Cry1*, or *Cry 2* (or a combination thereof), and *CK1ϵ*, thus demonstrating the relevance of these molecular-genetic mechanisms for expression of circadian rhythmicity at the behavioral level (Vitaterna et al., 1994; van der Horst et al., 1999; Lowrey et al., 2000; Albrecht et al., 2001).

Pacemaker Inputs

Entrainment of the SCN circadian pacemaker by environmental light-dark cycles is mediated by a direct retinohypothalamic (RHT) projection originating in a broadly scattered population of retinal ganglion cells and terminating mainly in the SCN (Moore and Lenn, 1972; Moore et al., 1995). Although other, less direct pathways exist by which photic signals may be conveyed to the SCN (discussed later), the RHT has been shown to be both necessary and sufficient for light entrainment (Johnson et al., 1988). Surprisingly, classical retinal rod and cone photoreceptors are not necessary for photic entrainment of the circadian pacemaker (Freedman et al., 1999). Instead, SCN-projecting retinal ganglion cells contain melanopsin, a previously unrecognized mammalian photopigment, conferring direct photosensitivity on this small subset of ganglion cells (Hannibal et al., 2002). Recent studies indicate that melanopsin knockout mice show attenuated phase-shift responses to light, but the responses are not entirely abolished, and melanopsin-null mice can still entrain to daily light-dark cycles, indicating the involvement of additional photoreceptive mechanisms (Panda et al., 2002; Ruby et al., 2002). Thus, photic entrainment of the circadian pacemaker is mediated, at least in part, by photoreceptors, ganglion cells, and retinal projections separate from those mediating visuoperceptual function.

RHT terminals release the excitatory amino acid neurotransmitter, glutamate, in response to photic stimulation, and glutamate acts through both N-methyl-D-aspartate (NMDA) and non-NMDA receptors and a variety of intracellular signaling molecules, including nitric oxide, protein kinase C and G (PKC, PKG), and CAMP responsive element binding protein (CREB) (Gillette, 1996), to activate *c-fos* and other immediate early genes (Kornhauser et al., 1996), leading to increased *Per* gene expression (Moriya et al., 2000), and thus to phase control of the molecular oscillator. In addition, RHT terminals release at least two peptide cotransmitters, substance P and pituitary adenyl cyclase activating peptide (PACAP), which appear to modulate the effects of glutamate (Hamada et al., 1999; Harrington et al., 1999).

In addition to the RHT, two other major SCN afferent systems have been characterized: (1) a projection arising from the intergeniculate leaflet (IGL), a discrete retinorecipient component of the lateral geniculate complex, and (2) a projection arising from the midbrain raphe, especially the median raphe nucleus (Moga and Moore, 1997; Moore, 1997). The projection from the IGL to the SCN, referred to as the geniculohypothalamic tract (GHT), releases both neuropeptide Y (NPY) and GABA, while the raphe projection to the SCN releases serotonin and probably other uncharacterized transmitters. These two projection systems form terminal fields that largely overlap with the RHT terminal field in the ventral (core) SCN. In general, IGL and raphe projections to the SCN have been characterized as subserving two distinct functions: (1) modulation of photic effects on the SCN pacemaker and (2) mediation of nonphotic, behavioral-state-related effects on the pacemaker. As examples of the latter, the period and phase of the mammalian circadian pacemaker may be affected by alterations in locomotor activity, arousal, and sleep (Mrosovsky, 1996; Mistlberger et al., 2000). Although the mechanisms underlying these effects are not as well understood as those mediating photic entrainment, extensive evidence indicates that experimental disruption of IGL or raphe projections, or pharmacological blockade of their transmitters, impairs the zeitgeber effects of behavioral-state-related cues, while disinhibiting photic effects on the circadian pacemaker. These interactions between photic and nonphotic entrainment signals are mediated at least in part by convergent effects of multiple pacemaker inputs on transcription of specific clock genes within individual SCN neurons (Maywood et al., 1999; Fukuhara et al., 2001).

Pacemaker Outputs

SCN efferents emerge from both core and shell divisions and project primarily to nearby diencephalic targets and, more sparsely, to mesencephalic and limbic forebrain areas (Watts et al., 1987; Watts, 1991; Morin et al., 1994; Buijs, 1996; Reuss, 1996). These efferents convey their output signals via the release of glutamate, GABA, and several peptide cotransmitters, including both vasopressin and VIP (Watts and Swanson, 1987; Cui et al., 2001). Within the diencephalon, SCN efferents project most densely to immediately surrounding regions, including the preoptic and retrochiasmatic areas, the dorsomedial hypothalamus (DMH), the subparaventricular zone (sPVZ), and, more sparsely, to the paraventricular nucleus (PVN) itself. Among first-order SCN efferents, special attention has focused on the dense SCN projections to the sPVZ: because the sPVZ itself projects to most other SCN targets, timing signals relayed through this area may serve to amplify or reinforce direct SCN output signals (Watts, 1991).

Given the wide range of autonomic, neuroendocrine, and behavioral systems displaying rhythmicity, perhaps the most surprising feature of first-order SCN efferents is their relatively limited distribution, indicating that circadian timing signals must be distributed throughout the brain by complex, multisynaptic pathways. For example, circadian control of the autonomic nervous system (ANS) is effected mainly via SCN efferents to the PVN (Reuss, 1996), which in turn projects to autonomic motor nuclei in the caudal brainstem and spinal cord (Swanson and Sawchenko, 1980). The first such pathway to be clearly identified was that underlying circadian control of pineal melatonin secretion, which includes projections from the SCN to the PVN, from the PVN to the intermediolateral column (IML) of the spinal cord, from the IML to the superior cervical ganglia (SCG), and, finally, from sympathetic SCG neurons to the pineal gland (Moore, 1996). More recently, the retrograde transneuronal tracer pseudorabies virus has been used to reveal SCN efferent circuitries through sympathetic projections to the adrenal cortex, pancreas, kidney, spleen, and brown and white adipose tissue, and through parasympathetic projections to the pancreas and liver (Bartness et al., 2001; Buijs et al., 1999, 2001). As examples, efferent pathways linking the SCN to the pineal gland and the adrenal cortex are depicted in Figure 7.2.

Progress has also been made in the identification of SCN output circuitry underlying other rhythmic functions. For example, SCN signals regulate the major hypothalamo-hypophyseal systems via projections to hypothalamic neurosecretory cells containing corticotrophin-releasing hormone (Vrang et al., 1995), gonadotropin-releasing hormone (Van der Beek et al., 1997),

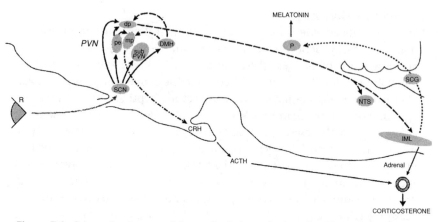

Figure 7.2: Schematic diagram of the rat brain in sagittal section illustrating pathways involved in the transmission of circadian timing information from the suprachiasmatic nucleus (SCN) to the pineal (p) and adrenal glands. The SCN projects to autonomic regions of the PVN, represented here by the dorsal (dp) and medial parvocellular (mp) parts, both directly and via the periventricular PVN (pe) and other nuclei of the medial hypothalamus, including the dorsomedial nucleus (DMH). Autonomic regions of the PVN project directly to the IML, which projects in turn to the adrenal gland and, via the superior cervical ganglion (SCG), to the pineal. In addition, neurons in the autonomic PVN release corticotropin-releasing hormone (CRH) into the portal system of the median eminence, stimulating the secretion of corticotropin (ACTH), which in turn stimulates the release of corticosterone from the adrenal cortex. R = retina; sub PVN = subparaventricular zone (abbreviated as sPVZ in the text); NTS = nucleus of the solitary tract. From Buijs et al. (1999).

and other hypothalamic releasing factors. With regard to sleep and arousal, SCN outputs relayed through the sPVZ (Lu et al., 2001) and the paraventricular thalamus (PVT; Buijs, 1996) convey circadian timing signals regulating behavioral and locomotor activity rhythms, while circadian modulation of arousal-promoting locus coeruleus neurons and sleep-promoting preoptic neurons is mediated by SCN outputs relayed through the dorsomedial hypothalamus (DMH) and other intermediate sites (Aston-Jones et al., 2001; Chou et al., 2002). Thus, even if considering only first- and second-order SCN efferents, plausible pathways may be constructed underlying the circadian control of sleep, activity, body temperature, affective arousal, sexual motivation, ingestive behavior, and autonomic and neuroendocrine systems (Buijs, 1996).

Finally, there is some evidence supporting a direct neurosecretory function for SCN neurons. Transplantion of fetal SCN neurons housed within a semipermeable capsule into arrhythmic SCN-lesioned hamsters has demonstrated that a diffusible signal secreted by the transplant is capable of driving a circadian rhythm in locomotor activity, even in the complete absence of neuronal outgrowth (Silver et al., 1996; LeSauter and Silver, 1998). More recently, several proteins that may mediate intrahypothalamic paracrine signaling by SCN neurons have been identified (Kramer et al., 2001; Cheng et al., 2002). Thus far, such effects have been linked only to behavioral activity rhythms and not to other SCN-dependent circadian rhythms, and the relative importance of neurocrine and paracrine SCN output signals under normal conditions remains to be determined.

Multiple Circadian Oscillators

Although the SCN output pathways just described could simply impose circadian timing on otherwise nonrhythmic effector systems, recent research indicates that several other brain regions as well as certain peripheral organs are capable of autonomous generation of at least damped circadian rhythmicity. Such findings have long been anticipated, given the extensive functional evidence for a multioscillatory circadian timing system (Rosenwasser and Adler, 1986). The initial demonstration of autonomous circadian rhythm generation in non-SCN mammalian tissue was the finding of circadian melatonin secretion by cultured hamster retina (Tosini and Menaker, 1996), but more recent studies have focused on the surprisingly widespread rhythmic expression of identified clock genes.

As mammalian clock genes were identified over the last few years, it quickly became apparent that these genes are expressed not only in the SCN, but also in a variety of other brain regions and in many peripheral tissues. By measuring light emission from microdissected brain tissue explants collected from transgenic rats carrying a *Per1-Luc* (*Luc*: firefly luciferase gene) construct, Abe and colleagues (2002) were able to demonstrate robust circadian rhythms of *Per* gene expression in the SCN that persisted with little reduction in amplitude for as long as they were measured (>11 days), as well as damped circadian rhythms in several other brain regions. The most robust non-SCN rhythms were seen in structures of the neuroendocrine hypothalamus and in explanted pineal and pituitary glands, but several other brain regions expressed either weaker cycling or no rhythmicity at all (see Table 7.2). Surprisingly, there was little correlation between the overall *level* and the *rhythmicity* of *Per* expression across different brain regions.

Abe and colleagues (2002) found that a 5-minute exposure of initially rhythmic cultures to forskolin after complete damping of circadian rhythmicity resulted in the reinstatement of rhythmicity, which damped out again over the next few days. In contrast, forskolin treatment failed to elicit circadian rhythmicity in tissues that were arrhythmic from the start. One possible explanation for these observations is that damping of circadian

Table 7.2: Brain structures assayed for circadian rhythmicity in *per-luc* bioluminescence (from Abe et al., 2002)

Rhythmic areas[a]	% Rhythmic	Damp rate[b] (Cycles)	Phase[c] (hour)	Arrhythmic areas	n
SCN	100 (42)	>10	9.7 ± 0.5	BNSTp	3
Pineal	100 (10)	1.3 ± 0.4	21.2 ± 0.2	CE	3
Pituitary	100 (16)	2.9 ± 0.3	14.3 ± 0.8	CP	5
RCH	100 (4)	1.8 ± 0.1	11.7 ± 0.1	DR	6
ME	100 (2)	1.3 ± 0.3	11.5 ± 1.8	HI	13
OB	100 (6)	1.6 ± 0.4	11.5 ± 0.3	M1p	3
AN	95 (19)	1.7 ± 0.2	14.3 ± 0.7	MR	3
PVN	92 (12)	1.0 ± 0.0	17.1 ± 0.6	MS	7
LH	75 (4)	0.8 ± 0.3	14.4 ± 3.4	NA	6
VOLT	50 (8)	0.8 ± 0.5	17.3 ± 2.3	PC	10
VLPO	43 (7)	1.1 ± 0.5	18.7 ± 1.8	S1	8
PVT	40 (5)	1.0 ± 0.0	22.4 ± 5.6	SN	7
SON	38 (8)	0.4 ± 0.2	13.0 ± 3.1	VTA	4
				Au 1	2

SCN = suprachiasmatic nucleus; RCH = retrochiasmatic nucleus; ME = median eminence; OB = olfactory bulb; AN = arcuate nucleus; PVN = paraventricular nucleus of the hypothalamus; LH = lateral hypothalamus; VOLT = vascular organ of the lamina terminalis; VLPO = ventrolateral preoptic nucleus; PVT = paraventricular nucleus of the thalamus; SON = supraoptic nucleus; BNSTp = bed nucleus of the stria terminalis, posterior portion; CE = cerebellum; DR = dorsal raphe; HI = hippocampus; M1p = primary motor cortex; MR = median raphe; MS = median septum; NA = nucleus accumbens; PC = piriform cortex; S1 = primary somatosensory cortex; SN = substantia nigra; VTA = ventral tegmental area; Au1 = primary auditory cortex.

[a] Cultures that showed at least two peaks at near-24-hour intervals. The total number of cultures is indicated in parentheses.

[b] Number of circadian cycles that exceeded 30% of the magnitude of the first cycle (mean ± SEM).

[c] Time of first circadian peak relative to the light-dark cycle underwhich the animal lived (lights on: 0 hour, lights off: 12 hours).

rhythms in *Per* expression is the result of rhythmic cells rapidly drifting out of phase from each other (Abe et al., 2002).

Autonomous but damped circadian rhythmicity in clock gene expression has also been demonstrated in liver, lung, kidney, heart, spleen, skeletal muscle, and testis (Yamazaki et al., 2000, 2002; Stokkan et al., 2001). In addition, clock genes are expressed in cultured fibroblast cell lines, and while such cell lines do not exhibit spontaneous rhythmicity in clock gene expression, damped rhythmicity can be induced by application of a serum or hormone pulse (Balsalobre et al., 1998, 2000). These results indicate that the SCN pacemaker may best be viewed as providing entraining and sustaining

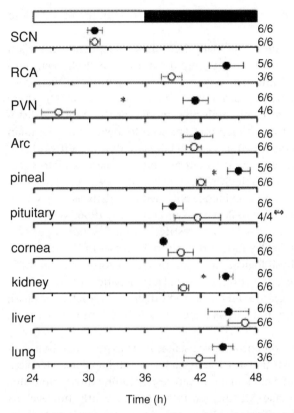

Figure 7.3: Phase map for central and peripheral circadian oscillators in young (*filled symbols*) and aged rats (*open symbols*). The data represent the mean (± SEM) times of the peaks of the *Per-luc* bioluminescence oscillations recorded during the interval between 12 hours and 36 hours in culture. The light-dark cycle that the animals had been exposed to prior to sacrifice is shown at the top. Sample sizes (number of rhythmic tissues/number of tissues tested) are indicated on the right. *Statistically significant difference ($p < .01$); **the pituitary glands from two aged rats had tumors and were excluded from the analysis. From Yamazaki et al. (2002).

signals to central and peripheral circadian oscillators located downstream from the pacemaker, rather than as imposing rhythmicity on passive, non-rhythmic targets.

Table 7.2 also shows that there are consistent phase differences between the rhythms expressed in non-SCN tissues. Recently, Yamazaki et al. (2002) found that the phases of some central and peripheral rhythms in *Per* expression (i.e., the times of their daily maxima relative to the light-dark cycle under which the animals had been housed) differed between young and old rats. As shown in Figure 7.3, the largest age-related difference was seen in the PVN, but significant differences were also observed in the pineal gland and kidney.

Following a phase-shift of the daily light-dark cycle, gene expression rhythms were found to reentrain most rapidly in the SCN itself; reentrainment rates in other rhythmic tissues, central as well as peripheral, were both slower and more variable (Yamazaki et al., 2000; Abe et al., 2002). These observations provide a plausible molecular model for the transient dissociation among different rhythmic functions seen following transmeridian travel (jet lag), and, as described in more detail later, during shift work.

The existence of one or more food-entrainable circadian oscillators separate from the light-entrainable SCN pacemaker has been recognized for some time, based on functional evidence obtained with daily feeding schedules (Boulos and Terman, 1980; Rosenwasser and Adler, 1986; Mistlberger, 1994). Thus, when food availability is restricted to a few hours per day, rats and other animals display food-anticipatory increases in locomotor activity and in several physiological processes, including corticosteroid secretion, body temperature, and various intestinal and hepatic functions. These food-anticipatory changes are only observed under feeding schedules with periods close to 24 hours and are not affected by SCN lesions (Boulos et al., 1980).

Localization of the food-entrainable oscillator(s) has proven to be a challenging problem. Food-anticipatory activity survives lesions of a wide variety of telencephalic and diencephalic structures, including those implicated in regulation of food intake (Mistlberger, 1994). More recently, however, restricted feeding schedules were shown to entrain the rhythmic expression of specific clock genes in the liver, but not in the SCN, suggesting a possible peripheral locus for at least one food-entrainable oscillator (Damiola et al., 2000; Stokkan et al., 2001).

Finally, although the relationship between the SCN circadian pacemaker and downstream central and peripheral oscillators is a hierarchical one, with the SCN at the top of the hierarchy, it is likely that peripheral circadian oscillators send signals back to central oscillators, including the SCN pacemaker, via their endocrine secretions as well as afferent autonomic axons (Buijs and Kalsbeek, 2001).

SLEEP MECHANISMS

Circadian and Homeostatic Regulation of Sleep

The regulation of sleep in humans and other mammals involves three distinct processes: a circadian process and a homeostatic process responsible for the global regulation of sleep and an ultradian process responsible for the cyclical alternation of the two basic sleep states, non–rapid eye movement (NREM) and rapid eye movement (REM) sleep, within the sleep episode (Borbély and Achermann, 2000; Pace-Schott and Hobson, 2002). Our focus here is on the circadian and homeostatic aspects of sleep regulation; for recent models of the NREM-REM sleep cycle and descriptions of the underlying neurophysiological mechanisms, see Pace-Schott and Hobson (2002).

Prominent circadian variations have been documented in several sleep parameters, including the propensity for sleep initiation, as indexed by sleep latency during multiple nap opportunities at different times of day (Richardson et al., 1982), sleep duration under entrained (Åkerstedt and Gillberg, 1981) and free-running conditions (Czeisler et al., 1980a; Zulley et al., 1981), and the probability of REM sleep within a sleep episode (Hume and Mills, 1977; Czeisler et al., 1980b). In both nocturnal rodents and diurnal monkeys, lesions of the SCN eliminate all circadian variations in sleep, leading to fragmented sleep bouts distributed evenly across the 24-hour day (Ibuka, 1979; Mistlberger et al., 1983; Edgar et al., 1993).

In human subjects living under normal entrained conditions, increasing wake duration by systematically delaying sleep onset from 23:00 to 11:00 results in a proportional decrease in sleep duration (Åkerstedt and Gillberg, 1981), a result that would appear to be at odds with a restorative or homeostatic process. A resolution of this apparent contradiction came with the realization that sleep varies in depth, or intensity, as well as in duration. Thus, sleep deprivation is followed by an increase in the deeper stages of NREM sleep (stages 3 and 4, known as slow-wave sleep, or SWS) at the expense of other sleep stages (Webb and Agnew, 1971), and daytime naps are followed by a selective decrease in SWS (Karacan et al., 1970).

A major criterion for classifying NREM sleep into stages is the percentage of slow waves in EEG recordings (Rechtschaffen and Kales, 1968). Thus, direct assessment of slow wave activity (SWA) by spectral analysis of the EEG signal in the delta frequency range (approximately 0.75–4.50 Hz) provides a more accurate measure of sleep depth than the somewhat arbitrary classification of NREM sleep into discrete stages. Quantitative EEG analysis has been applied to data from normal sleep, sleep deprivation, sleep displacement, and napping studies. The results indicate that SWA during NREM sleep is a function of prior sleep and wakefulness and shows little or no relation to circadian phase. Thus, SWA is typically highest at the start of the sleep episode, regardless of circadian phase, and decreases progressively over successive NREM-REM sleep cycles (Borbély et al., 1981). Furthermore, SWA increases monotonically with prior wake duration (Borbély et al., 1981; Dijk et al., 1987a; Daan et al., 1988) and decreases following daytime naps (Feinberg et al., 1985; Knowles et al., 1990; Werth et al., 1996). Finally, suppression of slow waves by acoustic stimulation during the first few hours of the sleep episode is followed by a substantial rise in SWA (Dijk et al., 1987b; Gillberg et al., 1991).

The circadian and homeostatic contributions to sleep are normally closely linked, as changes in prior wake duration, for example, generally entail a change in circadian phase as well. An effective method for separating the two factors, known as the forced desynchrony protocol, consists of maintaining subjects for many days on sleep-wake schedules with periods considerably longer or considerably shorter than 24 hours (e.g., 20 hours or 28 hours). Such periods fall outside the range of circadian entrainment, and the subjects' circadian rhythms therefore free-run at periods close to 24 hours. Thus, averaging the sleep data across the free-running period shows the circadian contribution to sleep, while averaging across the period of the sleep-wake schedule shows the homeostatic (sleep-dependent) contribution. Such data show clear circadian as well as sleep-dependent variations in wakefulness during the sleep episode, in REM and NREM sleep, and in EEG sigma activity (12.75–15.0 Hz, the frequency of sleep spindles, which are a hallmark of stage 2 sleep); in contrast, SWA decreases monotonically over the sleep episode but shows only weak dependence on circadian phase (Dijk and Czeisler, 1995).

The Two-Process Model of Sleep Regulation

The observations summarized in the previous section indicate that the timing of sleep is controlled primarily by a circadian process, while the intensity or depth of sleep is controlled by a homeostatic process and is allowed to vary in response to prior sleep and wakefulness. The two processes have been incorporated in a highly influential model of sleep regulation (Borbély, 1982; Daan et al., 1984), which postulates that a sleep-regulating variable, process S, accumulates during waking and dissipates during sleep. Sleep begins and ends when S reaches an upper and a lower threshold, respectively, and the two thresholds are modulated by the same circadian process, process C, controlled by the SCN pacemaker (Fig. 7.4).

In the original quantitative version of the two-process model (Daan et al., 1984), the time constants of process S were derived from the rate of change of SWA recorded during regular sleep and during recovery from different durations of sleep deprivation. Based on these data, both the rate of buildup of process S during wake and its rate of decline during sleep were modeled by saturating exponential functions, as shown in Figure 7.4. The model has undergone several revisions and refinements (see Borbély and Achermann, 2000), but even in its original version, it allows quantitative simulations of a number of well-known phenomena, including the

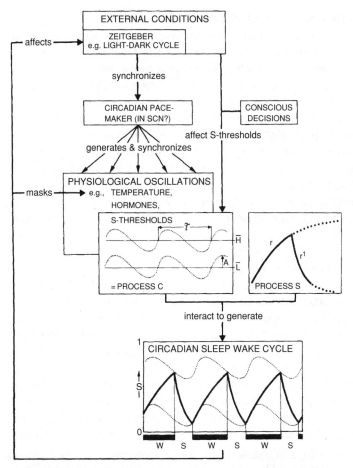

Figure 7.4: Schematic representation of the two-process model of sleep regulation. See text for explanations. SCN = suprachiasmatic nucleus. From Daan et al. (1984).

dependence of sleep duration on circadian phase in temporal isolation studies, during recovery from sleep deprivation, and during shift work. The model can also be used to generate the polyphasic sleep patterns of rodents and other small mammals by simply decreasing the distance between the upper and lower thresholds (Daan et al., 1984).

The success of the two-process model has prompted the search for a sleep-promoting neurochemical marker of process S, one that would increase during wakefulness and decrease during sleep. Although a number of peptides and hormones are involved in sleep regulation (Krueger et al., 1998), data summarized later suggest that the nucleoside, adenosine, is a strong candidate as a neurohumoral correlate of homeostatic sleep drive.

Neural and Neurohumoral Mechanisms

Historically, the search for neural mechanisms of sleep regulation has focused on two broadly defined brain regions: the preoptic basal forebrain and the rostral pole of the brainstem reticular formation. In the last several years, however, rapid advances have occurred in elucidating specific neural and neurohumoral sleep and arousal mechanisms in the basal forebrain, preoptic area, and anterior hypothalamus (Jones and Muhlethaler, 1999; Shiromani et al., 1999). At the same time, neural systems of the pons and midbrain have been closely linked to mechanisms of sleep-state selection, and in particular, with the segregation of REM sleep and waking states (Semba, 1999). In the sections that follow, we emphasize the role of forebrain sleep and arousal mechanisms, as well as possible anatomic and physiological bases for the integration of sleep-related and circadian timing signals.

Basal Forebrain and Anterior Hypothalamus

A broadly distributed swath of basal forebrain neurons (nucleus basalis, substantia innominata, and magnocellular preoptic area) provides widespread excitatory cholinergic input to the hippocampus and neocortex, and activation of these neurons during both waking and REM sleep states leads to cortical EEG activation and behavioral arousal (Jones and Muhlethaler, 1999; Shiromani et al., 1999; Vazquez and Baghdoyan, 2001). In contrast, an important concentration of sleep-active GABAergic neurons is found in the medial preoptic–anterior hypothalamic area (POAH; Shiromani et al., 1999). These neurons underlie the somnogenic effects of indirect GABA-A agonists, including benzodiazepines and ethanol (Mendelson, 2001) and provide a critical link between sleep-regulatory and thermoregulatory systems (Gong et al., 2000; Jha et al., 2001). Wake-active and sleep-active neuronal populations are not completely segregated anatomically, and some commingling occurs in both basal forebrain and preoptic areas. Furthermore, both populations are subject to complex afferent regulation by ascending brainstem cholinergic and monoaminergic projections and by both excitatory and inhibitory amino acid neurotransmitters (Jones and Muhlethaler, 1999; Shiromani et al., 1999).

Sherin et al. (1996) used functional autoradiography during spontaneous sleep to identify an anatomically discrete cluster of sleep-active neurons in the ventrolateral preoptic area (vLPO). Because vLPO neurons fire at relatively high rates during both SWS and REM states (Szymusiak et al., 1998), their activity appears to be correlated more closely with sleep per se than with cortical EEG. Furthermore, lesions restricted to the core of the vLPO dramatically reduce SWS, whereas slightly larger lesions also reduce

REM sleep (Lu et al., 2000). Neurons of the vLPO contain both GABA and the neuropeptide, galanin; because galanin is an important hypothalamic modulator of growth hormone (GH) secretion, these galanin-positive sleep-active neurons could be important for the sleep-dependent secretion of GH and other hormones (Shiromani et al., 1999).

Posterior and Lateral Hypothalamus

Histaminergic neurons of the tuberomammillary nucleus (TMN) of the posterior hypothalamus comprise an important wake-promoting neuronal system. TMN lesions result in hypersomnolence (McGinty, 1969), while TMN neurons show maximal discharge rates during wakefulness (and in some cells, during REM sleep) and are essentially silent during sleep (Steininger et al., 1999). While TMN neurons may influence cortical arousal in part via interactions with other sleep- or arousal-related forebrain systems, at least some TMN neurons project directly to the cortex (Saper, 1985).

Recently, an additional arousal-promoting neural system has been identified in the lateral hypothalamus (LH), specifically in the perifornical region. This cell group is defined anatomically by the presence of a newly recognized neuropeptide, hypocretin (Hcrt, also known as orexin). Indeed, hypocretin-positive neuronal somata are found only in the perifornical LH, and nowhere else in the brain (de Lecea et al., 1998). The clinical importance of Hcrt is highlighted by findings linking deficient expression of Hcrt peptide or receptors to human narcolepsy and to animal models of narcolepsy, respectively (Kilduff and Peyron, 2000; Sutcliffe and de Lecea, 2000). Because narcolepsy is often viewed as the inappropriate intrusion of REM sleep into ongoing periods of wakefulness, Hcrt may play a specific role in maintaining the separate integrities of these two states that are both associated with EEG arousal. Consistent with this hypothesis, Gerashchenko et al. (2001) employed a targeted immunotoxin (hypocretin-2-saporin) to show that selective destruction of LH Hcrt neurons results in increased SWS and REM sleep, as well as narcolepsylike sleep-onset REM episodes. In cats, functional autoradiographic studies (Torterolo et al., 2001) have revealed that Hcrt neurons are activated during behaviorally *active* but not behaviorally *quiet* wakefulness, and also that Hcrt neurons are nearly as activated during REM sleep as during active behavior. These results suggest that the function of the Hcrt system is more closely related to active behavior or REM sleep regulation than to wakefulness per se.

Role of Adenosine

Adenosine is a purine nucleoside that functions as a central neuromodulator (Fredholm and Hedqvist, 1980). Adenosine and adenosine agonists interact with distinct G-coupled postsynaptic receptors to produce generally inhibitory postsynaptic potentials, whereas adenosine heteroceptors exert presynaptic inhibition of neurotransmitter release in several transmitter systems. Several findings implicate a potential role for adenosine in sleep-wake regulation: (1) central adenosine administration increases SWS and REM sleep while decreasing wakefulness (Radulovacki, 1985; Ticho and Radulovacki, 1991; Portas et al., 1997); (2) adenosine antagonists such as caffeine and theophylline produce both subjective and EEG arousal (Virus et al., 1990; Landolt et al., 1995; Roehrs et al., 1995); and (3) extracellular adenosine levels accumulate during spontaneous or forced wakefulness and dissipate gradually during sleep (Porkka-Heiskanen et al., 1997). This last result is of considerable interest because the profile of adenosine levels during sleep-wake cycles is essentially identical to that of the hypothesized process S in human sleep regulation, as described earlier. Thus adenosine level may serve as a neurohumoral marker for homeostatic sleep drive.

A critical issue in understanding the role of adenosine in sleep regulation concerns the potential global and local actions of this modulator. Adenosine is a near-ubiquitous substance, and indeed, it is thought to be released as a by-product of neural metabolism throughout the brain. Because neurometabolic activity is relatively high during wakefulness and shows widespread reductions during SWS, adenosine could function globally to regulate brain energy expenditure, in part by triggering sleep (Benington and Heller, 1995). Alternatively, adenosine could regulate sleep and wakefulness directly, via local action within forebrain and brainstem state-control systems, including those discussed earlier. In support of this hypothesis, anatomically selective increases in adenosine levels within the basal forebrain or pontine cholinergic regions trigger sleep, whereas similar magnitude increases in the thalamus do not (Porkka-Heiskanen et al., 1997; Portas et al., 1997). Despite its generally inhibitory effects, central adenosine administration produced a selective *increase* in activity of the vLPO, consistent with the hypothesized role of this structure in promoting sleep (Scammell et al., 2001). Although these results do not rule out an additional global role for adenosine in the regulation of neuronal physiology and metabolism, they do indicate that adenosine acts specifically to increase activity in sleep-active regions while decreasing activity in wake-active regions of the forebrain and brainstem.

Connections with the Circadian Timing System

As described earlier, sleep is subject to both homeostatic and circadian regulation. Recent studies have begun to identify neural linkages that could account for functional interactions between these two regulatory systems. For example, efferents from the SCN circadian pacemaker reach the vLPO both directly (Novak and Nunez, 2000) and via several polysynaptic intrahypothalamic pathways (Deurveilher et al., 2002; Chou et al., 2002). In vitro electrophysiological studies indicate that SCN efferents exert both GABA-A-mediated inhibitory effects and NMDA- and non-NMDA-mediated excitatory effects on neurons of the vLPO and immediately surrounding area (Sun et al., 2000; 2001). However, most neurons within the core of the vLPO showed excitatory responses to SCN stimulation, consistent with the fact that SCN neurons are generally day-active, in association with daytime sleep in the nocturnal rat. In addition, the vLPO receives a direct projection from retinal ganglion cells (Lu et al., 1999), which could mediate the well-known masking effects of light on sleep and wakefulness in both animals and humans (Alfoldi et al., 1991; Campbell et al., 1995).

SCN efferents also project directly (Abrahamson et al., 2001) and indirectly (Deurveilher et al., 2001) to lateral hypothalamic Hcrt neurons, and Hcrt-positive fibers and Hcrt receptors are found in the SCN (McGranaghan and Piggins, 2001; Backberg et al., 2002). In addition, Hcrt fibers are seen in the IGL and midbrain raphe nuclei, which project in turn to the SCN (McGranaghan and Piggins, 2001). These anatomic observations provide a basis for reciprocal interaction between the SCN circadian pacemaker and this critical wake-promoting forebrain system.

In addition to its putative role as a sleep-promoting neuromodulator, adenosine contributes to circadian pacemaker regulation. Thus, the SCN displays robust circadian rhythms in adenosine levels, and these rhythms are phased oppositely to the day-active neural and metabolic rhythms seen within the SCN (Yamazaki et al., 1994). Activation of SCN adenosine receptors inhibits photic input via the RHT at night (Watanabe et al., 1996; Elliott et al., 2001) and modulates the phase of the circadian pacemaker during daytime (Antle et al. 2001). Both these effects are mimicked by enforced sleep deprivation (Mistlberger et al., 1997; Antle and Mistlberger, 2000), suggesting that the effects of sleep deprivation on the circadian pacemaker could be mediated in part by increased adenosine release due to prolonged wakefulness. A very similar pattern of results is also seen for serotonin, in that (1) serotonin neurons fire maximally during wakefulness (Jacobs and Fornal, 1999), (2) serotonin and serotonergic agonists act within the SCN to inhibit RHT input at night and to regulate pacemaker phase by day (Mistlberger et al., 2000; Rea and Pickard, 2000; Prosser, 2000), and (3) sleep deprivation increases firing rates and neurotransmitter turnover in serotonergic neurons (Asikainen et al., 1997; Gardner et al., 1997). Thus, several neurotransmitters and neuromodulators – including adenosine, serotonin,

GABA, and others – may underlie the coregulation and integration of sleep homeostasis and circadian timing.

SHIFT WORK AND ALLOSTASIS

Shift-work schedules, particularly those that include night work, entail a displacement of sleep toward the activity phase of the circadian rest-activity cycle and of wakefulness toward the rest phase. As a result, sleep quality and duration are adversely affected, as are work-time alertness, mood, and performance. Because they are required to function in a day-oriented society, shift workers also must contend with changes in their social patterns, including their domestic and family lives. Disruptions of circadian timing, sleep, and social patterns represent three sources of stress for shift workers, and there is now convincing evidence that, over time, these can lead to chronic deleterious effects on both physical and mental health (Moore-Ede and Richardson, 1985; U.S. Congress, Office of Technology Assessment, 1991; Härmä, 1998).

Shift work is an increasingly common feature of modern industrialized society and of many developing countries, with more and more men and women engaged in work outside the standard daytime hours. This trend is attributed to economic, technological, and social forces seeking to maximize the useful life of expensive industrial plants and facilities, especially as these become more highly automated, and to satisfy the rising demand for round-the-clock retail, entertainment, and other services (Mott et al., 1965; Kogi, 1985; Gordon et al., 1986). According to the Current Population Survey of 1991, 20.1% of the 100 million or so employed Americans aged 18 years or older worked nonstandard hours, with 8.3% on fixed evening shifts, 2.9% on fixed night shifts, 3.4% on rotating shifts, and a further 5.4% working irregular schedules (Presser, 1995).

Another important trend from a public health point of view is the aging of the population of industrial nations, which has meant an increase in the proportion of older workers engaged in shift work. Indeed, a recent survey across the European Union shows that 24.4% of male and 12.3% of female workers aged 45 years or more now work on schedules that include night shifts, only slightly less than the 28.7% of men and 14.8% of women under the age of 45 working on such schedules (Härmä and Ilmarinen, 1999). The ability to tolerate nonstandard work hours is subject to wide individual differences, and several potential factors underlying these differences have been identified. Of these, aging is one of the best documented, and it is generally recognized that the adverse effects of shift work on sleep and health are exacerbated in older workers (Härmä, 1993, 1996).

Research into the effects of shift work on health and well-being faces a number of difficulties, not the least of which is the sheer diversity of work schedules, thought to number in the hundreds in the United States alone (U.S. Congress, Office of Technology Assessment, 1991) and in the thousands worldwide (Smith et al., 1998). Distinguishing characteristics of work schedules include permanent versus rotating shifts, whether they include night work, the number of hours worked per shift, continuous (7 days/week) versus discontinuous shifts (Monday to Friday or Monday to Saturday), speed of rotation (i.e., number of consecutive shifts of each type), direction of rotation (forward: morning-afternoon/evening-night, or backward: morning-night-afternoon/evening), shift start and end times, distribution of days off within the shift cycle, length of the shift cycle, and regularity of the shift work system (Knauth and Rutenfranz, 1982).

Another difficulty is that many workers who experience serious problems adjusting to night work or rotating schedules – about 20% by some estimates (Rutenfranz et al., 1977; Costa, 1996) – change to day work. Surveys of former shift workers indicate that most of those who transfer to day jobs do so for health reasons, often at the advice of a physician (Frese and Okonek, 1984). Studies that fail to take this self-selection process, or healthy-worker effect, into account will therefore tend to underestimate the adverse effects of shift work, even more so if the comparison day worker population includes shift work dropouts.

Limited Adaptation of Circadian Rhythms during Shift Work

The extent to which circadian rhythms adjust, or reentrain, to the changes in sleep and wake time imposed by shift work has been the subject of numerous studies, dating as far back as the late 19th century (Aschoff, 1981). Many of the earlier studies involved body temperature measurements, in some cases recorded continuously over several days. In one such study, young male subjects with no prior shift work experience worked on an industrial assembly task for one or more day shifts followed by up to 21 consecutive night shifts (Knauth et al., 1978). Their rectal temperature profiles showed clear masking effects following the change from day to night shifts. For the first several days, however, temperature during daytime sleep remained higher than it had been during night sleep, whereas temperature at night remained lower than it was during daytime wake, resulting in a flattening of the daily temperature rhythm. The daily range of the temperature rhythm gradually increased over successive days, reaching normal levels by day 7. Even after 21 days of night work, however, the peak of the temperature rhythm still occurred in the evening, as it did during day work, indicating incomplete adjustment of the circadian component of the rhythm.

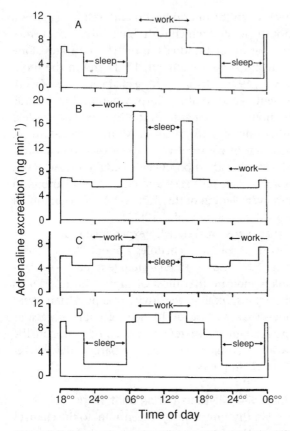

Figure 7.5: Mean urinary adrenaline excretion levels in shift workers during, **(A)** the last week of day work, **(B)** the first week of night work, **(C)** the third week of night work, and **(D)** the first week of return to day work. For clarity, data from the first 12 hours of each condition are double plotted. From Åkerstedt (1977).

A similar flattening and distortion of the daily temperature rhythm with, at best, only partial circadian adjustment, has been observed in several other laboratory studies (e.g., Colquhoun et al., 1968) as well as in field studies in experienced shift workers (e.g., Knauth et al., 1981). Indeed, in the latter case, circadian adjustment is even more limited, not only during rapidly rotating schedules with only 2 or 3 consecutive night shifts, but also in workers on permanent night shifts, because such workers usually revert to a normal day-active schedule during days off. Gradual, and partial, circadian adjustment has also been documented for several other rhythmic functions, including cortisol and melatonin secretion (e.g., Sack et al., 1992; Hennig et al., 1998).

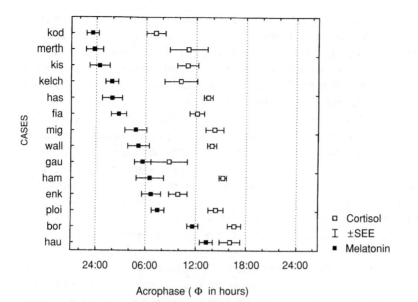

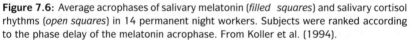

Figure 7.6: Average acrophases of salivary melatonin (*filled squares*) and salivary cortisol rhythms (*open squares*) in 14 permanent night workers. Subjects were ranked according to the phase delay of the melatonin acrophase. From Koller et al. (1994).

Figure 7.5 shows daily rhythms of urinary adrenaline excretion in a group of railroad workers during a week of day work, during the first and third of 3 consecutive weeks of night work, and during the first week back on a day work schedule (Åkerstedt, 1977). The data for the first week of night work show direct effects of sleep and activity on adrenaline levels, with no evidence of circadian adaptation. Note that the average daily level was higher than it was during day work, indicating greater stress. Daytime levels decreased during the third week, but the levels during work at night remained lower than they were during day work and the daily rhythm was flatter, indicating only partial adaptation. The return to day work resulted in a rapid restitution of the normal circadian pattern.

The flattening of daily rhythms in individual shift workers is attributable to masking effects superimposed on an incompletely adjusted circadian rhythm, but in group data such as those of Figure 7.5, a flattening of average daily rhythms can also reflect differences in the extent of circadian adjustment of individual workers, which are often considerable (e.g., Sack et al., 1992). Figure 7.6 shows individual acrophases (the times of the peaks of 24-hour sinusoids fitted to the data) of the circadian rhythms of salivary melatonin and salivary cortisol in 14 watchmen on permanent night duty, working 5 consecutive night shifts per week (Koller et al., 1994). The data, collected during the last 48 hours of the shift period, show wide individual differences in the acrophases of both hormones. In the case of melatonin, the acrophases spanned a 12-hour range, from around midnight, indicating no shifting of the melatonin rhythm, to around noon, indicating complete circadian realignment. The range of the individual cortisol acrophases was more limited, but still covered about 8 hours.

The fact that circadian reentrainment during shift work is rarely complete is generally attributed to the competing influences of external *zeitgebers,* particularly that of the light-dark cycle, although social cues may also play a role. In this respect, shift work differs from rapid travel across multiple time zones, which is accompanied by a shift in all external time cues. As a result, the circadian rhythms of transmeridian travelers fully reentrain to the new time zone, albeit gradually, at average rates of 1.5 hours/day for westward flights and 1 hour/day for eastward flights (Aschoff et al., 1975).

For night workers, exposure to bright light at the end of the work shift in the morning is especially important because light exposure at that time of day causes a phase advance, thereby opposing the phase delay required for realignment of the circadian timing system following a change from day or evening shifts to night shifts. The effect of morning light exposure was examined in the study depicted in Figure 7.6, which included continuous measurements of light intensity through light sensors mounted on eyeglass frames. A high negative correlation was obtained between maximum light exposure in the morning and the extent to which the melatonin acrophase shifted into the daytime (Koller et al., 1994).

Shift work studies that include recordings of multiple rhythmic variables show that different rhythms adjust at different rates (Chaumont et al., 1979; Touitou et al., 1990; Weibel et al. 1996; Costa et al., 1997). As a result, the phase relationships between the daily rhythms of night workers often differ from those in day workers. Such internal dissociations have generally been attributed to differences in the relative contributions of exogenous and endogenous factors to the daily pattern, with variables that are more susceptible to masking effects showing faster adjustment than those with a stronger circadian component. The recent evidence from animal studies described earlier, however, indicates that there are damped circadian oscillators in a number of brain regions and in various peripheral tissues and that these oscillators show different reentrainment rates following shifts of the light-dark cycle. These data thus suggest that the internal dissociations observed in shift workers may also reflect the different kinetics of multiple circadian oscillators.

Sleep in Shift Workers

Sleep disturbances are consistently associated with both rotating and permanent night work. Data from several studies involving a total of 5,766 workers, compiled by Rutenfranz et al. (1981), show that such disturbances are reported by about 15–20% of day workers, 5% of shift workers not engaged in night work, 60% of permanent night workers, and up to 80% of workers on rotating schedules that include night work. Indeed, the sleep disruptions experienced during shift work are sufficiently widespread as to be recognized as a sleep disorder, both in the International Classification of Sleep Disorders (ICSD) and in the Diagnostic and Statistical Manual of Mental Disorders (DSM).

The primary complaints of shift workers are of reductions in the quantity and quality of daytime sleep following night work, and of increased sleepiness and fatigue at work. Data from four samples of workers on permanent and rotating work schedules, compiled by Colligan and Tepas (1986), indicate that sleep duration is longest after afternoon-evening shifts and shortest after night shifts. The data also show that sleep duration after the night shift is shorter in rotating than in permanent night workers, and other studies indicate that workers on rapidly rotating schedules generally have shorter sleeps than workers on weekly rotations (Wilkinson, 1992). These observations indicate a certain degree of adaptation across consecutive night shifts, and there is some evidence that the improvement in sleep is associated with circadian adjustment (Dahlgren, 1981), but the adaptation is incomplete because sleep duration remains shorter than it is during day work.

The sleep disturbances experienced while on the night shift are due in large part to internal factors – namely, the timing of sleep at an inappropriate phase of the circadian cycle. The two curves in Figure 7.7 represent sleep duration as a function of the time of sleep onset in large groups of German and Japanese shift workers (Kogi, 1985). The two curves are virtually identical, both showing that sleep is longest when it is initiated in the late evening and early night, and shortest when it is initiated in the afternoon. In addition, however, shift workers frequently complain that their daytime sleep is disrupted by external factors, primarily noise from traffic, children, and telephones (Rutenfranz et al., 1981), and, in many developing countries, by high daytime temperatures (e.g., Khaleque, 1991).

Sleep duration is also abnormally short in workers on morning shifts with early start times (04:00–06:00), often even shorter than on the night shift (Kecklund and Åkerstedt, 1995). The shortening of sleep in this case

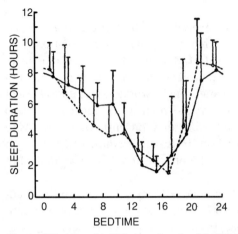

Figure 7.7: Mean (+ SD) sleep length of 2,332 German (*solid line and filled symbols*) and 3,240 Japanese (*dotted line and open symbols*) shift workers as a function of the time of sleep onset. From Kogi (1985).

is due to the failure of most workers to sufficiently advance their bedtimes to compensate for the early wake-up required on such schedules (Folkard and Barton, 1993). In part, this is because they may be reluctant to give up valuable leisure time in the evening, but there is a chronobiological reason as well – namely, that falling asleep at that time of day – about 2–3 hours before habitual bedtime – is more difficult than at any other. Indeed, this phase of the circadian sleep-wake cycle has been called the "forbidden zone" for sleep (Lavie, 1986) or, less ominously, the "wake maintenance zone" (Strogatz et al., 1987).

The data summarized in the preceding paragraphs refer to the duration of the main daily sleep episode and do not include naps taken before or after the work shift. Several studies have shown that such naps are longer and/or more frequent in connection with night (and early morning) shifts than with other shifts (Åkerstedt et al., 1989; Rosa, 1993). Night workers also show a greater increase in sleep duration on days off and when working other shifts, particularly evening shifts (Tepas et al., 1981). As a result, total daily sleep averaged across the entire shift cycle may be comparable to that of day workers (e.g., Tune, 1969), although in their study, Tepas et al. (1981) found that adjusted total sleep time remained somewhat lower in rotating workers (7.14 hours) than in day, afternoon-evening, and permanent night workers (7.53 hours, 7.86 hours, and 7.66 hours, respectively).

A number of studies have included EEG recordings of shift workers' sleep patterns, some in a sleep laboratory and others in the worker's home environment (Kripke et al., 1971; Bryden and Holdstock, 1973; Foret and

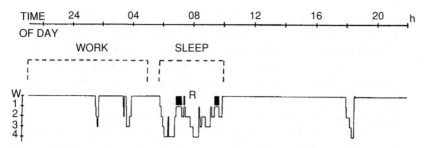

Figure 7.8: Hypnogram derived from 24-hour ambulatory electroencephalogram recorded from a rotating shift worker during a night shift. R = REM sleep. From Åkerstedt (1989).

Benoit, 1974, 1979; Dahlgren, 1981; Matsumoto, 1981; Walsh et al., 1981; Tilley et al., 1982). The results confirm the main conclusions derived from questionnaire and sleep diary data, including a shortening of sleep during night shifts, and a reduction in sleep quality, as indicated by an increase in wake and in stage 1 (a transitional stage between wake and sleep) during the sleep episode.

In some studies, ambulatory monitoring was used to obtain continuous 24-hour EEG recordings during different shift schedules (Torsvall et al., 1989; Åkerstedt et al., 1991). The example in Figure 7.8, recorded from a rotating three-shift worker during a night shift, shows a main daytime sleep episode lasting about 4 hours, followed by a brief nap later in the day. The record also shows that the worker fell asleep three times during the work shift. Indeed, such unintentional (and unauthorized) sleep episodes were observed in 5 of the 25 workers studied during the night shift, despite the fact that they were obviously aware of being monitored; in contrast, none of the workers fell asleep during the afternoon shift (Torsvall et al., 1989). Napping during work hours is allowed in some countries, notably Japan (Kogi, 1981), but not in most Western industrial nations. Nevertheless, in a survey of workers from various industrial plants in the United States, 53% reported falling asleep while on the night shift (Coleman and Dement, 1986). Similarly, Åkerstedt et al. (1983) found that 11% out of a total of 1,000 train drivers admitted to "dozing off" on most night trips, with a further 59% reporting that they had done so at least once.

Studies of former shift workers indicate that sleep disturbances are an important reason for their switching to day work. For example, Rutenfranz et al. (1981) found that about 90% of former shift workers reported that their sleep had been disturbed while they had been engaged in night work. That figure decreased dramatically following transfer to day work, but other data show that former shift workers still report more sleep disturbances than day workers who were never engaged in night work (Butat et al., 1993; Brugère et al., 1997). In a survey of 479 nurses on day or evening schedules but who had previously worked night shifts, Dumont and colleagues (1997) found that scores on an insomnia index were highest in those who had worked more than 5 nights per month for 4 to 10 years. EEG recordings obtained from a subset of 15 nurses confirmed that a high insomnia index score was associated with an increase in the number of awakenings.

Finally, there are indications that the sleep disturbances experienced during shift work worsen with age (Foret et al., 1981; Härmä, 1996). Some of the most convincing evidence was obtained in a field study involving two groups of locomotive engineers aged 25–35 and 50–60 years, respectively (Torsvall et al., 1981). EEG recordings were obtained in the subjects' homes during both night and day sleep (following day and night work, respectively). Sleep length did not differ between the two age groups, either during the day or at night, but the older subjects showed more awakenings and stage 1 sleep, greater diuresis, and higher noradrenaline excretion during day sleep. The worsening of sleep quality in older shift workers has been attributed to a decrease in phase tolerance, which refers to the capacity of an individual to sleep at abnormal circadian phases (Campbell et al., 1995). In addition, sleep becomes more fragile with age, that is, it is more easily disrupted by external factors, as indicated, for example, by a decrease in auditory awakening thresholds (McDonald et al., 1981).

In summary, the available evidence indicates that shift workers go through repeated periods of partial sleep deprivation while on night or early-morning shifts, alternating with periods of recovery when they are on day or evening shifts and on days off. In addition, the often poor quality of sleep while on night shifts suggests that shift workers may suffer from chronic sleep deprivation, even when total sleep duration averaged across the entire work cycle approaches normal, day worker levels. Finally, while the sleep disturbances experienced during shift work are attenuated following transfer to day schedules, they may still exceed levels reported by day workers for many years following such transfer.

Social and Domestic Effects of Shift Work

Shift work inevitably affects the social life of the worker, including his or her relations with family members, friends, and the community at large (Rutenfranz et al., 1977; Wedderburn, 1981; Walker, 1985). An early study by Mott et al. (1965) remains one of the most thorough investigations of the social aspects of shift work, involving more than 1,000 male workers from 5 manufacturing plants in the eastern and central United States. The workers surveyed were on fixed day, afternoon, or night shifts or on rotating shifts; they were further subdivided into those who were dissatisfied with their work hours and wished to change and those who did not, the former group comprising 66% of the workers on rotating schedules, and 44%, 38%, and 6% of those on fixed night, afternoon, and day schedules, respectively.

Compared with day workers, workers on other shifts, and especially those who wished to change their work hours, reported greater difficulty fulfilling their family roles as husbands and fathers. The specific problems reported by nonday workers depended on their work schedule. Thus, afternoon and rotating workers reported greater difficulty than night workers in providing companionship, diversion, and relaxation to their wives, and felt that their work hours interfered more with decision making regarding family issues. Rotating workers were less able to assist with housework than night or afternoon workers and felt that their schedule interfered more with their reaching mutual understanding with their wives, while night and rotating workers felt that their schedules interfered with their ability to protect their wives and with sexual relations. Shift workers also expressed greater dissatisfaction with their ability to provide companionship and teaching skills to their children and to maintain discipline. The dissatisfaction was greatest for afternoon workers followed by rotating workers, with night workers reporting the least dissatisfaction.

Mott et al. (1965) also found that shift work had a negative effect on organizational participation, particularly in workers on fixed afternoon shifts, who reported membership in fewer voluntary associations than day workers. The proportion of shift workers who reported being officers or committee members was also lower than that of day workers. One area in which shift work was felt to be an advantage was the opportunity to engage in solitary activities such as gardening, hunting, and fishing, and the availability of facilities that are only open on weekdays. On balance, however, the social disadvantages of shift work outweighed its advantages, especially in younger workers.

The night shift is also seen as an advantage by female workers with young children, as it allows them to be at home to take care of their children during the day. However, the ability to juggle child care by day and work at night carries a price: studies in nurses on night duty indicate that those with infants or young children sleep significantly less than other night workers, and spend less time on leisure activities (Gadbois, 1981; Kurumatani et al., 1994).

The social and domestic costs of shift work affect the partners of shift workers, who also report negative effects of their spouse's work schedule on marital happiness and family integration. In addition, the wives of workers on night or rotating shifts often have to postpone their household duties and prevent their children from making noise in order not to disturb their husbands who are asleep during the day; many also complain about the need to prepare meals for their husbands at unusual times of day (Mott et al., 1965). Similar complaints were reported in a more recent survey involving the partners of nuclear power shift workers in the United Kingdom, at least half of whom were unhappy with their spouse's work schedule and felt that their lives were substantially disrupted as a result (Smith and Folkard, 1993).

Finally, there are indications that the perceived social advantages and disadvantages of shift work schedules are related to the physical and mental health of the worker, although specific causal relationships have yet to be demonstrated (Zedeck et al., 1983). In a recent survey of shift workers from several different organizations, Taylor et al. (1997b) found that family advantages were negatively associated with psychological distress, cognitive and somatic anxiety, chronic fatigue, and digestive and cardiovascular problems, whereas social and domestic advantages were negatively related to psychological distress.

Health Effects of Shift Work
Epidemiological data on the effects of shift work on mortality are as yet very limited. Indeed, the only study with a relatively large sample size was reported 30 years ago (Taylor and Pocock, 1972). The study involved about 8,600 male manual workers from 10 organizations in England and Wales. The workers had been on fixed day shifts or on rotating shifts that included night work for at least 10 years; a group of day workers who had previously been on shift work for at least 6 months was also included. Age-specific rates of mortality from all causes were slightly lower than national rates in the day worker group, and slightly higher in both the shift worker and former shift worker groups, but the differences were not statistically significant. One criticism of this study, however, is the use of national mortality rates as the standard, rather than direct comparisons between the three groups (Kristensen, 1989). Clearly, more data will be required before this issue is resolved.

In contrast, the effects of shift work on morbidity have been extensively documented, and there is now clear evidence that shift work has detrimental effects on a wide range of physiological systems, comparable to the effects of chronic exposure to psychosocial stressors described in the allostasis literature.

Gastrointestinal System
Gastrointestinal disorders are among the most frequently reported health effects of shift work, with symptoms ranging from epigastric pain, irregularity

of bowel movements with frequent constipation, flatulence, heartburn, and lack of appetite, to enteritis and gastric and duodenal ulcers. A recent review shows that a higher incidence of such symptoms among shift workers was observed in 25 studies, with 10 studies finding no significant differences and one showing a higher incidence among day workers (Costa, 1996). Although there are discrepancies in this literature, these are not unexpected, given that the self-selection process described earlier was not always adequately taken into account, and given the wide differences in the type of data collected in the different studies.

A good example is the carefully conducted, large-scale study of Segawa et al. (1987), who compared the incidence of peptic ulcers among three categories of Japanese workers, based on X-ray and endoscopic examinations (Segawa et al., 1987). The incidences of gastric and duodenal ulcers were both about twice as high in shift workers than in day workers, whereas former shift workers showed a higher incidence of gastric ulcers. In some studies, the incidence of peptic ulcers was found to be highest in shift workers who had transferred to day work for health reasons (Angersbach et al., 1980), suggesting that gastrointestinal problems are a major component of shift work intolerance. Data from a retrospective cohort study that compared the medical records of workers at a chemical firm over an 11-year period indicate that the cumulative incidence of gastrointestinal disease in shift workers only starts to diverge from that in day workers after about 5 years of shift work (Angersbach et al., 1980).

The physiological mechanisms underlying the effects of shift work on gastrointestinal health have yet to be clarified, but they are likely to involve multiple factors (Rubin, 1988; Vener et al., 1989). Many digestive and postdigestive processes are known to exhibit daily rhythms controlled, to varying extents, by endogenous circadian mechanisms and exogenous factors, the latter including food intake and the sleep-wake cycle. The locus of circadian control of the different rhythms is also likely to differ, given the recent identification of circadian oscillators in several peripheral organs including the liver, kidney, and pancreas. Furthermore, these peripheral oscillators are differentially influenced by the timing of food intake (Damiola et al., 2000; Stokkan et al., 2001). As noted earlier, animal studies have shown that many rhythmic functions, including hepatic and intestinal enzyme activity and synthetic and transport processes, are entrained by food.

The changes in the daily food intake patterns of shift workers are of course not as drastic as those imposed on experimental animals in food entrainment studies. Indeed, shift workers generally try to maintain their normal three-meal schedule even when on the night shift. However, some changes in the times of the main meals are often unavoidable. In addition, workers on the night shift obtain a significant amount of their daily caloric intake – 20% or more in some studies (Reinberg et al., 1979; Romon-Rousseaux et al., 1986; Nikolova et al., 1990; Lennernäs et al., 1994) – in the form of snacks at night. There are indications that even relatively minor changes in meal timing can substantially alter the daily profile of some digestive functions. For example, Duroux et al. (1989) found that a 3-hour change in the time of the evening meal, from 18:00 to 21:00, increases gastric acidity across the entire night. Studies in shift workers also show changes in the levels or daily patterns (or both) of serum gastrin and pepsinogen following a change from day to night work (Åkerstedt and Theorell, 1976; Tarquini et al., 1986).

Disruptions of the daily rhythm of gastric pH are likely to play an important role in the higher incidence of peptic ulcers in shift workers. Moore (1988) has proposed a model of ulcerogenesis based on phase differences or dissociations between rhythms in aggressive factors that increase gastric acidity, including gastrin and pepsinogen, and protective factors that enable the gastric mucosa to resist injury. The latter would include the human trefoil peptide TFF2, which plays a major cytoprotective role in the stomach and was recently shown to exhibit pronounced diurnal variations in gastric juice (Semple et al., 2001). Other processes known to show circadian or diurnal variations in humans include plasma leptin levels (e.g., Sinha et al., 1996) and gastric emptying time (Goo et al., 1987), which are likely to be involved in the disturbances in appetite and bowel movement, respectively, experienced during shift work.

Finally, changes in the timing of sleep have been shown to affect a number of metabolic processes, independently of any changes in meal times. For example, a study by Van Cauter et al. (1994) measured plasma glucose level and insulin secretion rate (ISR) in a group of subjects receiving a constant glucose infusion over 57 hours. The subjects slept from 23:00 to 07:00 on the first night, but their sleep was delayed by 12 hours on the second. Both functions show a prominent increase on the first night, reflecting the well-documented nocturnal decrease in glucose tolerance. The nocturnal peak was blunted on the second night, when the subjects were kept awake, but plasma glucose level and ISR increased again during the subsequent daytime sleep episode, indicating control by both sleep-dependent and sleep-independent (circadian) processes.

Cardiovascular System

An increased risk for cardiovascular disease in shift workers has long been suspected, but more conclusive evidence was only obtained in the last several years (Bøggild and Knutsson, 1999). This evidence includes age-specific

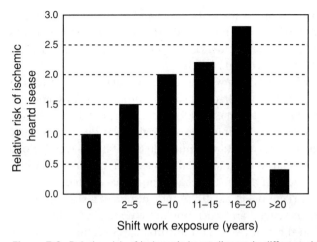

Figure 7.9: Relative risk of ischaemic heart disease in different shift work exposure categories. Plotted from data by Knutsson et al. (1986).

incidence data for ischaemic heart disease (IHD) from several case-referent and cohort studies with 4- to 15-year follow-up, and with appropriate controls for adverse lifestyle factors (Knutsson et al., 1986, 1999; Tüchsen, 1993; Kawachi et al., 1995; Tenkanen et al., 1997, 1998). Overall, Bøggild and Knutsson (1999) estimate that shift work increases the risk of cardiovascular disease by about 40%. That risk is considerably increased when shift work is combined with smoking or obesity, the joint effects showing a better fit to a multiplicative than to an additive model (Tenkanen et al., 1998).

The risk of cardiovascular disease increases gradually with exposure to shift work. In a prospective study of close to 80,000 female nurses who were free of diagnosed coronary heart disease and stroke at entry, the age-adjusted relative risk (RR) increased from 1.21 in nurses with less than 6 years of shift work exposure to 1.51 in those with 6 or more years (Kawachi et al., 1995). Similar results were obtained by Knutsson et al. (1986) in male blue-collar shift workers, as shown in Figure 7.9. The risk dropped dramatically for those with more than 20 years of shift work experience, presumably reflecting the healthy worker effect (Knuttson et al., 1986).

Shift work has also been associated with changes in various biological markers that precede, and predict, manifest cardiovascular disease (Bøggild and Knutsson, 1999). This includes increases in atherogenic lipids (e.g., Romon et al., 1992; Peter et al., 1999) and in the incidence of ventricular ectopic beats (Härenstam et al., 1987), a prolongation of the heart rate-adjusted QT interval (Murata et al., 1999), and a decrease in cardiac sympathetic modulation of the sinoatrial node and in sympathovagal balance (Furlan et al., 2000). Some, but not all, studies also report increases in blood pressure and a higher incidence of hypertension (Morikawa et al., 1999; Peter et al., 1999; Ohira et al., 2000).

Like other physiological systems, the cardiovascular system includes many rhythmic processes controlled, in varying proportions, by endogenous and exogenous factors (Lemmer, 1989). Indeed, the incidence of acute myocardial infarction is itself subject to diurnal variation and is three times higher at 09:00 than at 23:00 (Muller et al., 1985). Several blood fibrinolytic parameters show diurnal variations as well, but in shift workers, these rhythms were found to be flatter and somewhat shifted relative to those in day workers (Peternel et al., 1990). The results of a study in rotating three-shift workers suggest an association between the redistribution of food intake while on the night shift and disturbances in lipid metabolism (Lennernäs et al., 1994). In that study, total cholesterol, low-density lipoprotein cholesterol and the low-density : high-density lipoprotein ratio were all correlated with eating during the night shift, particularly with carbohydrate intake.

Women's Reproductive Health and Breast Cancer

Concern over the reproductive health of women shift workers is reflected in the Night Work Convention (No. 171) and Recommendation (No. 178), adopted in 1990 by the International Labour Organization (Kogi and Thurman, 1993). The former requires that women workers in industry be given an alternative to night work for a minimum of 16 weeks around the time of childbirth, of which 8 weeks should immediately precede the expected date of childbirth, while the latter recommends assignment to day work as far as possible at any point during pregnancy.

Data from a number of studies indicate that this concern is justified. A recent review by Nurminen (1998) shows that women engaged in work schedules that include night work are at increased risk for spontaneous abortion, a result obtained in 7 out of 9 studies that have examined this issue. Higher incidences of preterm birth and low birth weight have also been observed in 4 of 5 studies (Nurminen, 1998). More recently, Bodin et al. (1999) found that midwives who were engaged in night work in the second trimester of pregnancy had an odds ratio (OR) of 5.6 for preterm birth (< 37 weeks of gestation) and 1.9 for low birth weight (< 2.5 kg), whereas midwives on rotating three-shift schedules had an OR of 2.3 for preterm birth. A study by Hrubá and colleagues (1999) also shows an association between shift work and fetal intrauterine growth retardation, with ORs of 1.46 and 1.98 for nonsmokers and smokers, respectively.

In a cross-sectional study involving 5,388 women with singleton pregnancies, the prevalence of preeclampsia, recorded if the woman reported hospitalization for hypertension or if proteinuria and hypertension were diagnosed more than once during the pregnancy, was found to be slightly higher for all women who were engaged in shift work after the first trimester (OR adjusted for demographic and lifestyle factors = 1.3) and significantly higher for parous women engaged in such schedules (adjusted OR = 2.0; Wergeland and Strand, 1997). An association between rotating work schedules and prolonged waiting time to pregnancy has also been observed in two studies (Nurminen, 1998), and several studies have reported higher rates of irregular menstrual cycles in women engaged in night, rotating, or irregularly varying work schedules (Nurminen, 1998; Hatch et al., 1999).

Other recent data indicate that shift work may increase the risk of breast cancer (Hansen, 2001a; Davis et al., 2001; Schernhammer et al., 2001), confirming the results of three smaller cohort studies (reviewed by Hansen, 2001b). For example, Davis et al. (2001) obtained an OR for breast cancer of 1.6 in women on night shifts, with a trend for increased risk as a function of the duration and frequency of night work. The increased risk for breast cancer is attributed to the suppressive effect of nighttime light exposure on nocturnal melatonin production by the pineal gland. This interpretation is based on data from rodent studies and from an in vitro study on human breast cancer cells suggesting that melatonin has an oncostatic effect, either directly or through its effects on reproductive hormone levels (Hansen, 2001b).

Immune System

The effects of shift work on the immune system have thus far received little attention, but the results of two studies indicate that cellular immune function may be impaired in shift workers. Both studies examined mitogen-stimulated blastogenesis of lymphocytes. Curti et al. (1982) found a weaker lymphocyte response in weekly rotating three-shift workers during the afternoon shift than in permanent afternoon workers. Similarly, Nakano et al. (1982) found a lower T cell response in permanent night workers than in day workers at two times of day (morning and evening), while in a second experiment, rotating three-shift workers showed a lower response than day workers in the evening, both when the rotating workers were on the night shift and when they were on the morning shift. The fact that these differences were observed even on days when the rotating workers and the day worker comparison groups were on the same work schedule suggests that the suppression of lymphocyte function was a long-term effect of the weekly rotation rather than an immediate response to the work schedule on the day of the study.

Many components of the immune system, both cellular and humoral, show circadian variations (Haus and Smolensky, 1999), and the daily profiles of several of these components are directly affected by sleep, which is believed to play an important role in immune regulation (Krueger et al., 1999). Indeed, even a single night of partial sleep deprivation, with sleep restricted to the first or second half of the night, is followed by a reduction in natural killer (NK) cell activity, in lymphokine-activated killer cell activity, and in mitogen-stimulated interleukin-2 (IL-2) production (Irwin et al., 1994, 1996). Furthermore, while NK cell activity returned to baseline after a night of recovery sleep, IL-2 production remained suppressed (Irwin et al., 1996). More recently, subjects vaccinated for influenza after 4 nights of partial sleep deprivation showed reduced antibody titers 10 days later (Spiegel et al., 2002). These observations suggest that similar impairments in immune function may occur in shift workers, many of whom experience partial sleep deprivation while on night or early morning shifts.

Mental Health

Evidence of a negative impact of shift work on mental health consists mainly of data obtained through surveys and questionnaires, which often include questions aimed at assessing the psychological well-being of shift workers. Thus, a large-scale survey in the United States showed that, compared with workers on regular schedules, men and women on variable work schedules that include both day and night work were more likely to report severe emotional problems and extreme work stress, and the women were much more likely to use tranquilizers and sleeping pills (Gordon et al., 1986). A German questionnaire study involving male blue-collar workers showed that shift work was associated with higher levels of irritation and strain than day work, even more so in former shift workers who had transferred to day work for health reasons (Frese and Semmer, 1986). Prospective studies on student nurses with no prior shift work experience showed a decrease in psychological well-being (Bohle and Tilley, 1989) and increased feelings of depression and helplessness (Healy et al., 1993) a few months after starting night work.

Some studies also show higher rates of psychiatric disturbances among shift workers, based on such evidence as admission rates to psychiatric hospitals and psychiatric evaluation using standard instruments and criteria. For example, Costa et al. (1981) found that neurotic disorders, defined as symptoms of anxiety or depression requiring hospitalization or treatment with psychotropic drugs for more than 3 months, were much more frequent among Italian textile workers on permanent night or rotating shifts than in day workers. Psychoneurotic disorders, diagnosed according to the International Classification of Disease (ICD), were also more common in male blue-collar shift workers than in day workers at an Austrian oil refinery (Koller, 1983). Michel-Briand and colleagues (1981) found that symptoms of depression were more frequent in retired shift workers than in retired day workers, especially in former shift workers who had transferred to day work before retirement. Koller et al. (1981) found that shift workers represented 25.5% of all patients admitted to a psychiatric clinic over a 4-month period, but accounted for only 15.7% of the general population.

Scott et al. (1997) examined the prevalence of major depressive disorder (MDD) among current and former shift workers, based on DSM-III-R criteria. The prevalence of MDD increased as a function of exposure to shift work, up to 20 years, with the biggest change occurring after 5 years; the prevalence dropped substantially in those with more than 20 years of shift work exposure, possibly reflecting the healthy worker effect described earlier.

Recently, Cho (2001) computed the volumes of the left and right temporal lobes in two groups of flight attendants from coronal magnetic resonance imaging slices. One group had recovery intervals of less than 5 days between outbound transmeridian flights across at least 7 time zones, while the other group had recovery intervals of more than 14 days, during which they flew on shorter flights with only small time shifts. Although the total flight time was similar in the two groups, the volume of the right temporal lobe was significantly smaller in the short-recovery group. Furthermore, a significant negative correlation was obtained between temporal lobe volume and salivary cortisol levels (sampled twice a day within 12 hours after a transmeridian flight and averaged over 3 sampling days), but only in the short-recovery group. The latter group also showed longer reaction times and lower accuracy on a visuospatial performance task.

These results are noteworthy given recent evidence that repeated or chronic stress can result in neuronal atrophy and inhibition of neurogenesis in the hippocampus and other temporal lobe structures (McEwen, 2000b). Thus, hippocampal atrophy has been reported in recurrent major depression, schizophrenia, posttraumatic stress disorder, and Cushing's syndrome (Sapolsky, 1996; McEwen, 2000b). These conditions are all associated with abnormal glucocorticoid or corticotropin-releasing hormone activity, suggesting a key role for these neurohormones in stress-related neuronal atrophy.

Whether the temporal lobe atrophy observed by Cho (2001) is limited to flight attendants on transmeridian routes remains to be determined. Like many shift workers, flight crews have irregular schedules that frequently involve night work, and although they are also exposed to repeated changes in the daily light-dark cycle and other external conditions, flight crews experience disruptions of circadian timing, sleep, and social patterns similar to those faced by the more typical shift worker (e.g., Smolensky et al., 1982).

Accidents and Injuries

Workers on the night shift report greater fatigue than their day worker counterparts, and this is reflected in their performance at work, which is often poorer than that of day workers (Folkard and Monk, 1979; Monk, 1990). As noted earlier, night workers are required to function at the circadian nadir of their alertness rhythm, often after insufficient or less restful sleep. Furthermore, shift work involves a restructuring of daily activity, and the sleep-work-leisure sequence typical of day work becomes sleep-leisure-work during night (and evening) work (Tepas et al., 1981). Thus, shift workers generally start work several hours after the end of sleep, and their homeostatic sleep drive is therefore at a higher level than it is during day work.

Given these differences, it is not surprising that the risk of occupational accidents and injuries has been found to be higher during night and rotating work than during day work (for reviews, see Colquhoun, 1976; Dinges, 1995; Monk et al., 1996). For example, Smith and colleagues (1994) found that the risk of injuries in a large engineering company was 23% higher on the night shift than on the morning shift, and the risk of serious injury for self-paced work was 82% higher. Data from a recent analysis of occupational fatalities in the United States from 1980 to 1994, plotted by time of day in Figure 7.10, indicate that the average death rate at night and in the early morning hours was considerably higher than during the standard 09:00–17:00 work hours (Smith and Kushida, 2000). Similarly, a prospective study involving close to 48,000 individuals from a national sample in Sweden shows that nonday work was associated with a significantly higher risk of fatal occupational accidents (Åkerstedt et al., 2002).

Several studies have shown that shift workers are also at increased risk for motor vehicle accidents and near-misses while driving to and from work (Liddell, 1982; Richardson et al., 1990; Gold et al., 1992). The shift worker groups in two of these studies were also much more likely to report poor sleep quality, fatigue during work hours, and nodding off both at work and while driving to and from work (Richardson et al., 1990; Gold et al., 1992).

Shift Work, Stress, and Health

A number of conceptual models have been proposed to account for the effects of shift work on health, and although these models differ in several respects, most share the view that shift work is stressful to the worker,

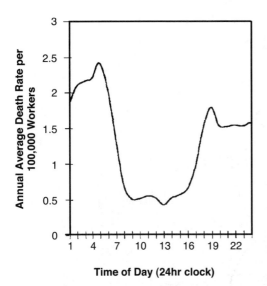

Figure 7.10: Average rate of fatal occupational injury in the United States as a function of time of day for the years 1980–94. From Smith and Kushida (2000).

and that the stress of shift work stems from three major sources, namely, disturbances in circadian timing, in sleep, and in social and family life (Fig. 7.11).

In general, however, the physiological mechanisms and pathways underlying the effects of these three types of stress on the health of shift workers have yet to be elucidated. The task is especially daunting in the case of disturbances in circadian timing, given that virtually all physiological parameters show circadian oscillations, and that, during shift work, many of these show changes in amplitude, waveform, or mean level, as well as changes in their phase relations, both to each other and to the external environment. Furthermore, the recent identification of circadian oscillators in several central and peripheral tissues adds a new level of complexity because it raises the possibility of interactions between the oscillators controlling different physiological outputs as well as between these oscillators and the SCN pacemaker, in addition to the many interactions between the outputs themselves. This complexity has hampered attempts at experimentally isolating and manipulating specific rhythmic variables, and the data implicating circadian disturbances in the health effects of shift work thus remain largely correlational.

The evidence linking sleep disturbances with specific health outcomes is more direct because it includes data from experimental studies, examples of which were described in the preceding sections, showing that changes in the duration and timing of sleep affect a number of specific physiological

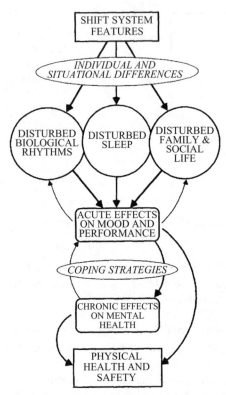

Figure 7.11: Conceptual model depicting the three major types of stress caused by shift work and their effects on mental and physical health. From Folkard (1993).

systems and parameters. In addition, there are epidemiological data showing that sleep disruptions in nonshift worker populations are associated with some of the same health risks as those observed in shift workers. For example, sleep complaints were recently shown to predict mortality from coronary artery disease in samples drawn from the general population (Nilsson et al., 2001; Mallon et al., 2002), and EEG-assessed sleep disturbances have been found to be associated with reduced NK cell activity in depressed patients (Irwin et al., 1992; Hall et al., 1998). Sleep disturbances are an important component of all diagnostic criteria for major depression (Thase, 2000), but there is also evidence that subjective insomnia is a risk factor for the later onset of depression, and there are indications that both insomnia and hypersomnia may be risk factors for the later development of depression, anxiety disorders, and substance abuse (Gillin, 1998).

Several investigators have recently commented on the relationships between stress, sleep, and health. There is substantial evidence for the view that primary insomnia is a disorder of generalized hyperarousal, characterized by increased corticotropin and cortisol levels, especially in the evening (Vgontzas et al., 2001), and by elevated high-frequency EEG activity, in the beta (14–35 Hz) and gamma (35–45 Hz) ranges, around the time of sleep onset and during NREM sleep (e.g., Perlis et al., 2000). Van Reeth et al. (2000) pointed out that shift workers, insomniacs, and patients suffering from mental disorders all show abnormal patterns of HPA secretory activity accompanied by sleep disturbances and that the changes in endocrine activity and sleep observed in these populations are similar to those seen in animals and humans exposed to chronic stress. Hall et al. (1998) noted that sleep disturbances may mediate the stress-immune relationship, while Van Cauter and Spiegel (1999) suggested that sleep loss increases allostatic load and contributes to the development of such chronic conditions as obesity, diabetes, and hypertension. Noting that these conditions are more common among individuals with low socioeconomic status (SES), Van Cauter and Spiegel hypothesized that the adverse impact of low SES on health is partly mediated by decrements in sleep quality and duration.

In tracing the historical development of shift work models, Taylor and colleagues (1997a) noted an increasing tendency for such models to incorporate concepts and frameworks from occupational stress theory to account for the relationship between shift work and health. In one model, for example, shift work is considered one of several occupational stressors and is listed alongside such factors as monotony and lack of autonomy (Olsson et al., 1990). Indeed, data from several studies indicate that shift work and other occupational stressors, including perceived job demand and decision latitude, have additive effects on various health outcomes (Frese and Semmer, 1986; Härenstam et al., 1987; Tenkanen et al., 1997; Parkes, 1999; Peter et al., 1999).

Finally, some shift work models, including the model depicted in Figure 7.11, share with occupational stress theory the view that the effects of work stressors on health are modulated by the coping strategies used by the worker. Thus, Olsson et al. (1990) found that active cognitive strategies in nurses and active physical strategies in paper workers attenuated the negative effects of shift work on health, whereas passive somatizing strategies were associated with more symptoms, especially gastrointestinal complaints and psychological symptoms.

CONCLUSIONS: TOWARD A DEFINITION OF ALLOSTASIS

As proposed originally by Sterling and Eyer (1988), the concept of allostasis encompasses two separate phenomena: changes in parameters of the

internal milieu resulting from chronic arousal, or stress, and daily variations in these parameters controlled directly or through its effects on the
sleep-wake cycle (or both) by the circadian timing system. Although these
two phenomena have a number of features in common, they also differ in
fundamental respects.

Both stress-induced and circadian variations in the body's internal environment are thought to reflect changes in physiological setpoints coordinated by the brain through multiple neural and neuroendocrine mechanisms. Indeed, there is even some overlap between the neural structures
and pathways involved, as in the case of the PVN, which plays a central role
in arousal pathology (Herman and Cullinan, 1997; Bhatnagar and Dallman,
1998; Schulkin et al., 1998) but is also the locus of a circadian oscillator and
an important relay of circadian timing signals from the SCN to both sympathetic and parasympathetic components of the autonomic nervous system.
Both stress-induced and circadian variations are considered to be adaptive,
allowing the organism to anticipate changes in need and to make appropriate adjustments in advance. But whereas the former depend on the specific
experiences and perceptions of an individual, the latter are obligatory and
common to all members of a species, preparing them for the regular, daily
changes that characterize various aspects of the external environment.

The most important distinction between stress-induced and circadian
variations, however, may lie in the notion of allostatic load, which refers
to the widespread pathological consequences of allostatic regulation. Thus,
while chronic stress has been associated with an increased risk for many diseases, there is at present no way to determine whether circadian variations
carry such a risk because there is no appropriate comparison group devoid
of all such variations. Nor are SCN lesions in animals likely to provide an adequate answer, even though such lesions abolish most circadian rhythms.
This is because along with circadian organization, different species have
evolved a host of sensory and behavioral adaptations specifically suited to
the temporal niches they occupy, and while animals bearing SCN lesions
survive and appear to function normally under laboratory conditions, they
are at a disadvantage in a more natural environment (DeCoursey et al.,
1997, 2000) or when competing with intact members of their species (Eskes,
1982).

We suggest that it is the chronic disruption of normal circadian patterns
in behavior and physiology, rather than the existence of such patterns, that
is stressful to the organism and entails the activation of allostatic mechanisms. Nowhere is such disruption more common than in shift workers, and
there are several arguments for including shift work among the psychosocial
factors that result in the accrual of allostatic load. Night and rotating work
schedules are widely considered to be stressful to the worker, and have been

associated with a higher incidence of several chronic diseases, as well as accidents and injuries. Furthermore, the patterns of morbidity documented in shift workers overlap substantially with those attributed to disruptions of intimate social relations. Indeed, disruptions of social and domestic patterns are one source of stress to shift workers and may contribute to the higher morbidity rates observed in this population.

But the health effects of shift work cannot be fully explained without taking into account the changes in circadian timing, including the timing of sleep and wakefulness, experienced by shift workers. The major difference between shift workers and their day worker counterparts is not in the activities they engage in but in the timing of these activities; the same activity can have different physiological consequences depending on the time of day at which it is performed. This is because the levels, or setpoints, around which many, and possibly all, physiological variables are regulated show daily oscillations, and a given value that falls within the regulated range at one time of day may represent a significant deviation at another. Indeed, the pervasiveness of circadian rhythmicity suggests that, in normal, healthy individuals, the traditional homeostatic model with fixed setpoints, if it applies at all, is the exception rather than the rule.

Circadian and homeostatic regulation, however, can be dissociated experimentally: several studies have shown that SCN lesions abolish circadian oscillations in such functions as body temperature, sleep, feeding and drinking, without interfering with the homeostatic regulation of these functions. Thus, animals bearing such lesions show mean daily levels of body temperature that are comparable to those of intact animals (Ruby et al., 1989; Edgar et al., 1993), even at low ambient temperature (Refinetti et al., 1994). Daily food and water intakes are similarly unchanged, and body weight is maintained within normal levels (Stephan and Zucker, 1972; Krieger et al., 1977; Boulos et al., 1980; Edgar et al., 1993). Similarly, sleep deprivation in SCN-lesioned rats is followed by compensatory increases in total sleep, in the deeper stages of NREM sleep, and in REM sleep, similar to those observed in intact rats (Mistlberger et al., 1983; Tobler et al., 1983).

These points are illustrated in Figure 7.12, where the same physiological functions appear in two compartments. The first compartment represents traditional homeostatic regulation with fixed setpoints and thresholds. This compartment receives inputs from the SCN pacemaker, directly and through several central and peripheral oscillators, and is influenced by various other factors, including chronic exposure to psychosocial stress and shift work schedules. The second compartment represents physiological regulation as observed in normal, intact organisms, where regulatory setpoints and thresholds show both rhythmic and nonrhythmic variations.

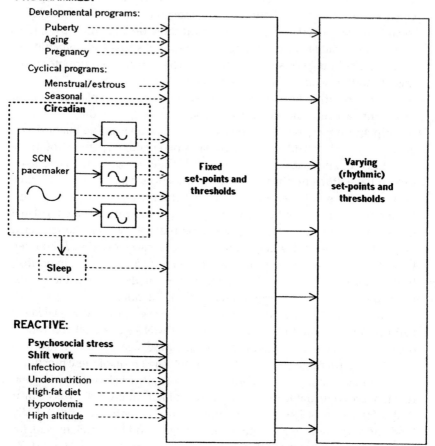

Figure 7.12: Schematic representation of the main arguments developed in this chapter. Under the influence of various factors, here divided into programmed and reactive, the fixed physiological setpoints and thresholds of the traditional homeostasis model show adaptive variations that enable the organism to anticipate changes in need. We suggest that the concept of allostasis be limited to those changes resulting from chronic exposure to psychosocial stress and from chronic disruption of circadian timing, as experienced during shift work (*solid arrows*), excluding all programmed changes, cyclical as well as developmental, and any reactive changes that are psychosocial in origin (*dashed arrows*).

Figure 7.12 also shows that circadian and stress-induced variations are each part of a larger category of physiological phenomena, best captured by Mrosovsky's concepts of programmed and reactive rheostasis, respectively. The "programmed" category, which includes both cyclical and developmental programs, represents obligatory phenomena, while the "reactive" category represents conditions that may or may not be encountered in the lifetime of an individual organism. As noted by Mrosovsky (1990), the distinction between the two categories is not absolute; pregnancy, for example, is not inevitable, but it is included in the "programmed" rather than the "reactive" category because reproduction is a normal part of the life cycles of all organisms. However, even if programmed and reactive events are thought of as belonging to a continuum, a useful distinction can still be made between the two ends of that continuum.

The importance of the allostasis paradigm lies in the conceptual framework it provides, linking a diverse set of social and psychological phenomena with an equally diverse set of negative health outcomes through common physiological regulatory mechanisms. By current criteria, however, all the phenomena listed on the left side of Figure 7.12 would qualify as allostatic regulation because they all result in changes in physiological setpoints that anticipate changes in need. Such an overly inclusive definition of allostatis can only detract from the role ascribed to it by Sterling and Eyer. We propose, therefore, that the criteria defining allostasis be modified so as to exclude all programmed changes in physiological setpoints, cyclic and developmental, as well as any reactive changes that are not psychosocial in origin, such as infection-induced changes in thermoregulatory setpoints, or changes in CO_2 chemostat thresholds brought about by high altitude.

REFERENCES

Abe, M., Herzog, E. D., Yamazaki, S., Straume, M., Tei, H., Sakaki, Y., Menaker, M., Block, G. D. (2002). Circadian rhythms in isolated brain regions. *J Neurosci* 22:350–6.

Abrahamson, E. E., Leak, R. K., Moore, R. Y. (2001). The suprachiasmatic nucleus projects to posterior hypothalamic arousal systems. *NeuroReport* 12:435–40.

Åkerstedt, T. (1977). Inversion of the sleep wakefulness pattern: Effects on circadian variations in psychophysiological activation. *Ergonomics* 20:459–74.

Åkerstedt, T., Fredlund, P., Gillberg, M., Jansson, B. (2002). A prospective study of fatal occupational accidents – relationship to sleeping difficulties and occupational factors. *J Sleep Res* 11:69–71.

Åkerstedt, T., Gillberg, M. (1981). The circadian variation of experimentally displaced sleep. *Sleep* 4:159–69.

Åkerstedt, T., Kecklund, G., Knutsson, A. (1991). Spectral analysis of sleep electroencephalography in rotating three-shift work. *Scand J Work, Environ Health* 17:330–6.

Åkerstedt, T., Theorell, T. (1976). Exposure to night work: Serum gastrin reactions, psychosomatic complaints and personality variables. *J Psychosom Res* 20:479–84.

Åkerstedt, T., Torsvall, L., Fröberg, J. E. (1983). A questionnaire study of sleep/wake disturbances and irregular work hours. *Sleep Res* 12:358.

Åkerstedt, T., Torsvall, L., Gillberg, M. (1989). Shift work and napping. In: *Sleep and Alertness: Chronobiological, Behavioral, and Medical Aspects of Napping* (ed. D. F. Dinges and R. J. Broughton), pp. 205–20. New York: Raven Press.

Albrecht, U., Zheng, B., Larkin, D., Sun, Z. S., Lee, C. C. (2001). MPer1 and MPer2 are essential for normal resetting of the circadian clock. *J Biol Rhythms* 16:100–4.

Alfoldi, P., Franken, P., Tobler, I., Borbély, A. A. (1991). Short light-dark cycles influence sleep stages and EEG power spectra in the rat. *Behav Brain Res* 43:125–31.

Angersbach, D., Knauth, P., Loskant, H., Karvonen, M. J., Undeutsch, K., Rutenfranz, J. (1980). A retrospective cohort study comparing complaints and diseases in day and shift workers. *Int Arch Occup Environ Health* 45:127–40.

Antle, M. C., Mistlberger, R. E. (2000). Circadian clock resetting by sleep deprivation without exercise in the Syrian hamster. *J Neurosci* 20:9326–32.

Antle, M. C., Steen, N. M., Mistlberger, R. E. (2001). Adenosine and caffeine modulate circadian rhythms in the Syrian hamster. *Neuroreport* 12:2901–5.

Aschoff, J. (1981). Circadian rhythms: Interference with and dependence on work-rest schedules. In: *Biological Rhythms, Sleep and Shift Work. Advances in Sleep Research, Volume 7* (ed. L. C. Johnson, D. I. Tepas, W. P. Colquhoun, M. J. Colligan), pp. 11–34. New York: Spectrum.

Aschoff, J. (1982). The circadian rhythm of body temperature as a function of body size. In: *A Companion to Animal Physiology* (ed. C. R. Taylor, K. Johansen and L. Bolis) pp. 173–88. London: Cambridge University Press.

Aschoff, J., Hoffmann, K., Pohl, H., Wever, R. (1975). Re-entrainment of circadian rhythms after phase-shifts of the Zeitgeber. *Chronobiologia* 2:23–78.

Aschoff, J. (1960). Exogenous and endogenous components in circadian rhythms. *Cold Spring Harbor Symp Quantitative Biol* 25:11–28.

Asikainen, M., Toppila, J., Alanko, L., Ward, D. J., Stenberg, D., Porkka-Heiskanen, T. (1997). Sleep deprivation increases brain serotonin turnover in the rat. *Neuroreport* 8:1577–82.

Aston-Jones, G., Chen, S., Zhu Y., Oshinsky, M. L. (2001). A neural circuit for circadian regulation of arousal. *Nat Neurosci* 4:732–8.

Backberg, M., Hervieu, G., Wilson, S., Meister, B. (2002). Orexin receptor-1 (OX-R1) immunoreactivity in chemically identified neurons of the hypothalamus: Focus on orexin targets involved in control of food and water intake. *Eur J Neurosci* 15:315–28.

Balsalobre, A., Brown, S. A., Marcacci, L., Tronche, F., Kellendonk, C., Reichardt, H. M., Schibler, U. (2000). Resetting of circadian time in peripheral tissues by glucocorticoid signaling. *Science* 289:2344–7.

Balsalobre, A., Damiola, F., Schibler, U. (1998). A serum shock induces circadian gene expression in mammalian tissue culture cells. *Cell* 93:929–37.

Bartness, T. J., Song, C. K., Demas, G. E. (2001). SCN efferents to peripheral tissues: Implications for biological rhythms. *J Biol Rhythms* 16:196–204.

Bauman, D. E., Currie, W. B. (1980). Partitioning of nutrients during pregnancy and lactation: A review of mechanisms involving homeostasis and homeorhesis. *J Dairy Sci* 63:1514–29.

Benington, J. H., Heller, H. C. (1995). Restoration of brain energy metabolism as the function of sleep. *Prog Neurobiol* 45:347–60.

Bhatnagar, S., Dallman, M. (1998). Neuroanatomical basis for facilitation of hypothalamic-pituitary-adrenal responses to a novel stressor after chronic stress. *Neuroscience* 84:1025–39.

Bligh, J., Johnson, K. G. (1973). Glossary of terms for thermal physiology. *J Appl Physiol* 35:942–61.

Bodin, L., Axelsson, G., Ahlborg, G., Jr. (1999). The association of shift work and nitrous oxide exposure in pregnancy with birth weight and gestational age. *Epidemiology* 10:429–36.

Bøggild, H., Knutsson, A. (1999). Shift work, risk factors and cardiovascular disease. *Scand J Work Environ Health* 25:85–99.

Bohle, P., Tilley, A. J. (1989). The impact of night work on psychological well-being. *Ergonomics* 32:1089–99.

Borbély, A. A. (1982). A two process model of sleep regulation. *Hum Neurobiol* 1:195–204.

Borbély, A. A., Achermann, P. (2000). Sleep homeostasis and models of sleep regulation. In: *Principles and Practice of Sleep Medicine,* 3d ed. (ed. M. H. Kryger, T. Roth, W. C. Dement), pp. 377–90. Philadelphia: W. B. Saunders.

Borbély, A. A., Baumann, F., Brandeis, D., Strauch, I., Lehmann, D. (1981). Sleep deprivation: Effect on sleep stages and EEG power density in man. *Electroencephalogr Clin Neurophysiol* 51:483–93.

Borbély, A. A., Neuhaus, H. U. (1978). Daily pattern of sleep, motor activity and feeding in the rat: Effects of regular and gradually extended photoperiods. *J Comp Physiol* 124:1–14.

Boulos, Z., Terman, M. (1980). Food availability and daily biological rhythms. *Neurosci Biobehav Rev* 4:119–31.

Boulos, Z., Rosenwasser, A. M., Terman, M. (1980). Feeding schedules and the circadian organization of behavior in the rat. *Behav Brain Res* 1:39–65.

Brugère, B., Barrit, J., Butat, C., Cosset, M., Volkoff, S. (1997). Shiftwork, age, and health: An epidemiologic investigation. *Int J Occup Environ Health* 3(suppl. 2): S15–9.

Bryden, G., Holdstock, T. L. (1973). Effects of night duty on sleep patterns of nurses. *Psychophysiology* 10:36–42.

Buijs, R. M. (1996). The anatomical basis for the expression of circadian rhythms: the efferent projections of the suprachiasmatic nucleus. In: *Hypothalamic Integration of Circadian Rhythms* (ed. R. M. Buijs, A. Kalsbeek, H. J. Romijn, C. M. A. Pennartz, M. Mirmiran), pp. 229–240. Amsterdam: Elsevier.

Buijs, R. M., Chun, S. J., Niijima, A., Romijn, H. J., Nagai, K. (2001). Parasympathetic and sympathetic control of the pancreas: A role for the suprachiasmatic nucleus and other hypothalamic centers that are involved in the regulation of food intake. *J Comp Neurol* 431:405–23.

Buijs, R. M., Kalsbeek, A. (2001). Hypothalamic integration of central and peripheral clocks. *Nat Rev Neurosci* 2:521–6.

Buijs, R. M., Wortel, J., van Heerikhuize, J. J., Feenstra, M. G. P., Ter Horst, G. J., Romijn, H. J., Kalsbeek, A. (1999). Anatomical and functional demonstration of a multisynaptic suprachiasmatic nucleus adrenal (cortex) pathway. *Eur J Neurosci* 11:1535–44.

Butat, C., Barrit, J., Brugère, D., Cosset, M., Touranchet, A., Volkoff, S. (1993). Troubles du sommeil en fonction de l'âge et des horaires du travail (enquête E.S.T.E.V.). *Arch Mal Prof Méd Travail Sécurité Sociale* 54:209–15.

Cabanac, M., Hildebrandt, G., Massonnet, B., Strempel, H. (1976). A study of the nycthemeral cycle of behavioural temperature regulation in man. *J Physiol* 257: 275–91.

Campbell, S. S. (1995). Effects of timed bright-light exposure on shift-work adaptation in middle-aged subjects. *Sleep* 18:408–16.

Campbell, S. S., Dijk, D.-J., Boulos, Z., Eastman, C. I., Lewy, A. J., Terman, M. (1995). Light treatment for sleep disorders: Consensus report. III. Alerting and activating effects. *J Biol Rhythms* 10:129–32.

Cannon, W. B. (1929). Organization for physiological homeostasis. *Physiol Rev* 9:399–431.

Cannon, W. B. (1930). The autonomic nervous system: An interpretation. *Lancet* 1109–15.

Chaumont, A.-J., Laporte, A., Nicolai, A., Reinberg, A. (1979). Adjustment of shift workers to a weekly rotation (Study 1). *Chronobiologia* (suppl. 1):27–34.

Cheng, M. Y., Bullock, C. M., Li, C., Lee, A. G., Bermak, J. C., Belluzi, J., Weaver, D. R., Leslie, F. M., Zhou, Q. Y. (2002). Prokineticin2 transmits the behavioral circadian rhythm of the suprachiasmatic nucleus. *Nature* 417:405–10.

Cho, K. (2001). Chronic "jet lag" produces temporal lobe atrophy and spatial cognitive deficits. *Nat Neurosci* 4:567–8.

Chou, T. C., Bjorkum, A. A., Gaus, S. E., Lu, J., Scammel, T. E., Saper, C. B. (2002). Afferents to the ventrolateral preoptic nucleus. *J Neurosci* 22:977–90.

Coleman, R. M., Dement, W. C. (1986). Falling asleep at work: A problem for continuous operations. *Sleep Res* 15:265.

Colquhoun, W. P. (1976). Accidents, injuries and shift work. In: *Shift Work and Health* (ed. P. G. Rentos, R. D. Shepard), pp. 160–175. HEW Publication No. National Institute for Occupational Safety and Health (NIOSH) 76–203, Washington, D.C.

Colquhoun, W. P., Blake, M. J. F., Edwards, R. S. (1968). Experimental studies of shift-work. II. Stabilized 8-hour shift systems. *Ergonomics* 11:527–46.

Colligan, M. J., Tepas, D. I. (1986). The stress of hours of work. *Am Industr Hygiene Assoc J* 47:686–95.

Costa, G. (1996). The impact of shift and night work on health. *Appl Ergonomics* 27:9–16.

Costa, G., Apostoli, P., d'Andrea, F., Gaffuri, E. (1981). Gastrointestinal and neurotic disorders in textile shift workers. In: *Night and Shiftwork: Biological and Social Aspects* (ed. A. Reinberg, N. Vieux, P. Andlauer), pp. 215–221. Oxford: Pergamon Press.

Costa, G., Bertoldi, A., Kovacic, M., Ghirlanda, G., Minors, D. S., Waterhouse, J. M. (1997). Hormonal secretion of nurses engaged in fast-rotating shift systems. *Int J Occup Environ Health* 3(suppl. 2):S35–9.

Cui, L.-N., Coderre, E., Renaud, L. P. (2001). Glutamate and GABA mediate suprachiasmatic nucleus inputs to spinal-projecting paraventricular neurons. *Am J Physiol* 281:R1283–9.

Curti, R., Radice, L., Cesana, G. C., Zanettini, R., Grieco, A. (1982). Work stress and immune system: Lymphocyte reactions during rotating shift work. Preliminary results. *Medicina Lavoro* 6:564–70.

Czeisler, C. A., Brown, E. N., Ronda, J. M., Kronauer, R. E., Richardson, G. S., Freitag, W. O. (1985). A clinical method to assess the endogenous circadian phase (ECP) of the deep circadian oscillator in man. *Sleep Res* 14:295.

Czeisler, C. A., Zimmerman, J. C., Ronda, J. M., Moore-Ede, M. C., Weitzman, E. D. (1980b). Timing of REM sleep is coupled to the circadian rhythm of body temperature in man. *Sleep* 2:329–46.

Czeisler, C. A., Khalsa, S. B. S. (2000). The human circadian timing system and sleep-wake regulation. In: *Principles and Practice of Sleep Medicine*, 3d ed. (ed. M. H. Kryger, T. Roth, W. C. Dement), pp. 353–75. Philadelphia: W. B. Saunders.

Czeisler, C. A., Weitzman, E. D., Moore-Ede, M. C., Zimmerman, J. C. Knauer, R. S. (1980a). Human sleep: Its duration and organization depend on its circadian phase. *Science* 210:1264–7.

Daan, S., Beersma, D. G. M., Borbély, A. A. (1984). Timing of human sleep: Recovery process gated by a circadian pacemaker. *Am J Physiol* 246:R161–78.

Daan, S., Beersma, D. G. M., Dijk, D.-J., Åkerstedt, T., Gillberg, M. (1988). Kinetics of an hourglass component involved in the regulation of human sleep and wakefulness. *Adv Biosci* 73:183–93.

Daan, S., Masman, D., Strijkstra, A., Verhulst, S. (1989). Intraspecific allometry of basal metabolic rate: Relations with body size, temperature, composition, and circadian phase in the kestrel, *Falco tinnunculus*. *J Biol Rhythms*, 4:267–83.

Dahlgren, K. (1981). Adjustment of circadian rhythms and EEG sleep functions to day and night sleep among permanent nightworkers and rotating shiftworkers. *Psychophysiology* 18:381–91.

Damiola, F., Le Minh, N., Preitner, N., Kornmann, B., Fleury-Olela, F., Schibler, U. (2000). Restricted feeding uncouples circadian oscillators in peripheral tissues from the central pacemaker in the suprachiasmatic nucleus. *Genes Dev* 14:2950–61.

Davis, S., Mirick, D. K., Stevens, R. G. (2001). Night shift work, light at night, and risk of breast cancer. *J Nat Cancer Inst* 93:1557–62.

DeCoursey, P. J., Krulas, J. R., Mele, G., Holley, D. C. (1997). Circadian performance of suprachiasmatic nuclei (SCN)-lesioned antelope ground squirrels in a desert enclosure. *Physiol Behav* 62:1099–108.

DeCoursey, P. J., Walker, J. K., Smith, S. A. (2000). A circadian pacemaker in free-living chipmunks: Essential for survival? *J Comp Physiol A* 186:169–80.

de Lecea, L., Kilduff, T. S., Peyron, C., Gao, X.-B., Foye, P. E., Danielson, P. E., Fukuhara, C., Battenberg, E. L. F., Gautvik, V. T., Bartlett, F. S., II, Frankel, W. N., van den Pol, A. N., Bloom, F. E., Gautvik, K. M., Sutcliffe, J. G. (1998). The hypocretins: Hypothalamus-specific peptides with neuroexcitatory activity. *Proc Nat Acad Sci U S A* 95:322–7.

Deurveilher, S., Burns, J., Semba, K. (2001). Indirect pathways from the suprachiasmatic nucleus to hypocretin/orexin-containing, monoaminergic and cholinergic cell groups in rat. *Soc Neurosci Abstracts* 27:1089.

Deurveilher, S., Burns, J., Semba, K. (2002). Indirect projections from the suprachiasmatic nucleus to the ventrolateral preoptic nucleus: A dual tract-tracing study in rat. *Eur J Neurosci* 16:1195–213.

Dijk, D.-J., Beersma, D. G. M., Daan, S. (1987a). EEG power density during nap sleep: Reflection of an hourglass measuring the duration of prior wakefulness. *J Biol Rhythms* 2:207–19.

Dijk, D.-J., Beersma, D. G. M., Daan, S., Bloem, G. M., Van den Hoofdakker, R. H. (1987b). Quantitative analysis of the effects of slow wave sleep deprivation during the first 3 h of sleep on subsequent EEG power density. *Eur Arch Psychiatry Neurol Sci* 236:323–8.

Dijk, D. J., Czeisler, C. A. (1995). Contribution of the circadian pacemaker and the sleep homeostat to sleep propensity, sleep structure, electroencephalographic slow waves, and sleep spindle activity in humans. *J Neurosci* 15:3526–38.

Dinges, D. F. (1995). An overview of sleepiness and accidents. *J Sleep Res* 4(suppl. 2):4–14.

Dumont, M., Montplaisir, J., Infante-Rivard, C. (1997). Sleep quality of former nightshift workers. *Int J Occup Environ Health* 3(suppl. 2):S10–4.

Dunlap, J. C. (1999). Molecular bases for circadian clocks. *Cell* 96:271–90.

Duroux, P., Bauerfeind, P., Emde, C., Koelz, H. R., Blum, A. L. (1989). Early dinner reduces nocturnal gastric acidity. *Gut* 30:1063–7.

Edgar, D. M., Dement, W. C., Fuller, C. A. (1993). Effect of SCN lesions on sleep in squirrel monkeys: Evidence for opponent processes in sleep-wake regulation. *J Neurosci* 13:1065–79.

Elliott, K. J., Weber, E., Rea, M. A. (2001). Adenosine A1 receptors regulate the response of the hamster circadian clock to light. *Eur J Pharmacol* 414:45–53.

Eskes, G. A. (1982). Functional significance of daily cycles in sexual behavior of the male golden hamster. In: *Vertebrate Circadian Systems* (ed. J. Aschoff, S. Daan and G. Groos), pp. 347–53. Berlin: Springer-Verlag.

Feinberg, I., March, J. D., Floyd, T. C., Jimison, R., Bossom-Demitrack, L., Katz, P. H. (1985). Homeostatic changes during post-nap sleep maintain baseline levels of delta EEG. *Electroencephalogr Clin Neurophysiol* 61:134–7.

Folkard, S. (1993). Editorial. *Ergonomics* 36:1–2.

Folkard, S., Barton, J. (1993). Does the "forbidden zone" for sleep onset influence morning shift sleep duration? *Ergonomics* 36:85–91.

Folkard, S., Monk, T. H. (1979). Shiftwork and performance. *Hum Factors* 21:483–92.

Foret, J., Benoit, O. (1974). Structure du sommeil chez des travailleurs à horaires alternants. *Electroencephalogr Clin Neurophysiol* 37:337–44.

Foret, J., Benoit, O. (1979). Sleep recordings of shift workers adhering to a three-to four-day rotation (Study 2). *Chronobiologia*, 6(suppl. 1):45–53.

Foret, J., Bensimon, G., Benoit, O., Vieux, N. (1981). Quality of sleep as a function of age and shift work. In: *Night and Shiftwork: Biological and Social Aspects* (ed. A. Reinberg, N. Vieux, P. Andlauer), pp. 149–54. Oxford: Pergamon Press.

Fredholm, B. B., Hedqvist, P. (1980). Modulation of neurotransmission by purine nucleotides and nucleosides. *Biochem Pharmacol* 29:1635–43.

Freedman, M. S., Lucas, R. J., Soni, B., von Schantz, M., Muñoz, M., David-Gray, Z., Foster, R. (1999). Regulation of mammalian circadian behavior by non-rod, non-cone, ocular photoreceptors. *Science* 284:502–4.

Frese, M., Okonek, K. (1984). Reasons to leave shiftwork and psychological and psychosomatic complaints of former shiftworkers. *J Appl Physiol* 69:509–14.

Frese, M., Semmer, N. (1986). Shiftwork, stress, and psychosomatic complaints: a comparison between workers in different shiftwork schedules, non-shiftworkers, and former shiftworkers. *Ergonomics* 29:99–114.

Fukuhara, C., Brewer, J. M., Dirden, J. C., Bittman, E. L., Tosini, G., Harrington, M. E. (2001). Neuropeptide Y rapidly reduces Period 1 and Period 2 mRNA levels in the hamster suprachiasmatic nucleus. *Neurosci Lett* 314:119–22.

Furlan, R., Barbic, F., Piazza, S., Tinelli, M., Seghizzi, P., Malliani, A. (2000). Modifications of cardiac autonomic profile associated with a shift schedule of work. *Circulation* 102:1912–16.

Gadbois, C. (1981). Women on night shift: Interdependence of sleep and off-the-job activities. In: *Night and Shift Work: Biological and Social Aspects* (ed. A. Reinberg, N. Vieux and P. Andlauer), pp. 223–7. Oxford: Pergamon Press.

Gardner, J. P., Fornal, C. A., Jacobs, B. L. (1997). Effects of sleep deprivation on serotonergic neural activity in the dorsal raphe nucleus of the freely moving cat. *Neuropsychopharmacology* 17:72–81.

Gerashchenko, D., Kohls, M. D., Greco, M., Waleh, N. S., Salin-Pascual, R., Kilduff, T. S., Lappi, D. A., Shiromani, P. J. (2001). Hypocretin-2-saprin lesions of the lateral hypothalamus produce narcoleptic-like sleep behavior in the rat. *J Neurosci* 21:7273–83.

Gillberg, M., Anderzén, I., Åkerstedt, T. (1991). Recovery within day-time sleep after slow wave sleep suppression. *Electroencephalogr Clin Neurophysiol* 78:267–73.

Gillette, M. U. (1996). Regulation of entrainment pathways by the suprachiasmatic circadian clock: Sensitivities to second messengers. In: *Hypothalamic Integration of Circadian Rhythms* (ed. R. M. Buijs, A. Kalsbeek, H. J. Romijn, C. M. A. Pennartz, M. Mirmiran), pp. 121–33. Amsterdam: Elsevier.

Gillin, J. C. (1998). Are sleep disturbances risk factors for anxiety, depressive and addictive disorders? *Acta Psychiatr* 393(suppl. 393):39–43.

Gold, D. R., Rogacz, S., Bock, N., Tosteson, T. D., Baum, T. M., Speizer, F. E., Czeisler, C. A. (1992). Rotating shift work, sleep, and accidents related to sleepiness in hospital nurses. *Am J Public Health* 82:1011–14.

Gong, H., Szymusiak, R., King, J., Steininger, T., McGinty, D. (2000). Sleep-related c-Fos protein expression in the preoptic hypothalamus: Effects of ambient warming. *Am J Physiol* 279:R2079–88.

Goo, R. H., Moore, J. G., Greenberg, E., Alazraki, N. P. (1987). Circadian variations in gastric emptying of meals in humans. *Gastroenterology* 93:515–18.

Gordon, N. P., Cleary, P. D., Parker, C. E., Czeisler, C. A. (1986). The prevalence and health impact of shiftwork. *Am J Public Health* 76:1225–8.

Graf, R., Krishna, S., Heller, H. C. (1989). Regulated nocturnal hypothermia induced in pigeons by food deprivation. *Am J Physiol* 256:R733–8.

Gronfier, C., Luthringer, R., Follenius, M., Schaltenbrand, N., Macher, J. P., Muzet, A., Brandenberger, G. (1997). Temporal relationships between pulsatile cortisol secretion and electroencephalographic activity during sleep. *Electroencephalogr Clin Neurophysiol* 103:405–8.

Halberg, F. (1953). Some physiological and clinical aspects of 24-hour periodicity. *Lancet* 73:20–32.

Hall, M., Baum, A., Buysse, D. J., Prigerson, H. G., Kupfer, D. J., Reynolds, C. F., III. (1998). Sleep as a mediator of the stress-immune relationship. *Psychosom Med* 60:48–51.

Hamada, T., Yamanouchi, S., Watanabe, A., Shibata, S., Watanabe, S. (1999). Involvement of glutamate release in substance P-induced phase delays of suprachiasmatic neuron activity rhythm in vitro. *Brain Res* 836:190–3.

Hamada, T., LeSauter, J., Venuti, J. M., Silver, R. (2001). Expression of Period genes: Rhythmic and nonrhythmic compartments of the suprachiasmatic nucleus. *J Neurosci* 21:7742–50.

Hannibal, J., Hindersson, P., Knudsen, S. M., Georg, B., Fahrenburg, J. (2002). The photopigment melanopsin is exclusively present in pituitary adenylate cyclase-activating polypeptide-containing retinal ganglion cells of the retinohypothalamic tract. *J Neurosci* 22:RC191(1–7).

Hansen, J. (2001a). Increased breast cancer risk among women who work predominantly at night. *Epidemiology* 12:74–7.

Hansen, J. (2001b). Light at night, shiftwork, and breast cancer risk. *J Nat Cancer Inst* 93:1513–15.

Härenstam, A., Theorell, T., Orth-Gomér, K., Palm, U.-B., Unden, A.-L. (1987). Shift work, decision latitude and ventricular ectopic activity: A study of 24-hour electrocardiograms in Swedish prison personnel. *Work Stress* 1:341–50.

Härmä, M. (1993). Individual differences in tolerance to shiftwork: A review. *Ergonomics* 36:101–9.

Härmä, M. (1996). Ageing, physical fitness and shiftwork tolerance. *Appl Ergonomics* 27:25–9.

Härmä, M. (1998). New work times are here – are we ready? *Scand J Work Environ Health* 24(Suppl. 3):3–6.

Härmä, M., Ilmarinen, J. E. (1999). Towards the 24-hour society – new approaches for aging shift workers? *Scand J Work Environ Health* 25:610–15.

Harrington, M. E., Hoque, S., Hall, A., Golombek, D., Biello, S. (1999). Pituitary adenylate cyclase activating peptide phase shifts circadian rhythms in a manner similar to light. *J Neurosci* 19:6637–42.

Hastings, J. W., Rusak, B., Boulos, Z. (1991). Circadian rhythms: The physiology of biological timing. In: *Comparative Animal Physiology, Fourth Edition, Neural and Integrative Animal Physiology* (ed. C. L. Prosser), pp. 435–546. New York: Wiley-Liss.

Hatch, M. C., Figa-Talamanca, I., Salerno, S. (1999). Work stress and menstrual patterns among American and Italian nurses. *Scand J Work Environ Health* 25:144–50.

Haus, E., Smolensky, M. H. (1999). Biologic rhythms in the immune system. *Chronobiol Int* 16:581–622.

Healy, D., Minors, D. S., Waterhouse, J. M. (1993). Shiftwork, helplessness and depression. *J Affect Disord* 29:17–25.

Hennig, J., Kieferdorf, P., Moritz, C., Huwe, S., Netter, P. (1998). Changes in cortisol secretion during shiftwork: Implications for tolerance to shiftwork? *Ergonomics* 41:610–21.

Herman, J. P., Cullinan, W. E. (1997). Neurocircuitry of stress: Central control of the hypothalamo-pituitary-adrenocortical axis. *Trends Neurosci* 20:78–84.

Hildebrandt, G. (1974). Circadian variations of thermoregulatory response in man. In: *Chronobiology* (ed. L. E. Scheving, F. Halberg, J. E. Pauly), pp. 234–240. Tokyo: Igaku Shoin.

Holl, R. W., Hartman, M. I., Veldhuis, J. D., Taylor, W. M., Thorner, M. O. (1991). Thirty-second sampling of plasma growth hormone in man: correlation with sleep stages. *J Clin Endocrinol Metab* 72:854–61.

Hrubá, D., Kukla, L., Tyrlík, M. (1999). Occupational risks for human reproduction: ELSPAC study. *Central Eur J Public Health* 7:210–15.

Hume, K. I., Mills, J. N. (1977). Rhythms of REM and slow-wave sleep in subjects living on abnormal time schedules. *Waking Sleeping* 1:291–6.

Ibuka, N. (1979). Suprachiasmatic nucleus and sleep-wakefulness rhythms. In: *Biological Rhythms and their Central Mechanism. The Naito Symposium* (ed. M. Suda, O. Hayaishi, H. Nakagawa), pp. 325–34. Amsterdam: Elsevier/North Holland Biomedical Press.

Inouye, S. T., Shibata, S. (1994). Neurochemical organization of circadian rhythm in the suprachiasmatic nucleus. *Neurosci Res* 20:109–30.

Irwin, M., Mascovich, A., Gillin, J. C., Willoughby, R., Pike, J., Smith, T. L. (1994). Partial sleep deprivation reduces natural killer cell activity in humans. *Psychosom Med* 56:593–8.

Irwin, M., McClintick, J., Costlow, C., Fortner, M., White, J., Gillin, J. C. (1996). Partial night sleep deprivation reduces natural killer and cellular immune responses in humans. *FASEB J* 10:643–53.

Irwin, M., Smith, T. L., Gillin, J. C. (1992). Electroencephalographic sleep and natural killer activity in depressed patients and control subjects. *Psychosom Med* 54: 10–21.

Jacobs, B. L., Fornal, C. A. (1999). An integrative role for serotonin in the central nervous system. In: *Handbook of Behavioral State Control* (ed. R. Lydic, H. A. Baghdoyan), pp. 181–93. Boca Raton: CRC Press.

Jha, S. K., Yadav, V., Mallick, B. N. (2001). GABA-A receptors in mPOAH simultaneously regulate sleep and body temperature in freely moving rats. *Pharmacol Biochem Behav* 70:115–21.

Johnson, C. H., Golden, S. S., Ishiura, M., Kondo, T. (1996). Circadian clocks in prokaryotes. *Mol Microbiol* 21:5–11.

Johnson, R. F., Moore, R. Y., Morin, L. P. (1988). Loss of entrainment and anatomical plasticity after lesions of the hamster retinohypothalamic tract. *Brain Res* 460: 297–313.

Jones, B. E., Muhlethaler, M. (1999). Cholinergic and GABAergic neurons of the basal forebrain: Role in cortical activation. In: *Handbook of Behavioral State Control* (ed. R. Lydic, H. A. Baghdoyan), pp. 213–33. Boca Raton: CRC Press.

Julius, S. (1988). The blood pressure seeking properties of the central nervous system. *J Hyperten* 6:177–85.

Karacan, I., Williams, R. L., Finley, W. W., Hursh, C. J. (1970). The effects of naps on nocturnal sleep: Influence on the need for stage-1 REM and stage 4 sleep. *Biol Psychiatry* 2:391–9.

Kawachi, I., Colditz, G. A., Stampfer, M. J., Willett, W. C., Manson, J. E., Speizer, F. E., Hennekens, C. H. (1995). Prospective study of shift work and risk of coronary heart disease in women. *Circulation* 92:3178–82.

Khaleque, A. (1991). Effects of diurnal and seasonal sleep deficiency on work effort and fatigue of shift workers. *Int Arch Occup Environ Health* 62:591–3.

Kecklund, G., Åkerstedt, T. (1995). Effects of timing of shifts on sleepiness and sleep duration. *J Sleep Res* 4(suppl. 2):47–50.

Kerkhof, G. A., Van Dongen H. P. A., Bobbert, A. C. (1998). Absence of endogenous circadian rhythmicity in blood pressure. *Am J Hyperten* 11:373–7.

Kilduff, T. S., Peyron, C. (2000). The hypocretin/orexin ligand-receptor system: implications for sleep and sleep disorders. *Trends Neurosci* 23:359–65.

Klein, D. C., Moore, R. Y., Reppert, S. M. (1991). *Suprachiasmatic Nucleus: the Mind's Clock*. New York: Oxford University Press.

Knauth, P., Emde, E., Rutenfranz, J., Kiesswetter, E., Smith, P. (1981). Re-entrainment of body temperature in field studies of shiftwork. *Int Arch Occup Environ Health* 49:137–49.

Knauth, P., Rutenfranz, J. (1982). Development of criteria for the design of shiftwork systems. *J Hum Ergol* 11(suppl.):337–67.

Knauth, P., Rutenfranz, J., Herrmann, G., Poeppl, S. J. (1978). Re-entrainment of body temperature in experimental shift-work studies. *Ergonomics* 21:775–83.

Knowles, J. B., Coulter, M., Wahnon, S., Reitz, W., MacLean, A. W. (1990). Variation in Process S: Effects on sleep continuity and architecture. *Sleep* 13:97–107.

Knutsson, A., Åkerstedt, T., Jonsson, B. G., Orth-Gomer, K. (1986). Increased risk of ischaemic heart disease in shift workers. *Lancet*: 89–92.

Knutsson, A., Hallquist, J., Reuterwall, C., Theorell, T., Åkerstedt, T. (1999). Shiftwork and myocardial infarction: A case-control study. *Occup Environ Med* 56:46–50.

Kogi, K. (1981). Comparison of resting conditions between various shift rotation systems for industrial workers. In: *Night and Shift Work: Biological and Social Aspects* (ed. A. Reinberg, N. Vieux, P. Andlauer), pp. 417–24. Oxford: Pergamon Press.

Kogi, K. (1985). Introduction to the problems of shiftwork. In: *Hours of Work* (ed. S. Folkard, T. H. Monk), pp. 165–84. New York: Wiley.

Kogi, K., Thurman, J. E. (1993). Trends in approaches to night and shiftwork and new international standards. *Ergonomics* 36:3–13.

Koller, M. (1983). Health risks related to shift work. *Int Arch Occup Environ Health* 53:59–75.

Koller, M., Haider, M., Kundi, M., Cervinka, R., Katschnig, H., Küfferle, B. (1981). Possible relations of irregular working hours to psychiatric psychosomatic disorders. In: *Night and Shiftwork: Biological and Social Aspects* (ed. A. Reinberg, N. Vieux, P. Andlauer), pp. 465–72. Oxford: Pergamon Press.

Koller, M., Härmä, M., Laitinen, J. T., Kundi, M., Piegler, B., Haider, M. (1994). Different patterns of light exposure in relation to melatonin and cortisol rhythms and sleep of night workers. *J Pineal Res* 16:127–35.

Koob, G. F., Le Moal, M. (1997). Drug abuse: Hedonic homeostatic dysregulation. *Science* 278:52–8.

Koob, G. F., Le Moal, M. (2001). Drug addiction, dysregulation of reward, and allostasis. *Neuropsychopharmacology* 24:97–129.

Kornhauser, J. M., Ginty, D. D., Greenberg, M. E., Mayo, K. E., Takahashi, J. S. (1996). Light entrainment and activation of signal transduction pathways in the SCN. In: *Hypothalamic Integration of Circadian Rhythms* (ed. R. M. Buijs, A. Kalsbeek, H. J. Romijn, C. M. A. Pennartz, M. Mirmiran), pp. 133–46. Amsterdam: Elsevier.

Kramer, A., Yang, F.-C., Snodgrass, P., Li, X., Scammell, T. E., Davis, F. C., Weitz, C. J. (2001). Regulation of daily locomotor activity and sleep by hypothalamic EGF receptor signaling. *Science* 294:2511–15.

Krieger, D. T., Hauser, H., Krey, L. C. (1977). Suprachiasmatic nuclear lesions do not abolish food-shifted circadian adrenal and temperature rhythmicity. *Science* 197:398–9.

Kripke, D. F., Cook, B., Lewis, O. F. (1971). Sleep of night workers: EEG recordings. *Psychophysiology* 7:377–84.

Kristensen, T. S. (1989). Cardiovascular diseases and the work environment: A critical review of the epidemiologic literature on nonchemical factors. *Scand J Work Environ Health* 20:165–79.

Krueger, J. M., Fang, J., Floyd, R. A. (1999). Relationships between sleep and immune function. In: *Regulation of Sleep and Circadian Rhythms* (ed. F. W. Turek, P. C. Zee), pp. 427–64. New York: Marcel Dekker.

Krueger, J. M., Fang, J., Hansen, M. K., Zhang, J., Obál, F., Jr. (1998). Humoral regulation of sleep. *News Physiol Sci* 13:189–94.

Kurumatani, N., Koda, S., Nakagiri, S., Hisashige, A., Sakai, K., Saito, Y., Aoyama, H., Dejima, M., Moriyama, T. (1994). The effects of frequently rotating shiftwork on sleep and the family life of hospital nurses. *Ergonomics* 37:995–1007.

Landolt, H. P., Werth, E., Borbély, A. A., Dijk, D. J. (1995). Caffeine intake (200 mg) in the morning affects human sleep and EEG power spectra at night. *Brain Res* 675:67–74.

Lavie, P. (1986). Ultrashort sleep-waking schedule. III. "Gates" and "forbidden zones" for sleep. *Electroencephalogr Clin Neurophysiol* 63:414–25.

Lemmer, B. (1989). Circadian rhythms in the cardiovascular system. In: *Biological Rhythms in Clinical Practice* (ed. J. Arendt, D. S. Minors, J. M. Waterhouse), pp. 51–70. London: Wright.

Lennernäs, M., Åkerstedt, T., Hambræus, L. (1994). Nocturnal eating and serum cholesterol of three-shift workers. *Scand J Work Environ Health* 20:401–6.

LeSauter, J., Silver, R. (1998). Output signals of the SCN. *Chronobiol Int* 15:535–50.

Liddell, F. D. K. (1982). Motor vehicle accidents (1973–6) in a cohort of Montreal drivers. *J Epidemiol Comm Health* 36:140–5.

Lowrey, P. L., Shimomura, K., Antoch, M. P., Yamazaki, S., Zemenides, P. D., Ralph, M. R., Menaker, M., Takahashi, J. S. (2000). Positional syntenic cloning and functional characterization of the mammalian circadian mutation *tau*. *Science* 288: 483–91.

Lu, J., Greco, M. A., Shiromani, P., Saper, C. B. (2000). Effect of lesions of the ventrolateral preoptic nucleus on NREM and REM sleep. *J Neuroscience* 20:3830–42.

Lu, J., Shiromani, P., Saper, C. B. (1999). Retinal input to the sleep-active ventrolateral preoptic nucleus in the rat. *Neuroscience* 93:209–14.

Lu, J., Zhang, Y.-H., Chou, T. C., Gaus, S. E., Elmquist, J. K., Shiromani, P., Saper, C. B. (2001). Contrasting effects of ibotenate lesions of the paraventricular nucleus and subparaventricular zone on sleep-wake cycle and temperature regulation. *J Neurosci* 21:4864–74.

Mallon, L., Broman, J.-E., Hetta, J. (2002). Sleep complaints predict coronary artery disease mortality in males: A 12-year follow-up study of a middle-aged Swedish population. *J Int Med* 251:207–16.

Matsumoto, K. (1978). Sleep patterns in hospital nurses due to shift work: An EEG study. *Waking Sleeping* 2:169–73.

Matsumoto, K. (1981). Effects of nighttime naps on body temperature changes, sleep patterns, and self-evaluation of sleep. *J Hum Ergology* 10:173–84.

Maywood, E. S., Mrosovsky, N., Field, M. D., Hastings, M. H. (1999). Rapid down-regulation of mammalian period genes during behavioral resetting of the circadian clock. *Proc Nat Acad Sci U S A* 96:15211–16.

McDonald, C. S., Zepelin, H., Zammit, G. K. (1981). Age and sex patterns in auditory awakening thresholds. *Sleep Res* 10:115.

McEwen, B. S. (1998a). Protective and damaging effects of stress mediators. *New Engl J Med* 338:171–9.

McEwen, B. S. (1998b). Stress, adaptation, and disease: Allostasis and allostatic load. *Ann N Y Acad Sci* 840:33–44.

McEwen, B. S. (2000a). Allostasis and allostatic load: Implications for neuropsychopharmacology. *Neuropsychopharmacology* 22:108–24.

McEwen, B. S. (2000b). The neurobiology of stress: From serendipity to clinical relevance. *Brain Res* 886:172–89.

McEwen, B. S., Seeman, T. (1999). Protective and damaging effects of mediators of stress: elaborating and testing the concepts of allostasis and allostatic load. *Ann N Y Acad Sci* 896:30–47.

McEwen, B. S., Stellar, E. (1993). Stress and the individual: Mechanisms leading to disease. *Arch Int Med* 153:2093–101.

McGinty, D. J. (1969). Somnolence, recovery and hyposomnia following ventro-medial diencephalic lesions in the rat. *Electroencephalogr Clin Neurophysiol* 26: 70–9.

McGranaghan, P. A., Piggins, H. D. (2001). Orexin A-like immunoreactivity in the hypothalamus and thalamus of the Syrian hamster (*Mesocricetus auratus*) and Siberian hamster (*Phodopus sungorus*), with special reference to circadian structures. *Brain Res* 904:234–44.

Mendelson, W. B. (2001). The sleep-inducing effect of ethanol microinjection into the medial preoptic area is blocked by flumazenil. *Brain Res* 892:118–21.

Michel-Briand, C., Chopard, J. L., Guiot, A., Paulmier, M., Studer, G. (1981). The pathological consequences of shift work in retired workers. In: *Night and Shiftwork: Biological and Social Aspects* (ed. A. Reinberg, N. Vieux, P. Andlauer), pp. 399–407. Oxford: Pergamon Press.

Mills, J. N., Minors, D. S., Waterhouse, J. M. (1978). Adaptation to abrupt time shifts of the oscillator(s) controlling human circadian rhythms. *J Physiol* 285:455–70.

Mistlberger, R. E. (1994). Circadian food-anticipatory activity: Formal models and physiological mechanisms. *Neurosci Biobehav Revi* 18:171–95.

Mistlberger, R. E., Antle, M. C., Glass, J. D., Miller, J. D. (2000). Behavioral and serotonergic regulation of circadian rhythms. *Biol Rhythm Res* 31:240–83.

Mistlberger, R. E., Bergmann, B. M., Waldenar, W., Rechtschaffen, A. (1983). Recovery sleep following sleep deprivation in intact and suprachiasmatic nuclei-lesioned rats. *Sleep* 6:217–33.

Mistlberger, R. E., Landry, G. J., Marchant, E. G. (1997). Sleep deprivation can attenuate light-induced phase shifts of circadian rhythms in hamsters. *Neurosci Lett* 238:5–8.

Moga, M. M., Moore, R. Y. (1997). Organization of neural inputs to the suprachiasmatic nucleus in the rat. *J Comp Neurol* 389:508–34.

Monk, T. H. (1990). Shiftworker performance. *Occup Med* 5:183–98.

Monk, T. H., Folkard, S., Wedderburn, A. I. (1996). Maintaining safety and high performance on shiftwork. *App Ergonomics* 27:17–23.

Moore, J. G. (1988). Chronobiology of gastric function: Relevance to human ulcerogenesis. In: *Trends in Chronobiology, Advances in the Biosciences, Volume 73* (ed. W. T. J. M. Hekkens, G. A. Kerkhof, W. J. Rietveld), pp. 295–305. Oxford: Pergamon Press.

Moore, R. Y. (1996). Neural control of the pineal gland. *Behav Brain Res* 73:125–30.

Moore, R. Y. (1997). Circadian rhythms: Basic neurobiology and clinical applications. *Ann Rev Med* 48:253–66.

Moore, R. Y., Eichler, V. B. (1972). Loss of a circadian adrenal corticosterone rhythm following suprachiasmatic lesions in the rat. *Brain Res* 42:201–6.

Moore, R. Y., Lenn, N. J. (1972). A retinohypothalamic projection in the rat. *J Comp Neurol* 146:1–14.

Moore, R. Y., Silver, R. (1998). Suprachiasmatic nucleus organization. *Chronobiol Int* 15:475–87.

Moore, R. Y., Speh, J. C., Card, J. P. (1995). The retinohypothalamic tract originates from a distinct subset of retinal ganglion cells. *J Comp Neurol* 352:351–66.

Moore-Ede, M. C. (1986). Physiology of the circadian timing system: Predictive versus reactive homeostasis. *Am J Physiol* 250:R735–52.

Moore-Ede, M. C., Richardson, G. S. (1985). Medical implications of shift-work. *Ann Rev Med* 36:607–17.

Morikawa, Y., Nakagawa, H., Miura, K., Ishizaki, M., Tabata, M., Nishijo, M., Higashiguchi, K., Yoshita, K., Sagara, T., Kido, T., Naruse, Y., Nogawa, K. (1999). Relationship between shift work and onset of hypertension in a cohort of manual workers. *Scand J Work Environ Health* 25:100–4.

Morin, L. P., Goodless-Sanchez, N., Smale, L., Moore, R. Y. (1994). Projections of the suprachiasmatic nuclei, subparaventricular zone and retrochiasmatic area in the golden hamster. *Neuroscience* 61:391–410.

Moriya, T., Horikawa, K., Akiyama, M., Shibata, S. (2000). Correlative association between N-methyl-D-aspartate receptor-mediated expression of period genes in the suprachiasmatic nucleus and phase shifts in behavior with photic entrainment of clock in hamsters. *Mol Pharmacol* 58:1554–62.

Mott, P. E., Mann, F. C., McLoughlin, Q., Warwick D. P. (1965). *Shift Work – the Social, Psychological, and Physical Consequences*. Ann Arbor: University of Michigan Press.

Mrosovsky, N. (1990). *Rheostasis: The Physiology of Change*. New York: Oxford University Press.

Mrosovsky, N. (1996). Locomotor activity and non-photic influences on circadian clocks. *Biol Rev* 71:343–72.

Mrosovsky, N. (1999). Masking: History, definitions, and measurement. *Chronobiol Int* 16:415–29.

Muller, J. E., Stone, P. H., Turi, Z. G., Rutherford, J. D., Czeisler, C. A., Parker, C., Poole, K., Passamani, E., Roberts, R., Robertson, T., Sobel, B. E., Willerson, J. T., Braunwald, E., and the MILIS Study Group. (1985). Circadian variation in the frequency of onset of acute myocardial infarction. *New Eng J Med* 313:1315–22.

Murata, K., Yano, E., Shinozaki, T. (1999). Impact of shift work on cardiovascular functions in a 10-year follow-up study. *Scand J Work Environ Health* 25:272–7.

Nakano, Y., Miura, T., Hara, I., Aono, H., Miyano, N., Miyajima, K., Tabuchi, T., Kosaka, H. (1982). The effect of shift work on cellular immune function. *J Hum Ergol* 11(suppl.):131–7.

Nikolova, N., Handjiev, S., Angelova, K. (1990). Nutrition of night and shiftworkers in transports. In: *Shiftwork: Health, Sleep and Performance* (ed. G. Costa, G. Cesana, K. Kogi, A. Wedderburn), pp. 583–7. Frankfurt am Main: Verlag Peter Lang GmbH.

Nilsson, P. M., Nilsson, J.-Å, Hedblad, B., Berglund, G. (2001). Sleep disturbance in association with elevated pulse rate for prediction of mortality – consequences of mental strain? *J Int Med* 250:521–9.

Novak, C. M., Nunez, A. A. (2000). A sparse projection from the suprachiasmatic nucleus to the sleep active ventrolateral preoptic area in the rat. *Neuroreport* 11:93–6.

Nurminen, T. (1998). Shift work and reproductive health. *Scand J Work Environ Health* 24(suppl. 3):28–34.

Ohira, T., Tanigawa, T., Iso, H., Odagiri, Y., Takamiya, T., Shimomitsu, T., Hayano, J., Shimamoto, T. (2000). Effects of shift work on 24-hour ambulatory blood pressure and its variability among Japanese workers. *Scand J Work Environ Health* 26:421–6.

Olsson, K., Kandolin, I., Kauppinen-Toropainen, K. (1990). Stress and coping strategies of three-shift workers. *Travail Humain* 53:175–88.

Pace-Schott, E. F., Hobson, J. A. (2002). The neurobiol sleep: Genetics, cellular physiology and subcortical networks. *Nature Rev Neurosci* 3:591–605.

Panda, S., Sato, T. K., Castrucci, A. M., Rollag, M. D., DeGrip, W. J., Hogenesch, J. B., Provencio, I., Kay, S. A. (2002). Melanopsin (*Opn4*) requirement for normal light-induced circadian phase shifting. *Science* 298:2213–16.

Parkes, K. R. (1999). Shiftwork, job type, and the work environment as joint predictors of health-related outcomes. *J Occup Health Psychol* 4256–68.

Perlis, M. L., Smith, M. T., Andrews, P. J., Orff, H., Giles, D. E. (2000). Beta/gamma EEG activity in patients with primary and secondary insomnia and good sleeper controls. *Sleep* 24:110–26.

Peter, R., Alfredsson, L., Knutsson, A., Siegrist, J., Westerholm, P. (1999). Does a stressful psychosocial work environment mediate the effects of shift work on cardiovascular risk factors? *Scand J Work Environ Health* 25:376–81.

Peternel, P., Stegnar, M., Salobir, U., Salobir, B., Keber, D., Vene, N. (1990). Shift work and circadian rhythm of blood fibrinolytic parameters. *Fibrinolysis* 4(suppl. 2):113–5.

Pittendrigh, C. S. (1993). Temporal organization: Reflections of a Darwinian clock-watcher. *Ann Rev Physiol* 55:17–54.

Porkka-Heiskanen, T., Strecker, R. E., Thakker, M.; Bjørkum, A. A., Greene, R. W., McCarley, R. W. (1997). Adenosine: A mediator of the sleep-inducing effects of prolonged wakefulness. *Science* 276:1265–8.

Portas, C. M., Thakkar, M., Rainnie, D. G., Greene, R. W., McCarley, R. W. (1997). Role of adenosine in behavioral state modulation: A microdialysis study in the freely moving cat. *Neuroscience* 79:225–35.

Presser, H. B. (1995). Job, family, and gender: Determinants of nonstandard work schedules among employed Americans in 1991. *Demography* 32:577–98.

Prosser, R. A. (2000). Serotonergic actions and interactions on the SCN circadian pacemaker: In vitro investigations. *Biol Rhythm Res* 31:315–39.

Radulovacki, M. (1985). Role of adenosine in sleep in rats. *Rev Clin Basic Pharmacol* 5:327–39.

Rea, M. A., Pickard, G. E. (2000). Serotonergic modulation of photic entrainment in the Syrian hamster. *Biol Rhythm Res* 31:284–314.

Rechtschaffen, A. Kales, A. (1968). *A Manual of Standardized Terminology, Techniques and Scoring System for Sleep Stages of Human Subjects.* Los Angeles: University of California.

Refinetti, R., Kaufman, C. M., Menaker, M. (1994). Complete suprachiasmatic lesions eliminate circadian rhythmicity of body temperature and locomotor activity in golden hamsters. *J Comp Physiol A* 175:223–32.

Reinberg, A., Migraine, C., Apfelbaum, M., Brigant, L., Ghata, J., Vieux, N., Laporte, A., Nicolai, A. (1979). Circadian and ultradian rhythms in the feeding behaviour and nutrient intakes of oil refinery operators with shift-work every 3–4 days. *Diabète Metab* 5:33–41.

Reuss, S. (1996). Components and connections of the circadian timing system in mammals. *Cell Tissue Res* 285:353–78.

Richardson, G. S., Carskadon, M. A., Orav, E. J., Dement, W. C. (1982). Circadian variation of sleep tendency in elderly and young adult subjects. *Sleep* 5(suppl. 2):S82–94.

Richardson, G. S., Miner, J. D., Czeisler, C. A. (1990). Impaired driving performance in shiftworkers: The role of the circadian system in a multifactorial model. *Alcohol Drugs Driving* 5/6:265–73.

Richter, C. P. (1967). Sleep and activity: Their relation to the 24-hour clock. *Proc Assoc Res Nerv Ment Dis* 45:8–27.

Roberts, A. J., Heyser, C. J., Cole, M., Griffin, P., Koob, G. F. (2000). Excessive ethanol drinking following a history of dependence: Animal model of allostasis. *Neuropsychopharmacology* 22:581–94.

Roehrs, T., Merlotti, L., Halpin, D., Rosenthal, L., Roth, T. (1995). Effects of theophylline on nocturnal sleep and daytime sleepiness/alertness. *Chest* 108:382–7.

Romon, M., Nuttens, M.-C., Fievet, C., Pot, P., Bard, J. M., Furon, D., Fruchart, J. C. (1992). Increased triglyceride levels in shift workers. *Am J Med* 93:259–62.

Romon-Rousseaux, M., Beuscart, R., Thuilliez, J. C., Frimat, P., Furon, D. (1986). Influence of different shift schedules on eating behaviour and weight gain in edible oil refinery workers. In: *Night and Shiftwork: Longterm Effects and their Prevention* (ed. M. Haider, M. Koller, R. Cervinka), pp. 433–40. New York: Verlag.

Rosa, R. R. (1993). Napping at home and alertness on the job in rotating shift workers. *Sleep* 16:727–35.

Rosenwasser, A. M., Adler, N. T. (1986). Structure and function in circadian timing systems: Evidence for multiple coupled circadian oscillators. *Neurosci Biobehav Rev* 10:431–48.

Rubin, N. H. (1988). Chronobiology of the gastrointestinal system. In: *Trends in Chronobiology, Advances in the Biosciences, Volume 73* (ed. W. T. J. M. Hekkens, G. A. Kerkhof, W. J. Rietveld), pp. 307–18. Oxford: Pergamon Press.

Ruby, N. F., Brennan, T. J., Xie, X., Cao, V., Franken, P., Heller, H. C., O'Hara, B. F. (2002). Role of melanopsin in circadian responses to light. *Science* 298:2211–13.

Ruby, N. F., Ibuka, N., Barnes, B. M., Zucker, I. (1989). Suprachiasmatic nuclei influence torpor and circadian temperature rhythms in hamsters. *Am J Physiol* 257:R210–15.

Rutenfranz, J., Colquhoun, W. P., Knauth, P., Ghata, J. N. (1977). Biomedical and psychosocial aspects of shift work. *Scand J Work Environ Health* 3:165–82.

Rutenfranz, J., Knauth, P., Angersbach, D. (1981). Shift work research issues. In: *Biological Rhythms, Sleep Shift Work. Advances in Sleep Research, Volume 7* (ed. L. C. Johnson, D. I. Tepas, W. P. Colquhoun, M. J. Colligan), pp. 165–96. New York: Spectrum.

Sack, R. L., Blood, M. L., Lewy, A. J. (1992). Melatonin rhythms in night shift workers. *Sleep* 15:434–41.

Saper, C. B. (1985). Organization of cerebral cortical afferent systems in the rat. II. Hypothalamocortical projections. *J Comp Neurol* 237;21–46.

Sapolsky, R. M. (1996). Why stress is bad for your brain. *Science* 273:749–50.

Satinoff, E. (1978). Neural organization and evolution of thermal regulation in mammals. *Science* 201:16–22.

Satinoff, E. (1983). A reevaluation of the concept of the homeostatic organization of temperature regulation. In: *Handbook of Behavioral Neurobiology, Vol. 6, Motivation* (ed. E. Satinoff, P. Teitelbaum), pp. 443–72. New York: Plenum.

Scammell, T. E., Gerashchenko, D. Y., Mochizuki, T., McCarthy, M. T., Estabrooke, I. V., Sears, C. A., Saper, C. B., Urade, Y., Hayaishi, O. (2001). An adenosine A2a agonist increases sleep and induces Fos in ventrolateral preoptic neurons. *Neuroscience* 107:653–63.

Schernhammer, E. S., Laden, F., Speizer, F. E., Willett, W. C., Hunter, D. J., Kawachi, I., Colditz, G. A. (2001). Rotating night shifts and risk of breast cancer in women participating in the Nurses' Health Study. *J Nat Cancer Inst* 93:1563–8.

Schmidt-Nielsen, K., Schmidt-Nielsen, B., Jarnum, S. A., Houpt, T. R. (1957). Body temperature of the camel and its relation to water economy. *Am J Physiol* 188:103–12.

Schulkin J., Gold, P. W., McEwen, B. S. (1998). Induction of corticotropin-releasing hormone gene expression by glucocorticoids: Implication for understanding the states of fear and anxiety and allostatic load. *Psychoneuroendocrinology* 23: 219–43.

Schulkin, J., McEwen, B. S., Gold, P. W. (1994). Allostasis, amygdala, and anticipatory angst. *Neurosci Biobehav Rev* 18:385–96.

Scott, A. J., Monk, T. H., Brink, L. L. (1997). Shiftwork as a risk factor for depression: a pilot study. *Int J Occup Environ Health* 3(suppl. 2):S2–S9.

Segawa, K., Nakazawa, S., Tsukamoto, Y., Kurita, Y., Goto, H., Fukui, A., Takano, K. (1987). Peptic ulcer is prevalent among shift workers. *Dig Dis Sci* 32:449–53.

Semba, K. (1999). The mesopontine cholinergic system: A dual role in REM sleep and wakefulness. In: *Handbook of Behavioral State Control* (ed. R. Lydic, H. A. Baghdoyan), pp. 161–80. Boca Raton: CRC Press.

Semple, J. I., Newton, J. L., Westley, B. R., May, F. E. B. (2001). Dramatic diurnal variation in the concentration of the human trefoil peptide TFF2 in gastric juice. *Gut* 48:648–55.

Shearman, L. P., Sriram, S., Weaver, D. R., Maywood, E. S., Chaves, I., Zheng, B., Kume, K., Lee, C. C., Van der Horst, G. T. J., Hastings, M. H., Reppert, S. M.

(2000). Interacting molecular loops in the mammalian circadian clock. *Science* 288:1013–19.

Sherin, J. E., Shiromani, P. J., McCarley, R. W., Saper, C. B. (1996). Activation of ventrolateral preoptic neurons during sleep. *Science* 271:216–19.

Shiromani, P. J., Scammell, T. E., Sherin, J. E., Saper, C. B. (1999). Hypothalamic regulation of sleep. In: *Handbook of Behavioral State Control* (ed. R. Lydic, H. A. Baghdoyan), pp. 311–25. Boca Raton: CRC Press.

Silver, R., LeSauter, J., Tresco, P. A., Lehman, M. N. (1996). A diffusible coupling signal from the transplanted suprachiasmatic nucleus controlling circadian locomotor rhythms. *Nature* 382:810–13.

Simpson, S., Galbraith, J. J. (1906). Observations on the normal temperature of the monkey and its diurnal variation, and on the effect of changes in the daily routine on this variation. *Trans Royal Soc Edinburgh* 45:65–106.

Sinha, M. K., Ohanessian, J. P., Heiman, M. L., Kriauciunas, A., Stephens, T. W., Magosin, S., Marco, C., Caro, J. F. (1996). Nocturnal rise of leptin in lean, obese, and non-insulin-dependent diabetes mellitus subjects. *J Clin Invest* 97: 1344–7.

Smith, L., Folkard, S. (1993). The perceptions and feelings of shiftworkers' partners. *Ergonomics* 36:299–305.

Smith, L., Folkard, S., Poole, C. J. M. (1994). Increased injuries on night shift. *Lancet* 344:1137–9.

Smith, L., Macdonald, I., Folkard, S., Tucker, P. (1998). Industrial shift systems. *Appl Ergonomics* 29:273–80.

Smith, R. S. Kushida, C. A. (2000). Risk of fatal occupational injury by time of day. *Sleep* 23(suppl. 2):A110–11.

Smolensky, M. H., Lee, E., Mott, D., Colligan, M. (1982). A health profile of American flight attendants (FA). *J Hum Ergol* 11(suppl.):103–19.

Spiegel, K., Sheridan, J. F., Van Cauter, E. (2002). Effect of sleep deprivation on response to immunization. *JAMA* 288:1471–2.

Steininger, T. L., Alam, M. N., Gong, H., Szymusiak, R., McGinty, D. (1999). Sleep-waking discharge of neurons in the posterior lateral hypothalamus of the albino rat. *Brain Res* 840:138–47.

Stephan, F. K., Zucker, I. (1972). Circadian rhythms in drinking behavior and locomotor activity of rats are eliminated by hypothalamic lesions. *Proc Nat Acad Sci U S A* 69:1583–6.

Stephenson, L. A., Kolka, M. A. (1985). Menstrual cycle phase and time of day alter reference signal controlling arm blood flow and sweating. *Am J Physiol* 249:R186–91.

Stephenson, L. A., Wenger, C. B., O'Donovan, B. H., Nadel, E. R. (1984). Circadian rhythm in sweating and cutaneous blood flow. *Am J Physiol* 246:R321–4.

Sterling, P., Eyer, J. (1981). Biological basis of stress-related mortality. *Social Sci Med* 15E:3–42.

Sterling, P., Eyer, J. (1988). Allostasis: A new paradigm to explain arousal pathology. In: *Handbook of Life Stress. Cognition and Health* (ed. S. Fisher, J. Reason), pp. 629–49. New York: John Wiley & Sons.

Stokkan, K.-A., Yamazaki, S., Tei, H., Sakaki, Y., Menaker, M. (2001). Entrainment of the circadian clock in the liver by feeding. *Science* 291:490–3.

Strogatz, S. H., Kronauer, R. E., Czeisler, C. A. (1987). Circadian pacemaker interferes with sleep onset at specific times each day: Role in insomnia. *Am J Physiol* 253:R172–8.

Sumova, A., Travnickova, Z., Mikkelsen, J. D., Illnerova, H. (1998). Spontaneous rhythm in c-Fos immunoreactivity in the dorsomedial part of the rat suprachiasmatic nucleus. *Brain Res* 801:254–8.

Sun, X., Whitefield, S., Rusak, B., Semba, K. (2001). Electrophysiological analysis of suprachiasmatic nucleus projections to the ventrolateral preoptic area in the rat. *Eur J Neurosci* 14:1257–74.

Sutcliffe, J. G., de Lecea, L. (2000). The hypocretins: Excitatory neuromodulatory peptides for multiple homeostatic systems, including sleep and feeding. *J Neurosci Res* 62:161–8.

Swanson, L. W., Sawchenko, P. E. (1980). Paraventricular nucleus: A site for the integration of neuroendocrine and autonomic mechanisms. *Neuroendocrinology* 31:410–17.

Szymusiak, R., Alam, N., Steininger, T. L., McGinty, D. (1998). Sleep-waking discharge patterns of ventrolateral preoptic/anterior hypothalamic neurons in rats. *Brain Res* 803:178–88.

Tarquini, B., Cecchettin, M., Cariddi, A. (1986). Serum gastrin and pepsinogen in shift workers. *Int Arch Occup Environ Health* 58:99–103.

Taylor, E., Briner, R. B., Folkard, S. (1997a). Models of shiftwork and health: An examination of the influence of stress on shiftwork theory. *Hum Factors* 39:67–82.

Taylor, E., Folkard, S., Shapiro, D. A. (1997b). Shiftwork advantages as predictors of health. *Int J Occup Environ Health* 3(Suppl.2):S15,S20, S29.

Taylor, P. J., Pocock, S. J. (1972). Mortality of shift and day workers 1956–68. *Br J Industr Med* 29:201–7.

Tenkanen, L., Sjöblom, T., Härmä, M. (1998). Joint effect of shift work and adverse life-style factors on the risk of coronary heart disease. *Scand J Work Environ Health* 24:351–7.

Tenkanen, L., Sjöblom, T., Kalimo, R., Alikoski, T., Härmä, M. (1997). Shift work, occupation and coronary heart disease over 6 years of follow-up in the Helsinki Heart Study. *Scand J Work Environ Health* 23:257–65.

Tepas, D. I., Walsh, J., Armstrong, D. (1981). Comprehensive study of the sleep of shift workers. In: *Biological Rhythms, Sleep and Shift Work. Advances in Sleep Research, Volume 7* (ed. L. C. Johnson, D. I. Tepas, W. P. Colquhoun, M. J. Colligan), pp. 347–56. New York: Spectrum Publications.

Thase, M. E. (2000). Treatment issues related to sleep and depression. *J Clin Psychiatry* 61(suppl.11):46–50.

Ticho, S. R., Radulovacki, M. (1991). Role of adenosine in sleep and temperature regulation in the preoptic area of rats. *Pharmacol Biochem Behav* 40:33–40.

Tilley, A. J., Wilkinson, R. T., Warren, P. S. G., Watson, B., Drud, M. (1982). The sleep and performance of shift workers. *Hum Factors* 24:629–41.

Tobler, I., Borbély, A. A., Groos, G. (1983). The effect of sleep deprivation on sleep in rats with suprachiasmatic lesions. *Neurosci Lett* 42:49–54.

Torsvall, L., Åkerstedt, T., Gillberg, M. (1981). Age, sleep and irregular work hours. *Scand J Work Environ Health* 7:196–203.

Torsvall, L., Åkerstedt, T., Gillander K., Knutsson, A. (1989). Sleep on the night shift: 24-hour EEG monitoring of spontaneous sleep/wake behavior. *Psychophysiology* 26:352–8.

Torterolo, P., Yamuy, J., Sampogna, S., Morales, F. R., Chase, M. H. (2001). Hypothalamic neurons that contain hypocretin (orexin) express c-fos during active wakefulness and carbachol-induced active sleep. *Sleep Res Online* 4:25–32.

Tosini, G., Menaker, M. (1996). Circadian rhythms in cultured mammalian retina. *Science* 272:419–21.

Touitou, Y., Motohashi, Y., Reinberg, A., Touitou, C., Bourdeleau, P., Bogdan, A., Auzéby, A. (1990). Effect of shift work on the night-time secretory patterns of melatonin, prolactin, cortisol and testosterone. *Eur J Appl Physiol* 60:288–92.

Tüchsen, F. (1993). Working hours and ischaemic heart disease in Danish men: A 4-year cohort study of hospitalization. *Int J Epidemiol* 22:215–21.

Tune, G. S. (1969). Sleep and wakefulness in a group of shift workers. *Br J Industr Med* 26:54–8.

U.S. Congress, Office of Technology Assessment. (1991, September). *Biological Rhythms: Implications for the Worker* (OTA-BA-463). Washington, DC: U.S. Government Printing Office.

Van Cauter, E., Blackman, J. D., Roland, D., Spire, J. P. Refetoff, S., Polonsky, K. S. (1991). Modulation of glucose regulation and insulin secretion by circadian rhythmicity and sleep. *J Clin Invest* 88:934–42.

Van Cauter, E., Spiegel, K. (1999). Sleep as a mediator of the relationship between socioeconomic status and health: A hypothesis. *Ann N Y Acad Sci* 896:254–61.

Van Cauter, E. V., Polonsky, K. S., Blackman, J. D., Roland, D., Sturis, J., Byrne, M. M., Scheen, A. J. (1994). Abnormal temporal patterns of glucose tolerance in obesity: relationship to sleep-related growth hormone secretion and circadian cortisol rhythmicity. *J Clin Endocr Metab* 79:1797–805.

Van der Beek, E. M., Horvath, T. L., Wiegant, V. M., Van den Hurk, R., Buijs, R. M. (1997). Evidence for a direct neuronal pathway from the suprachiasmatic nucleus to the gonadotropin-releasing hormone system: Combined tracing and light and electron microscopic immunocytochemical studies. *J Comp Neurol* 384:569–79.

Van der Horst, G. T. J., Muijtjens, M., Kobayashi, K., Takano, R., Kanno, S.-I., Takao, M., de Wit, J., Verkerk, A., Eker, A. P. M., van Leenen, D., Buijs, R., Bootsma, D., Hoeijmakers, J. H. J., Yasui, A. (1999). Mammalian Cry1 and Cry2 are essential for maintenance of circadian rhythms. *Nature* 398:627–30.

Van Dongen, H. P. A., Maislin, G., Kerkhof, G. A. (2001). Repeated assessment of the endogenous 24-hour profile of blood pressure under constant routine. *Chronobiol Int* 18:85–98.

Van Reeth, O., Weibel, L., Spiegel, K., Leproult, R., Dugovic, C., Maccari, S. (2000). Interactions between stress and sleep: From basic research to clinical situations. *Sleep Med Rev* 4:201–19.

Vazquez, J., Baghdoyan, H. A. (2001). Basal forebrain acetylcholine release during REM sleep is significantly greater than during waking. *Am J Physiol* 280:R598–601.

Vener, K. J., Szabo, S., Moore, J. G. (1989). The effect of shift work on gastrointestinal (GI) function: A review. *Chronobiologia* 16:421–39.

Vgontzas, A. N., Bixler, E. O., Lin, H.-M., Prolo, P., Mastorakos, G., Vela-Bueno, A., Kales, A., Chrousos, G. P. (2001). Chronic insomnia is associated with nyctohemeral activation of the hypothalamic-pituitary-adrenal axis: Clinical implications. *J Clin Endocrinol Metab* 86:3787–94.

Virus, R. M., Ticho, S., Pilditch, M., Radulovacki, M. (1990). A comparison of the effects of caffeine, 8-cyclopentyltheophylline, and alloxazine on sleep in rats: Possible roles of central nervous system adenosine receptors. *Neuropsychopharmacology* 3:243–9.

Vitaterna, M. H., King, D. P., Chang, A.-M., Kornhauser, J. M., Lowrey, P. L., McDonald, J. D., Dove, W.F., Pinto, L. H., Turek, F. W., Takahashi, J. S. (1994). Mutagenesis and mapping of a mouse gene, *Clock*, essential for circadian behavior. *Science* 264:719–25.

Vrang, N., Larsen, P. J., Mikkelsen, J. D. (1995). Direct projection from the suprachiasmatic nucleus to hypophysiotropic corticotropin-releasing factor immunoreactive cells in the paraventricular nucleus of the hypothalamus demonstrated by means of *Phaseolus vulgaris*-leucoagglutinin tract tracing. *Brain Res* 684:61–9.

Walker, J. Social problems of shiftwork (1985). In: *Hours of Work* (ed. S. Folkard, T. H. Monk), pp. 211–25. New York: Wiley.

Walsh, J., Tepas, D. I., Moss, P. (1981). The EEG sleep of night and rotating shift workers. In: *Biological Rhythms, Sleep and Shift Work. Advances in Sleep Research, Volume 7* (ed. L. C. Johnson, D. I. Tepas, W. P. Colquhoun, M. J. Colligan), pp. 371–81. New York: Spectrum.

Watanabe, A., Moriya, T., Nisikawa, Y., Araki, T., Hamada, T., Shibata, S., Watanabe, S. (1996). Adenosine A1-receptor agonist attenuates the light-induced phase shifts and fos expression in vivo and optic nerve stimulation-evoked field potentials in the suprachiasmatic nucleus in vitro. *Brain Res* 740:329–36.

Watts, A. G. (1991). The efferent projections of the suprachiasmatic nucleus: Anatomical insights into the control of circadian rhythms. In: *Suprachiasmatic Nucleus: The Mind's Clock* (ed. D. C. Klein, R. Y. Moore, S. M. Reppert), pp. 77–106. New York: Oxford University Press.

Watts, A. G., Swanson, L. W. (1987). Efferent projections of the suprachiasmatic nucleus: II. Studies using retrograde transport of fluorescent dyes and simultaneous peptide immunohistochemistry in the rat. *J Comp Neurol* 258:230–52.

Watts, A. G., Swanson, L. W., Sanchez-Watts, G. (1987). Efferent projections of the suprachiasmatic nucleus: I. Studies using anterograde transport of Phaseolus vulgaris leucoagglutinin in the rat. *J Comp Neurol* 258:204–29.

Webb, W. B., Agnew, H. W., Jr. (1971). Stage 4 sleep: Influence of time course variables. *Science* 174:1354–6.

Wedderburn, A. A. I. (1981). How important are the social effects of shiftwork? In: *Biological Rhythms, Sleep and Shift Work. Advances in Sleep Research, Volume 7* (ed. L. C. Johnson, D. I. Tepas, W. P. Colquhoun, M. J. Colligan), pp. 257–69. New York: Spectrum.

Weibel, L., Spiegel, K., Follenius, M., Ehrhart, J., Brandenberger, G. (1996). Internal dissociation of the circadian markers of the cortisol rhythm in night workers. *Am J Physiol* 270:E608–13.

Welsh, D. K., Logothetis, D. E., Meister, M., Reppert, S. M. (1995). Individual neurons dissociated from rat suprachiasmatic nucleus express independently phased circadian firing rhythms. *Neuron* 14:697–706.

Wenger, C. B., Roberts, M. F., Stolwijk, J. A. J., Nadel, E. R. (1976). Nocturnal lowering of thresholds for sweating and vasodilation. *J Appl Physiol* 41:15–19.

Wergeland, E., Strand, K. (1997). Working conditions and prevalence of preeclampsia, Norway 1989. *Int J Gynecol Obstetr* 58:189–96.

Werth, E., Dijk, D.-J., Achermann, P., Borbély, A. A. (1996). Dynamics of the sleep EEG after an early evening nap: Experimental data and simulations. *Am J Physiol* 271:R501–10.

Wever, R. (1979). *The Circadian System of Man: Experiments under Temporal Isolation.* New York: Springer-Verlag.

Wilkinson, R. T. (1992). How fast should the night shift rotate? *Ergonomics* 35:1425–46.

Yamazaki, S., Ishida, Y., Inouye, S. (1994). Circadian rhythms of adenosine triphosphate contents in the suprachiasmatic nucleus, anterior hypothalamic area and caudate putamen of the rat – negative correlation with electrical activity. *Brain Res* 664:237–40.

Yamazaki, S., Numano, R., Abe, M., Hida, A., Takahashi, R., Ueda, M., Block, G. D., Sakaki, Y., Menaker, M., Tei, H. (2000). Resetting central and peripheral circadian oscillators in transgenic rats. *Science* 288:682–5.

Yamazaki, S., Straume, M., Tei, H., Sakaki, Y., Menaker, M., Block, G. D. (2002). Effects of aging on central and peripheral mammalian clocks. *Proc Nat Acad Sci U S A* 99:10801–6.

Young, M. W. (2000). Life's 24-hour clock: Molecular control of circadian rhythms in animal cells. *Trends Biochem Sci* 25:601–6.

Zedeck, S., Jackson, S. E., Summers, E. (1983). Shift work schedules and their relationship to health, adaptation, satisfaction, and turnover intention. *Acad Manage Inst* 26:297–310.

Zulley, J., Wever, R., Aschoff, J. (1981). The dependence of onset and duration of sleep on the circadian rhythm of rectal temperature. *Pflügers Arch* 391:314–18.

8 Allostatic Load and Life Cycles: Implications for Neuroendocrine Control Mechanisms

John C. Wingfield

INTRODUCTION

Animals live in environments that change predictably over time, and individuals adjust their life cycles in anticipation of those changes. Habitat variation is most predictable at middle and high latitudes where short days of autumn and winter are accompanied by low primary productivity (plant growth) resulting in decreasing food resources for most animals. Lengthening days of spring and early summer bring a resurgence of growth and increasing food abundance. It is not surprising then that many vertebrates breed in spring when resources are increasing. This maximizes the chances for reproductive success. Even those animals that mate or give birth in autumn and winter time the period of maximum food requirements for feeding young with spring and summer (Lack, 1968; Bronson, 1989). In tropical regions, resources for breeding also fluctuate but often on a more variable time scale. The most well-known seasonal change in the tropics is rainfall, but the onset of the rainy season can vary in timing and amount from year to year. Thus, animals time breeding to coincide with maximum plant growth and subsequent fruit and insect production – or with the end of the rainy season when densities of young animals are great, in the case of many predators and vultures (Brown and Britton, 1980). Virtually all of these vertebrate organisms must initiate gonadal development, and many migrate to favorable breeding areas in anticipation of the onset of environmental conditions that are conducive to breeding. Equally important is termination of the reproductive period and gonadal regression as, or

Preparation of this chapter and the formulation of ideas presented in it were supported by a John Simon Guggenheim Fellowship and a Benjamin Meaker Fellowship. I am also grateful to the Division of Integrative Biology and Neurobiology and the Office of Polar Programs, National Science Foundation, for many years of generous grant support. I acknowledge a Shannon Award from the National Institutes of Health and the Russell F. Stark University Professorship from the University of Washington. Both awards have had a major influence on development of ideas presented in this review.

before, resources for breeding decline. Many species then undergo a molt, and migratory populations move back to wintering areas. These major changes in life history stages (LHSs; Jacobs and Wingfield, 2000) require profound adjustments of physiology, morphology, and behavior.

Although predictable cycles of morphology, physiology, and behavior are easily studied at the population level (e.g., daily, seasonal, tidal rhythms), individuals must also regulate their life cycles differently according to social status, body condition, parasite load, and so on. For example, resources for breeding are rarely distributed uniformly through the environment. Dominant individuals complete for the best-quality territories on which to breed and may gain greatest reproductive success. Females in the best body condition can begin breeding earlier, produce more young, and gain higher reproductive success. Males with high endogenous reserves of fat and protein are able to compete more successfully for mates and also have higher reproductive success than males in poor condition (e.g., Drent and Daan, 1980; Sapolsky, 1987). Parasite load often goes hand-in-hand with body condition, and there is now extensive evidence that individuals with parasite infections do not compete as well for breeding resources, are discriminated against in mate choice experiments, and are more susceptible to stress (Hamilton and Zuk, 1982; Dunlap and Schall, 1995). Clearly, many aspects of an individual's phenotype must be integrated with environmental changes.

There is also an unpredictable component to all environments that can have further influences on the life cycle of an individual. Responses of populations to unpredictable events, called labile perturbation factors (LPFs; Wingfield et al., 1998; Wingfield and Ramenofsky, 1999; Wingfield and Romero, 2000) involve a characteristic spectrum of processes that also can vary at the individual level. Examples of LPFs include severe storms that disrupt breeding, migration, or force an animal out of its winter home range. They can also include an increase in predators, change in social status (and thus access to resources), or a more long-term change in habitat such as an El Niño Southern Oscillation event, that can disrupt climate for months or even years. Over the next decades we will likely see global climate changes, human disturbance and pollution that will have permanent impact on the environment and its inhabitants – including ourselves. We know that neuroendocrine and endocrine mechanisms are in place to cope with temporary disruption of the environment (LPFs), but one disturbing unknown is how organisms will deal with permanent changes.

This chapter focuses on a framework that integrates physiological function in relation to complex individual responses to all levels of environmental change, internal and external, social and physical. Using this framework, some examples are used to illustrate how various hormones act at these different levels. Many hormones also may have actions at two or more levels of function, posing particularly fascinating problems for control mechanisms. Hormonal control at the population level on broad predictable schedules and also at the individual level as an organism deals with unpredictable local conditions and social status are emphasized. An understanding of how organisms adjust "homeostasis" to cope with seasonal environments, unpredictable events, and the uncertainties of social interactions will be critical if we are to deal with the major changes that will accompany global climate change. Our own future is inexorably intertwined with these processes.

THE ROLE OF HORMONES – LINKS AMONG ENVIRONMENT, MORPHOLOGY, PHYSIOLOGY, AND BEHAVIOR

Neuroendocrine and endocrine secretions regulate many aspects of vertebrate life cycles, including transitions in morphology, physiology, and behavior in relation to predictable changes in the environment. Because most vertebrates live in fluctuating environments, they adjust their state to maximize survival at different times of year. These changes can be summarized as a finite state machine (FSM) consisting of a temporal sequence of LHSs. The overall benefit of these transitions is to maximize lifetime fitness (Jacobs, 1996; Jacobs and Wingfield, 2000). Hormone secretions have several roles in the orchestration of LHSs in the life cycle. They are involved in the regulation of developmental trajectories and transitions of LHS; they activate and deactivate physiological and behavioral states within a LHS; and they drive facultative responses to unpredictable events in the environment (e.g., Gorbman et al., 1983; Nelson, 1999; Wingfield et al., 1999; Jacobs and Wingfield, 2000). Most life history trajectories involve ontogenetic stages followed by a transition into cyclic adult LHSs (Fig. 8.1). During ontogeny, hormones regulate growth and differentiation including determination of sex. It is during this process that morphological, physiological, and behavioral phenotypes are developed. These characters are largely irreversible (Arnold and Breedlove, 1985) or can be expressed in a fixed number of ways regulated by hormone action in the adult (relative plasticity model of Moore, 1991). Adult life history cycles also involve growth and differentiation as the individual changes morphology, physiology, and behavior from one stage to the next. Unlike ontogeny, developmental changes associated with adult life history cycles are reversible and usually cyclic. These are hormone dependent, but the mechanisms may be markedly different from ontogeny (Jacobs and Wingfield, 2000).

The patterns of ontogenetic and adult life history stages can be complex. Semelparous species such as some Pacific salmon of the genus *Oncorhynchus*

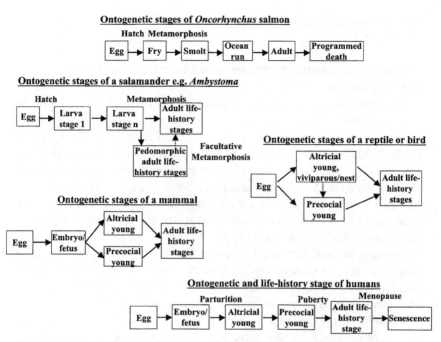

Figure 8.1: Vertebrate life cycles consist of ontogenetic stages that are essentially irreversible followed by a transition to the adult life cycle. The latter consists of distinct life history stages that are cyclic, usually on an annual basis. In some cases animals, become senescent. Semelparous species have a single trajectory from ontogeny to adult. They reproduce once and die. Iteroparous species undergo more than one cycle of adult life history stages.

appear to have one ontogenetic sequence culminating in reproduction, senescence, and death (Fig. 8.1). Iteroparous species also show a single ontogenetic sequence (examples given in Fig. 8.1) but then make a transition into adult life history cycles that may be repeated many times. For example, many amphibians have a larval stage that is aquatic. As growth progresses, metamorphosis to the terrestrial form occurs. The animal then enters the cycle of adult LHSs including breeding (Fig. 8.1). There are some exceptions in which larval forms persist and enter the adult sequence of LHSs (e.g., pedomorphosis in salamanders). In some cases, the adult "larval" form is permanent, and in others facultative metamorphosis can occur if the pond dries up (Fig. 8.1, see Denver, 1997, 1998; Hayes, 1997). In other tetrapods, such as birds, reptiles, and mammals, the young are born, or hatch, with a similar body plan to the adult but at varying stages of development (Fig. 8.1). The transition from ontogeny to adult life history cycles often, but not always, occurs at puberty. Compare, for example, the sequences for an amphibian, reptile, and bird with those of mammals, including humans, in Figure 8.1.

FINITE STATE MACHINE THEORY AND CYCLES OF ADULT LIFE HISTORY STAGES

Once mature, most vertebrates express a series of LHSs, each with a unique set of substages. Because different combinations of substages can be expressed depending on local environmental conditions, they define state of the individual at that time (Wingfield and Jacobs, 1999; Jacobs and Wingfield, 2000). They may be altered further by, for example, changes in social status, body condition, disease, and so on. Some examples of LHSs are given in Figure 8.2. Each box represents a distinct LHS, and in species with more than two LHSs, the temporal sequence cannot be reversed. For example, in a migratory bird it is not possible to revert to vernal migration after the breeding season; the sequence must move on to the next stage – prebasic molt (Fig. 8.2). All LHSs must be expressed in the correct sequence before vernal migration is again attained. Each individual, then, is a finite state machine of characteristic LHSs that occur in a fixed sequence usually on a schedule determined by the changing seasons. Substages expressed within each LHS are determined more by local conditions, social status, and so on (Wingfield and Jacobs, 1999). A number of potential states can thus be manifest within each LHS. These not only represent the actual LHS and the combination of substages, but include factors in the extended phenotype such as territory quality, presence of a mate, or social status, within a group (Dawkins, 1982) as well as body condition, disease, pollution, and other factors. Therefore, state varies with time associated with adjustments in morphology, physiology, and behavior of the organism as it acclimates to fluctuations in its environment, as well as with "extended phenotypic factors." Changes in state are triggered by environmental cues and regulated by neuroendocrine and endocrine mechanisms that must occur in specific sequences and at different levels (Jacobs and Wingfield, 2000; Wingfield and Ramenofsky, 1999; Wingfield et al., 2000).

It is important to bear in mind that there are three phases in the expression of a LHS. Each has a development phase followed by a "mature capability" in which a number of substages can then be activated. The LHS is terminated at an appropriate time, although there may be varying degrees of overlap with other LHSs. Hormones play a major role in development of each LHS (and its termination) as well as activation of substages (Wingfield and Jacobs, 1999; Jacobs and Wingfield, 2000). Some species may have more LHSs than others and, within an individual, some LHSs may have a more complex set of substages than others. For example, a coral reef fish has only

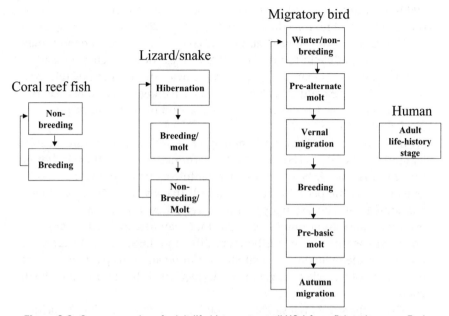

Figure 8.2: Some examples of adult life history stages (LHSs) from fish to humans. Each stage has a development phase, mature capability (when onset of the LHS can begin), and a termination phase. Each LHS also has a unique repertoire of substages (Wingfield and Jacobs, 1999). Note that when there are more than two life history stages, the temporal sequence through each LHS is one way – it cannot be reversed. Expression of a LHS can be suppressed (e.g., reproductively mature but not actually breeding), and there may be varying degrees of overlap of LHSs. Thus the life cycle of an organism is a finite state machine of LHSs and the substages expressed at any point. State is determined by the combination of LHSs and substages (Wingfield and Jacobs, 1999; Jacobs and Wingfield, 2000). Because of the development and termination phases of each LHS, there is a limit to how many can be expressed in an organism in 1 year. Furthermore, the more LHSs expressed, the less flexibility in timing of those stages. Humans appear to be unique in having only one LHS and thus have great flexibility in the life cycle.

two obvious LHSs, breeding and nonbreeding, whereas many temperate-zone lizards and snakes have three (Fig. 8.2). In contrast, a migratory bird has as many as six distinct LHSs (Wingfield and Jacobs, 1999; Jacobs and Wingfield, 2000). At the opposite end of the spectrum, humans have only one adult LHS (Fig. 8.2).

The boundaries of the temporal sequence of LHSs can be defined at many levels such as day, month, tidal fluctuation, length of wet-dry seasons, a year, or longer (Jacobs and Wingfield, 2000). Duration of each LHS can be measured directly from field observations. In migratory birds (Fig. 8.2), individuals may be in breeding condition for 1 to 3 months, whereas prebasic molt may be restricted to 1 month (Wingfield and Jacobs,

1999). In lizards and snakes, it may not be possible to separate molt from the breeding and nonbreeding stages at all (Fig. 8.2). The degree of overlap of LHSs varies as well. In general, some LHSs (such as winter and molt) may be more compatible in terms of overlap, whereas others such as migration and breeding may be mutually exclusive. Obviously, it is not possible for a bird to build a nest and incubate eggs while covering long distances on migration, but it may be possible for some fish and mammals to be pregnant during migration.

Because each LHS has a development phase, followed by mature capability (and onset of that LHS), and finally a termination phase, there is a lower time limit to each LHS. Given the natural history data from vertebrates in general, this time period is about a month (Jacobs and Wingfield, 2000), and some LHSs may last many months. Given these time constraints, there must be a maximum number of LHSs that may be expressed by any one phenotype within a year. Furthermore, the more LHSs that are expressed, the greater the reduction of flexibility in timing and thus plasticity of that phenotype's life cycle (Wingfield and Jacobs, 1999; Jacobs and Wingfield, 2000).

The simplest FSM is a single adult LHS, although this appears to be rare in vertebrates. For humans with one LHS, there will be maximum plasticity in the life cycle because we are not constrained by transitions among several LHSs. Virtually all other vertebrates have life cycles with much less plasticity. Evidence that humans have only one LHS is as follows: once adult, reproduction can be constant, we do not have a molt period, and we do not show regular migration patterns that exclude other functions. It is well known that seasonal patterns in human births occur for virtually every population worldwide, and they tend to be more pronounced in regions with summer heat or at extreme latitude. This suggests at first glance that there may be breeding and nonbreeding stages in humans. Demographic data from Louisiana indicate a 45% difference in high and low birth rates over the year. This annual shift has declined in recent years, however, possibly due to the advent of air-conditioning, suggesting the differences are adjustments of substages and not separate LHSs (Lam and Miron, 1994).

It is well known that women menstruate at all seasons, and although the length of the cycle may vary, this is not significant (Rosetta, 1993). Men's semen quality and testosterone levels also change, but only by $\pm 10\%$ over a year (Dabbs, 1990; Levine, 1994). These authors suggest a biological clock is involved that may be reset by photoperiod, but the evidence is weak. There is no doubt that humans can respond to day length to entrain circadian rhythms and other functions, but there appears to be little if any effect on reproduction (Ingram and Dauncey, 1993).

The evidence indicates that although temperature, humidity, nutrition, and other factors all influence human reproductive function, these effects may be on the mature capability phase (of reproduction) rather than separate breeding and nonbreeding LHSs. It has been suggested that because our human ancestors were tropical and only recently colonized temperate and polar regions, a large brain and social and communication skills allowed us to solve problems of hostile environment by developing clothing, houses, heating, food storage etc. We could then maintain tropical-like breeding patterns (largely aseasonal) and devoid of strong photoperiodism (Short, 1994). Humans thus have the unique ability to create their own microenvironment and maintain it anywhere, thus allowing us to colonize virtually all regions on earth (Farhi, 1987). It is through these largely artificial means that we are able to survive with a single LHS in extremely variable environments. This does not mean, however, that we are not vulnerable to unpredictable stressful events. Indeed, LPFs may temporarily interrupt reproductive function in human adults.

THE EMERGENCY LIFE HISTORY STAGE AND UNPREDICTABLE ENVIRONMENTAL EVENTS

The emergency life history stage (ELHS) can be expressed at any time in the life cycle. The LPFs that trigger an ELHS are diverse, but the substages within the ELHS are remarkably constant in all vertebrates studied to date. They serve to direct the individual into a survival mode and then allow it to return to the normal LHS once the LPF passes. Behavioral and physiological components that make up the substages of an ELHS are as follows:

1 "Leave-it" strategy – movements away from LPFs
2 "Take-it" strategy – switch to an alternate set of energy-conserving behavioral and physiological traits
3 "Take at first and then leave-it" strategy – switch to energy conserving mode first and then move away if conditions do not improve

Once a "strategy" has been adopted, mobilization of stored energy sources such as fat and protein to fuel movement away from the source of the LPF or to provide energy while sheltering in a refuge becomes critical. Finally, as the LPF passes or the individual has moved away, then it must settle in alternate habitat once an appropriate site is identified or return to the original site and resume the normal sequence of LHSs. These dramatic changes in behavior and physiology can occur within minutes or hours of exposure to an LPF and have been the subject of many experiments to determine the hormonal mechanisms underlying them (Wingfield and Ramenofsky, 1999; Wingfield and Kitaysky, 2002).

It is possible to look at many levels of LHSs and the ELHS down to regulation of single genes, and the states of all cells within the individual. This could generate an apparent infinite number of microstates for an organism, although many of these will be essentially identical and can be grouped into a finite number of states expressed by an underlying set of LHSs. Note that individuals within a state will not be completely identical, but FSM theory predicts that their interactions with the environment in similar social contexts and in relation to body condition and so on will be largely indistinguishable (Jacobs and Wingfield, 2000). We can now explore using FSM theory to construct a framework to assess neuroendocrine mechanisms by which environmental cues, internal and external, influence the state of an individual. The attractive feature of this approach is that it takes into account LHSs and substages, extended phenotypic factors and unpredictable components of environmental change (LPFs) as a whole. In other words, FSM theory has the potential to address all of the external and internal factors that impinge on an organism's life and make predictions on how that individual can make neuroendocrine and endocrine adjustments to maintain homeostasis in sometimes dramatically different conditions.

THE CONCEPT OF ALLOSTASIS AND ALLOSTATIC LOAD

Homeostasis in its purest form is difficult to reconcile with adaptation and acclimation of individuals to a changing environment and the cyclic nature of LHSs. Regulation over a wide range of physiological and developmental conditions is required (Bauman, 2000). This has been debated at length throughout the 20th century, and numerous new terms have been coined to refer to changing regulatory processes relating to homeostasis overall (Bauman, 2000). "Homeorhesis" is a term used to refer to regulatory mechanisms of nutrient partitioning and physiological processes in different stages of a LHS. For example, a pregnant or lactating mammal is in a different homeostatic state from an estrus female (Bauman, 2000). A similar concept involves the changes in regulation of state during the day (rheostasis), circadian rhythms and so on (Mrosovsky, 1990). At any one instant in an individual's life cycle (i.e., state in the FSM model), homeostatic processes are in effect, but they change predictably over time. Kuenzel et al. (1999) introduced the term "poikilostasis" to refer to changes in regulatory processes from one LHS to another (e.g., nonbreeding, migration, breeding, molt, etc.).

These terms suggest different levels of physiology, including appropriate adjustments of morphology and behavior. Basic homeostasis could be described as the fundamental mechanisms required to maintain life (Fig. 8.3, level A). For example, glucose transporters, Na/K ATPase must

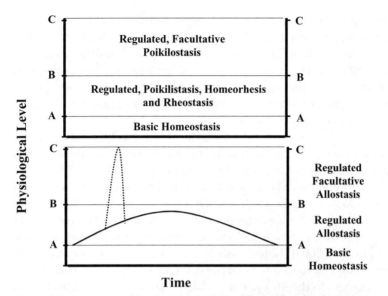

Figure 8.3: A summary of the three levels of physiological function. In the top panel, level A represents basic homeostatic mechanisms required for simple existence of an organism. As the organism goes about its daily routine or changes life history stage (LHS) on a seasonal basis, physiological function varies. These processes require change in homeostatic mechanisms as environmental needs and LHS requirements change. These occur on a predictable basis and can be regulated in anticipation of environmental change within level B. There have been several terms coined to describe these changes, including poikilostasis (Kuenzel et al., 1999), homeorhesis (Bauman, 2000), and rheostasis (Mrosovsky, 1990). Superimposed on this predictable component of the life cycle is the unpredictable. The latter includes storms, predator pressure, human disturbance, disease, injury, and other challenges. These have been termed "labile perturbation factors" (LPFs; Wingfield et al., 1998, 2000) and can occur at any time in the life cycle and during any LHS. In response to LPFs is a complex suite of physiological and behavioral events that allow the individual to adjust to the challenge. This could be termed "regulated, facultative poikilostasis" operating from levels B to level C. Following the concepts of Sterling and Eyer (1988), McEwen (1998), and Schulkin (2003), this complex terminology can be unified as allostasis – stability through change (lower panel). Basic homeostasis is still important for simple existence (level A), and regulated allostasis includes the predictable life cycle and changes in physiological function within level B. Unpredictable events in the environment trigger regulated, facultative allostasis from levels B to C. See text for further explanation.

be expressed at some basal level otherwise cells die. Basic homeostasis is required at all times. Beyond this level, homeostatic mechanisms must change with time of day or season (regulated rheostasis), with substage in a LHS (e.g., estrous, pregnancy, lactation; regulated homeorhesis), or with changing LHS (regulated poikilostasis). These combine to regulate homeostasis within the predictable life cycle (Fig. 8.3, level B). Superimposed on this predictable environment are transitory responses to LPFs. These trigger emergency adjustments of physiology and behavior to maintain homeostasis in the face of challenges. This is level C (Fig. 8.3) and represents the ELHS (Wingfield et al., 1998; Wingfield and Ramenofsky, 1999; Wingfield and Romero, 2000) that maximizes fitness in the face of an unexpected challenge. Because this involves a change in stage, we can call it regulated, facultative poikilostasis (Fig. 8.3).

The term "allostasis" was introduced by Sterling and Eyer (1988) to refer to changing regulatory systems (including setpoints). Allostasis can be considered to be the process of maximizing fitness (ensure viability) in the face of environmental change as well as unpredictable challenges (see also Schulkin, 2003). In other words, regulatory mechanisms must change to maintain or achieve a state appropriate for the time of day or year and in response to perturbations. It includes anticipatory as well as reactive processes (Schulkin, 2003). Here I wish to extend this idea to include physiological, morphological and behavioral responses by individuals to predictable changes in the life cycle (i.e., daily, tidal, annual rhythms, etc.) as well as unpredictable perturbations (Wingfield et al., 1998, 2000). Allostasis can be regarded as encompassing all the levels of function (lower panel of Fig. 8.3), including the unpredictable events that can trigger an ELHS. Regulated allostasis covers the changes in state to accommodate the changing environment in the predictable life cycle and regulated, facultative allostasis includes all the emergency reactions to unpredictable events (lower panel of Fig. 8.3). Once again, at any point in the predictable life cycle or when in an ELHS, homeostasis is maintained. The concept also comprises individual differences in response to environmental change and includes a highly coordinated suite of physiological and behavioral responses to maintain internal stability. These reactions may anticipate overload of regulatory capacities so that chronic stress can be avoided (McEwen and Stellar, 1993; McEwen, 1998; Sterling and Eyer, 1988; Schulkin, 2003; McEwen and Wingfield, 2003).

McEwen and Stellar (1993) and McEwen (1998) developed the concept further and coined the term "allostatic load" to describe potential permanent taxing of homeostatic processes in humans by disease, deleterious lifestyle, and so on. Given the levels of allostasis outlined in Figure 8.3, it is possible to extend this idea further so that overall allostatic load = basic homeostasis + regulated homeorhesis + regulated rheostasis + regulated poikilostasis + facultative poikilostasis (Fig. 8.4). Allostasis is then the process of adjusting morphology, physiology, and behavior to cope with allostatic load, and allostatic state is the hormonal state required to regulate that process (McEwen, 1998; Schulkin, 2003; McEwen and Wingfield,

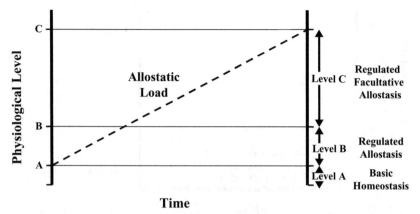

Figure 8.4: The concept of allostasis and levels A–C of physiological function extended to allostatic load (cf. McEwen, 1998; McEwen and Wingfield, 2003). Allostatic load increases as a function of daily and seasonal needs (level B) and of responses to unpredictable events (level C). Factors such as disease, injuries that result in permanent disability, habitat change due to human disturbance, and so on can result in permanent cumulative allostatic load in addition to those accrued through levels A–C.

2003; Goldstein, this volume). One can imagine allostatic load increasing as a function of rising energetic costs of fueling regulatory processes in the face of environmental changes and challenges from unpredictable sources (Fig. 8.4). Information about habitat locations, how to acquire resources, and at what time of day a particular nutrient might be available is also important (Schulkin, 2003; McEwen and Wingfield, 2003) to maximize fitness, especially in the face of a LPF or when arriving from migration, dispersal, or a facultative movement away from a perturbation. Behavioral flexibility, a suite of physiological responses, and appropriate morphological changes are essential to allostatic regulation (Schulkin, 2003; McEwen and Wingfield, 2003).

Energetic and social demands on the individual can be further exacerbated by injury resulting in permanent disability, disease, or habitat degradation that could provide cumulative allostatic load (McEwen, 1998) before all of the other energetic costs associated with the life cycle are added on. Such a cumulative, permanent allostatic load would be particularly debilitating for an organism with many LHSs and little flexibility in control of timing of the life cycle (McEwen and Wingfield, 2003).

Implications of Allostatic Load for Hormone Control Mechanisms

Given the three levels of physiological and behavioral functions in Figures 8.3 and 8.4, it is possible to extrapolate to levels of endocrine secretion and

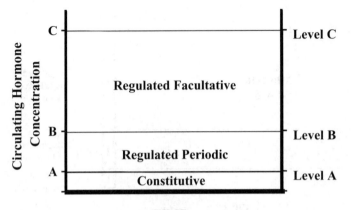

Figure 8.5: The concept of levels A–C can be extended to levels of hormone secretion that actually regulate different physiological levels in basic homeostasis and regulated and facultative allostasis. At level A (hormone levels regulating basic homeostasis), secretion may be constitutive. Some regulation may be possible, however. At level B, hormone secretion is regulated according to needs as a function of endogenous rhythms (e.g., circadian, episodic) or in response to environmental signals transduced through neural and neuroendocrine pathways. From level B to C, regulated facultative secretion is controlled exclusively by environmental signals. Depending on environmental and social contexts, different hormones may operate at each level. In many cases, the same hormone may act at two or at all three levels. The latter scenario raises important issues in terms of mechanisms of action of a hormone at more than one level. Examples are given in the text.

control on same scheme (Fig. 8.5, Wingfield et al., 1997a, 2001). This is appropriate because the three levels can be distinguished throughout the animal's life cycle and thus are likely to be regulated differently. Using the same scheme as in Figures 8.3 and 8.4, the following levels are proposed (Fig. 8.5):

Level A – Absolute baseline of hormone secretion to maintain basic functions critical for cell survival and for negative feedback signals. This probably involves constitutive secretion of the appropriate hormones although some regulation may also be in effect.

Level B – Regulated changes in baseline levels of hormones in relation to predictable changes in environment (diel and circadian rhythms, seasonal and tidal changes, etc.). These probably involve periodic secretion of hormones regulated by endogenous rhythms or environmental factors perceived by the individual and transduced into neuroendocrine secretions that triggers cascades of responses.

Level C – Facultative surges in hormone secretion to deal with unpredictable events (LPFs) in the environment. These are always regulated by environmental signals and include facultative hormone responses to social interactions, stress responses, and so on.

Initiation of hormone responses to environmental change occurs in the brain as environmental signals (external, social, and internal) are transduced into neuroendocrine secretion (signalized pathway, or directly in response to environmental stimuli, such as hormones of the gastrointestinal tract and many others in ectotherms) (see also de Wilde, 1978). Hormonal cascades follow that customize the response to the particular context and LHS. In many cases, various hormones may act at levels A–C depending on context; in other cases, the same hormone may have actions at two or all three levels (Wingfield et al., 1990, 1997a, 1997b). For example, interactions of glucocorticosteroids, insulin, glucagon, and other hormones regulate glucose levels through much of the life cycle. In response to LPFs, however, epinephrine and glucocorticosteroids are important (e.g., Gorbman et al., 1983; Norris, 1997). In other examples, such as the breeding LHS of birds, testosterone is active at level A because a nonbreeding baseline is essential to maintain negative feedback on gonadotropin and gonadotropin-releasing hormone secretion. Level B is the breeding LHS baseline of testosterone required for male reproductive morphology, physiology, and behavior. Transitory surges above level B to level C induced by aggressive challenges from other males, or the sexual behavior of females provide facultative behavioral and physiological responses to social demands (Wingfield et al., 1990, 2001). Another example is given in Figure 8.6. In passerine birds, an absolute baseline of circulating corticosteroid levels is required (level A) because adrenalectomized individuals will die (e.g., Gorbman et al., 1983; Norris, 1999). However, glucocorticosteroid levels do change with time of day and season within the range of level B. On the other hand, LPFs result in transitory surges of glucocorticosteroids to level C, often triggering an ELHS (Fig. 8.6, Wingfield and Romero, 2000).

An important concept emerging from this framework is the need for tissues and cells to respond appropriately to various levels of the same hormone. Multiple receptor types are obvious ways by which this could be achieved. In the case of glucocorticosteroids, it is well known that there are two genomic receptor types with different binding affinities (e.g., McEwen, 1988; Sapolsky et al., 2000), and a putative fast-acting membrane receptor (Breuner et al., 1998; Orchinik et al., 2002). The concept of multiple receptor types is, of course, not new, but linking them as mechanisms for different responses to changing environments as a function of allostatic load is important.

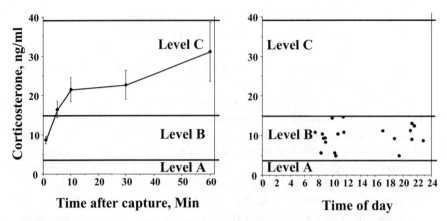

Figure 8.6: An example at levels A–C secretion of a hormone, corticosterone, in a free-living bird, the snow bunting, *Plectrophenax nivalis* (from Wingfield et al., 1994). Level A represents the baseline circulating level of corticosterone required for basic homeostasis. Corticosterone levels increase following capture, handling, and restraint of the bird (a potent stress that results in increased glucocorticosteroid secretion in virtually all vertebrates studied to date) to maximum levels at about 30–60 minutes after capture. Plasma levels of corticosterone at capture represent prestress levels and can be plotted as a function of time of day (level B, right-hand panel). Stress-induced high levels of corticosterone represent level C (left-hand panel). Further examples are given in Wingfield et al., 1997a.

A MODEL FOR LIFE CYCLES, ENDOCRINE REGULATION, AND ALLOSTATIC LOAD

Allostatic mechanisms drive the individual into a state appropriate to environmental context, challenges from social interactions and other factors. At the facultative level (level C, Figs. 8.4–8.6), these states are high energy, short term, and potentially deleterious in the long term, but they are necessary for viability (Schulkin, 2003). How, then, can we think of allostatic load and the responses of individuals to that load as a function of environmental conditions and LHSs? Some simple energy models have been proposed to explain these issues in a common framework. Energy (E) is assumed here to be the sum of all the requirements (including protein, essential fatty acids, vitamins, minerals, and so on) of an individual in any given state (Wingfield et al., 1998; Wingfield and Ramenofsky, 1999; McEwen and Wingfield, 2003). Of course, it is possible to model many other nutritional requirements, but for simplicity this discussion will assume E as all encompassing. Using Figure 8.7 as a beginning, the following energy model definitions are proposed:

EE = Existence energy required for basic homeostasis (equivalent to level A, Figs. 8.4–8.6)

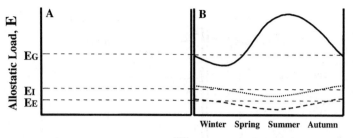

Figure 8.7: A framework for modeling energetic requirements (E) of organisms during their life cycle. This energy requirement, E, includes all potential nutritional requirements including energy per se. They are grouped together here for convenience, although essential components of nutrition could also be modeled. EE represents the energy required for basic homeostasis (i.e., level A). EI represents the extra energy required for the individual to go out, find, process, and assimilate food under ideal conditions (i.e., level B). EG represents the amount of energy (in food) available in the environment (from Wingfield et al., 1998; Wingfield and Ramenofsky, 1999; McEwen and Wingfield, 2003). In panel A, these requirements are represented as straight lines. In B, the changes in energy levels have been adjusted to represent probable changes in relation to seasons. EG would be expected to rise dramatically in spring and summer and then decline through autumn and winter, when primary productivity is low. EE would be lowest in summer when ambient temperatures are highest. EI should be fairly constant (under ideal conditions) and varies in parallel with EE. Bearing in mind that energy levels will vary in potentially complex ways, EE and EI will be held as straight lines for simplicity.

EI = Energy required to obtain food and process and assimilate it under ideal conditions (i.e., regulated allostasis, equivalent to level B, Figs. 8.4–8.6)

EG = Energy to be gained from food in the environment (i.e., what nutritional resources are out there for the individual to go out and procure)

Figure 8.7A shows EE, EI, and EG as straight lines. In reality, all will fluctuate with time, particularly seasons (Fig. 8.7B). For example, food supplies (EG) will increase in spring as a direct function of primary productivity and decline through autumn and winter. Similarly, EE will be lower in summer when ambient temperatures are high and, conversely, higher in winter when temperatures decrease. EI will also fluctuate, probably in parallel with EE (Fig. 8.7B). Although these changes are important and can be modeled as part of the unique experience an individual gains throughout its life cycle, EE and EI will be modeled as straight lines for simplicity.

Effects of Labile Perturbation Factors on Allostatic Load

Ideal conditions under which EE + EI remain less than EG rarely exist for extended periods. Physical and social events in the environment increase the energy required to forage, process, and assimilate food. It is possible to

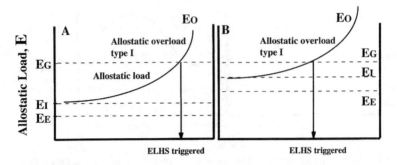

Labile Perturbation Factors (snow, wind speed, rain, temperature, predator pressure, human disturbance, etc.)

Figure 8.8: The effects of labile perturbation factors (LPFs) such as storms, predators, and human disturbance on energy levels in free-living organisms. In this case, EG is held constant, but as environmental conditions deteriorate, the energy required to go out, obtain food, and process and assimilate it increases, and a new line, EO, can be inserted. Allostatic load (Type 1) increases as a function of EE + EI + EO. As the LPF persists (panel A), EO increases until EE + EI + EO > EG. At this point, the individual goes into negative energy balance (allostatic overload), and secretion of glucocorticosteroids increases dramatically, thus driving the transition to an emergency life history stage (ELHS; see Wingfield et al., 1998, 2000). EE can be increased because of low ambient temperature, changing basic homeostasis, or because cumulative allostatic load following infection, injury resulting in permanent disability, and so on (panel B). In this case, the same LPF would result in a similar rise in EO, but allostatic overload triggers an ELHS more quickly. In this way, the energy model can explain individual differences in coping with the environment but keeping the potential mechanisms on a common framework. Note also that when in negative energy balance, the triggering of an ELHS is designed to reduce allostatic overload (EE + EI + EO) so that the individual can regain positive energy balance and survive the LPF in the best condition possible. If allostatic overload is so high that even an ELHS cannot reduce it sufficiently to gain positive energy balance, chronic stress will ensue.

represent this by another line, EO, which is the additional energy required to obtain food under nonideal conditions such as a storm, predator pressure, human disturbance, and so on (Fig. 8.8). This would be regulated and facultative allostasis (equivalent to the transition from level B to level C, Figs. 8.4–8.6). As long as EE + EI + EO < EG, then the individual has the resources to fuel allostasis. If EE + EI + EO > EG, then negative energy balance results in allostatic overload type I, a transition to level C (McEwen and Wingfield, 2003). This state cannot be sustained for long, even if endogenous stores of fat and protein are mobilized to supplement EG. It is at this point that an ELHS must be triggered to reduce allostatic load to a level that can be sustained by available resources (Wingfield and Ramenofsky, 1999; Wingfield and Romero, 2000; McEwen and Wingfield, 2003). If EO remains below EG, but is a permanent contribution to allostatic load, such as injury, disease and low social status, then this is allostatic overload type II. Glucocorticoids also tend to be elevated but may not trigger an ELHS (McEwen and Wingfield, 2003).

Allostatic load increases as a function of EO and would be accompanied by neuroendocrine and endocrine secretions that would allow allostatic processes to operate. For example, humoral signals occur in proportion to body energy stores and provide a negative feedback to brain areas that control food intake and energy expenditure. Insulin and leptin appear to be the mediators of this signal (Schwartz, 2001), and glucocorticoids and epinephrine also interact at many levels (Dallman and Bhatnagar, 2000). In cases of allostatic overload, neuroendocrine secretions act in synergy with glucocorticosteroids to orchestrate the ELHS in response to challenges from the environment (discussed later).

An example of EO (includes labile perturbation factors) is the effect of a storm or global climate change that can have dramatic effects on food supply or increase allostatic load as organisms need to search further afield for food and may have to compete more intensely to maintain food intake. Winter snowstorms that bury food supplies (i.e., increase EO so that EG was exceeded, Fig. 8.8a) for free-living dark-eyed juncos, *Junco hyemalis*, in eastern North America, resulted in abandonment of the home range. The allostatic overload type I triggered an ELHS that was accompanied by high circulating levels of corticosterone (Rogers et al., 1993; Wingfield and Ramenofsky, 1997, 1999). Very severe oceanic storms near the Antarctic convergence of the South Georgia Islands increased EO for large numbers of common diving petrels, *Pelecanoides urinatrix*, foraging at sea in this area. As EE + EI + EO increased and exceeded EG (allostatic overload type I), this triggered an ELHS. The petrels were captured and sampled as they flew toward their breeding islands to shelter. They had high baseline levels of corticosterone that did not increase further with capture and handling, suggesting that they were indeed stressed by the storm (Smith et al., 1994). Such examples are consistent with the models outlined in Figure 8.8.

It is also important to note that EE + EI can be higher if environmental conditions are more severe (see also Fig. 8.7). In winter, lower environmental changes increase EE, and because EI is the same the sum of the two will be higher than in summer (i.e., higher allostatic load, Fig. 8.8B). Thus, even if EG does not change, the same perturbation (EO) can exceed EG more quickly resulting in allostatic overload (Figs. 8.8A and 8.8B). Similarly, individuals experiencing allostatic overload type II are more susceptible to LPFs and reach allostatic overload type I more quickly (McEwen and Wingfield, 2003).

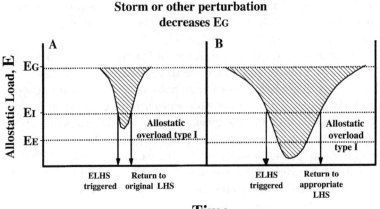

Figure 8.9: The effects of labile perturbation factors (LPFs) such as storms, predators, and human disturbance on energy levels in free-living organisms can be modeled in different ways. In this scenario, EE and EI remain the same, but the LPFs reduce EG. In panel A, a LPF reduces EG (shaded area), it briefly drops and below the sum of EE and EI. Thus allostatic overload type I can result because EG decreases and not because allostatic load per se increases. An animal may be able to withstand such a brief period of negative energy balance, or it may have to trigger an emergency life history stage (ELHS). As EG increases again following the passing of the LPF, the individual will return to the original life history stage (LHS). In panel B, a more severe and long-lasting LPF is modeled (shaded area). Here an ELHS will be maintained longer; once the LPF passes, EG increases and the individual will return to the original LHS, or perhaps the next LHS in the temporal sequence. As in Figure 8.8, the ELHS allows the individual to move away and find better resources (EG) or shelter in an attempt to reduce EE + EI. In both cases, an attempt is made to reduce allostatic overload type I and regain positive energy balance.

In some environmental scenarios, a storm or other LPF may decrease EG without EO being applied (Fig. 8.9). In these cases allostatic load may not change during the perturbation, but as EG decreases (shaded areas of Fig. 8.9) and crosses EI so that EE + EI < EG, then allostatic overload type I occurs and an ELHS is triggered. The time in an ELHS varies as a function of severity of the LPF. Some last longer and are more intense than others (Figs. 8.9A and 8.9B). These types of LPFs can be extremely severe because the only way to reduce allostatic load is to reduce EE + EI. This may require torpor (hibernation-like strategy to reduce EE drastically), or the individual must leave the area to find a habitat with higher EG. An example of the former is the male rufous hummingbird, *Selasphorus rufus*, that enters torpor at night when EG declines and the individual is unable to store enough fat to survive the night at normal body temperature. Onset of torpor was accompanied by elevated urinary corticosterone (Hiebert et al., 2000a, 2000b).

The worst-case scenario would be when a LPF results in a simultaneous decrease of EG and an increase in EO, that may result in rapid allostatic overload type I. Examples exist in nature; the effects of oceanic storms on seabird populations may be one example. The diving petrels mentioned earlier have to forage more intensively because the storm disturbs the ocean surface so that their food, euphausids, move down in the water column and may be out of reach (decreased EG). Thus these birds must forage more intensively (dive deeper) while at the same time they are exposed to lower temperatures and high winds (increase EE + EI + EO). Under such conditions, allostatic overload type I may result quickly; plasma levels of glucocorticosteroids increase rapidly, triggering the ELHS (Smith et al., 1994). This model is flexible and, using the same framework, can explain many scenarios characteristic of the life cycle of individuals.

The Effects of the Reproductive Life History Stage on Allostatic Load
Different LHSs can introduce varying degrees of allostatic load. Two of the most energetically expensive LHSs are migration and reproduction. For the latter we can introduce another term, EY, that represents the additional energy required to obtain food to raise young to independence (Fig. 8.10).

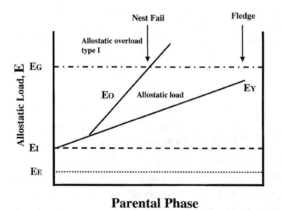

Parental Phase

Figure 8.10: Using the energy models of Figures 8.8 and 8.9, it is possible to add further lines contributing to allostatic load. In this case, EY represents increasing energetic requirements to breed successfully. In the case of birds, this would be the costs of incubating eggs and then feeding the young until they reach independence (or at least fledging state), increasing allostatic load. The number of young raised is adjusted so that the cost of rearing them (allostatic load) always remains below EG. If a labile perturbation factor (LPF) were to increase these costs, however, and thus allostatic load (EO), then EE + EI + EY + EO > EG. At this point, allostatic overload type I sets in, and the reproductive effort would fail (young would die). Further reproduction would be held in abeyance until the LPF passed.

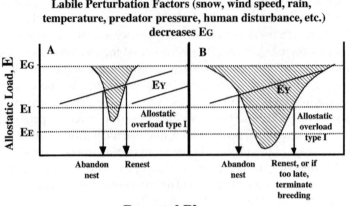

Labile Perturbation Factors (snow, wind speed, rain, temperature, predator pressure, human disturbance, etc.)
decreases EG

Parental Phase

Figure 8.11: Here we use the energy models of Figures 8.8 and 8.9 again to add further lines contributing to allostatic load, this time modeling the effects of a labile perturbation factor (LPF) on EG. As in Figure 8.10, EY represents increasing energetic requirements (allostatic load) to breed successfully. In the case of birds, this would be the costs of incubating eggs and then feeding the young until they reach independence (or at least fledging state). The number of young raised is adjusted so that the cost of rearing them always remains below EG. If a LPF were to decrease EG (panel A) so that EG < EE + EI + EY, however allostatic overload type I sets in, the nest would be abandoned, and reproductive function would be inhibited until the LPF passes. By eliminating EY, allostatic load is reduced to levels sustainable by EG. When the LPF passes and EG > EE + EI + EY, then a renest attempt will follow, and EY will rise as before. If the LPF is more severe and prolonged (panel B), then EG may not rise until it is too late to renest and the breeding season is nearly ended. In this case, the individual would make an early transition into the next LHS.

When breeding, allostatic load could be expressed as EE + EI + EY. Even if EE + EI remain the same, EY increases as the young grow and require more food to be supplied by parents. As long as EE + EI + EY remain below EG, then breeding can progress to successful completion. Many organisms regulate the number of young they produce so that EY (allostatic load due to breeding) does not exceed EG (e.g., Lack, 1968). Breeding animals are, however, vulnerable to perturbations because allostatic load is already increased. Thus, in response to a LPF, EO could increase allostatic load dramatically. If EG > EE + EI + EO + EY, then breeding continues. If EG < EE + EI + EO + EY, allostatic overload type I sets in, and breeding is interrupted (Fig. 8.10). This will reduce allostatic load (because EY is eliminated), and the individual can return to an energy-positive state. However, if EO continued to increase and remain in excess of EG, then repeated allostatic overload type I would trigger an ELHS.

Alternatively, one can model a LPF when in the breeding LHS with a decrease in EG (Fig. 8.11). As EG decreases and becomes less than EE +

EI + EY, then the breeding attempt is abandoned. As the storm passes and EG rises again to exceed EE + EI, then a second breeding attempt can begin (Fig. 8.11A). If the storm depresses EG for longer, then a second breeding attempt (renest in birds) may not be possible, and the individual makes an early transition into the next LHS (Fig. 8.11B). These scenarios are particularly severe because EG can drop so low that an ELHS is triggered (Fig. 8.11).

There are several examples of the effects of LPFs on the breeding LHS of birds. Severe storms during the breeding season of white-crowned sparrows, *Zonotrichia leucophrys pugetensis,* resulted in abandonment of the breeding territory (including young in the nest) and formation of winter-like flocks. This type of ELHS was accompanied by high circulating levels of corticosertone. After the storms passed, corticosterone levels declined, and the birds returned to their territories and renested (Wingfield et al., 1983). These data are consistent with allostatic overload type I in Figures 8.10 and 8.11. Note also that the ELHS (abandonment of breeding) reduced allostatic load so that EG remained greater than EE + EI + EO (EY was eliminated). Chronic stress was thus avoided, and the reproductive system remained in a near functional state so that when the storm passed, renesting could begin immediately. If the animals had become chronically stressed (by remaining in allostatic overload), then the reproductive system would have been suppressed probably delaying renesting until recrudescence occurred. By this time the breeding season would probably be over and reproductive success zero. Thus, by triggering an ELHS and inhibiting reproductive behavior but not the reproductive system, individuals retain the ability to renest, thus maximizing reproductive success (Wingfield et al., 1983, 1998, 2000).

Neuroendocrine and Endocrine Bases of the Energy Model Framework and Allostatic Load

As allostatic load increases energetic costs through levels A–C, then combinations of neuroendocrine and endocrine secretions are regulated according to context. For example, in situations that require osmotic adjustments, one would expect appropriate changes in arginine vasopressin (AVP) or ariginine vasotocin (AVT) in non-mammalian tetrapods, aldosterone, atrial natriuretic factor, and so on. If in a breeding LHS, then changes in gonadotropin-releasing factor and associated neuromodulators would regulate the hypothalamo-pituitary gonad axis (see Wingfield et al., 1990, for discussion in relation levels A–C). In the winter LHS, a small bird may show regular changes in thyroid hormones epinephrine and glucocorticosteroids to maintain body temperature. Despite the many different contexts owing to LHS (social status, disease, human disturbance etc.), there is one common response – a gradual increase in glucocorticosteroid secretion resulting from activation of the hypothalamic-pituitary-adrenal (HPA) axis. In other words, glucocorticosteroid secretion increases as a function of allostatic load (McEwen and Wingfield, 2003). Changes in glucocorticosteroid levels within level B probably trigger changes in metabolism and behavior associated with the predictable life cycle or when resources (EG) are sufficient to fuel allostatic load. If allostatic load exceeds EG (overload type I, energy output exceeds input), then glucocorticosteroid levels increase further above level B to level C and an ELHS is triggered.

Glucocorticosteroids are elevated under conditions in which hunger (or at least negative energy balance) is a central state (Dallman and Bhatnagar, 2000; Schulkin, 2003). Triggering an ELHS acts to reduce allostatic load once it gets beyond what local resources and stored energy can sustain. The ELHS directs the individual away from the current LHS thus reducing allostatic load to a level manageable with the current resources available. In other words, by relieving allostatic overload type I, all energy can be channeled into survival. That is, an ELHS avoids stress. If the ELHS does not allow the individual to reduce allostatic load, especially for individuals in which cumulative allostatic load (disease etc.) is high (allostatic overload type II), then chronic stress can result (McEwen and Wingfield, 2003).

Important additional components of this process are increases in sympathetic nervous system activity and the actions of other hormones in HPA axis to promote adaptation (McEwen and Stellar, 1993; McEwen, 1998) to adjust internal stability as internal and external environmental conditions change. Secretion of glucocorticosteroids has complex effects on neuropeptide systems depending on the ecological context and biological needs of the individual (Sapolsky et al., 2000; Wingfield and Romero, 2000; Schulkin, 2003). These interrelationships play an essential role in the expression of motivational states and provide a suite of "feed-forward" allostatic mechanisms in the regulation of life cycles (Schulkin, 2003). Steroids, by facilitating neuropeptides, neurotransmitters, or receptor sites, can influence state of the individual in relation to functional requirements. These states are dependent on environment, and other cognitive and physiological events (Schulkin, 2003). Glucocorticosteroids have a diverse array of effects on many regulatory systems in the brain and can be described in terms of long- and short-term effects, permissive, suppressive, stimulatory, or preparative (see Sapolsky et al., 2000, for a useful classification of these complex effects). These all may have environmental contexts, especially in terms of the predictable life cycle and unpredictable perturbations (Wingfield and Romero, 2000).

Motivational states represent readiness to behave appropriately in suitable environments or in response to unpredictable environmental events.

Glucocorticosteroids can facilitate these motivational behaviors (preparatory role of Sapolsky et al., 2000; see also Wingfield and Romero, 2000, for comparative aspects). Because animals anticipate environmental events and adjust their morphology physiology and behavior accordingly (Jacobs and Wingfield, 2000), hormonal and physiological mechanisms maintain stability and regulate energy expenditure to match resources available to the individual. Glucocorticosteroids and their interactions with brain peptides and other hormones maintain internal stability in changing environments with regard to nutrient availability and energy homeostasis (see Dallman, 2000; Schulkin, 2003). Additionally, the individual must deal with unpredictable events in the environment, such as LPFs. These can add dramatically to allostatic load, resulting in overload type I that requires immediate remediation. Schulkin (2003) identified three subtypes of allostatic overload type II: (1) overstimulation by frequent LPFs resulting in excessive stress hormone exposure (i.e., early responses to stress – the ELHS designed to reduce allostatic load); (2) failure to inhibit allostatic responses when they are not needed or an inability to habituate (get away from) the LPF resulting in overexposure to stress hormones; (3) an inability to stimulate allostatic responses when needed, in which case other systems (e.g., immune system) may become hyperactive and produce other types of wear and tear (McEwen, 1998; Schulkin, 2003). In free-living animals, the ELHS reduces allostatic overload type I and allows the individual to regain positive energy balance. Failure to do so results in chronic stress and death. However, in human society and also in agricultural animals, chronic stress often persists, resulting in the well-known pathological effects.

Interactions of Corticosterone and Other Hormones in the Orchestration of the Emergency Life History Stage

Once allostatic overload type I has set in, the ELHS is a mechanism to redirect the individual away from the normal LHS into a survival mode. This is intended to reduce allostatic load to a level that EG can sustain. There are many other hormonal responses to stress in addition to glucocorticosteroids, and this has been reviewed several times in relation to comparative aspects of stress biology (Wingfield, 1994; Wingfield et al., 1998, 2000; Wingfield and Romero, 2000). Here the focus is on neuroendocrine aspects of corticosterone actions in relation to components of the ELHS (see also Wingfield and Ramenofsky, 1999; Wingfield and Kitaysky, 2002).

Suppression of Reproductive and Territorial Behavior

Subcutaneous implants of corticosterone reduced territorial aggression in free-living male song sparrows, *Melospiza melodia,* compared to controls (Wingfield and Silverin, 1986), consistent with abandonment of a breeding territory or winter range. In side-blotched lizards, *Uta stansburiana*, similar corticosterone treatment reduced home range (DeNardo and Sinervo, 1994a), despite simultaneous implants of testosterone. Corticosterone may thus act directly on territoriality and not by inhibition of testosterone secretion (DeNardo and Licht, 1993; DeNardo and Sinervo, 1994b). High corticosterone levels also suppressed sexual and parental behavior in ducks (*Anas platyrhychos*) (Deviche et al., 1979; Deviche and Delius, 1981). In breeding pied flycatchers, *Ficedula hypoleuca*, corticosterone treatment decreased parental behavior in both sexes. They fed young less frequently, their nestlings weighed less, and fewer fledged than in control birds. Higher doses of corticosterone resulted in abandonment of nests and territories. Reproductive success was zero for these birds (Silverin, 1986). In the rough-skinned newt, *Taricha granulosa*, sexual behavior such as male clasping was inhibited by corticosterone injection (Moore and Miller, 1984). This response was within 8 minutes, suggesting a nongenomic action of glucocorticosteroids (Orchinik et al., 1991).

Intracerebro ventricular injection of both corticotropin-releasing factor (CRF) and beta-endorphin, suppressed estrogen-induced courtship displays in female white-crowned sparrows (Maney and Wingfield, 1998a) possibly involving interactions with AVT and Chicken GnRH-I and II (Maney et al., 1997a, 1997b). Furthermore, injections of CRF to lateral brain ventricles of male white-crowned sparrows in the field inhibited territorial aggression. AVT was without effect (Romero et al., 1998). How peptides of the HPA axis interact with corticosterone secreted peripherally remains largely unknown.

Increase Foraging Behavior
Elevated circulating levels of corticosteroids are widely known to facilitate food intake (Tempel and Liebowitz, 1989), circadian activity patterns (Iuvone and van Hartesveldt, 1977; Micco et al., 1980), exploratory behavior (Veldhuis et al., 1982), avoidance behavior (Moyer and Leshner, 1976), and learning retention (Bohus and de Kloet, 1981). In white-crowned sparrows and song sparrows corticosterone treatment may play a permissive role in facilitation of foraging behavior (Gray et al., 1990; Astheimer et al., 1992). Corticosterone in males also decreases growth rate, results in involution of the bursa of Fabricius, increases liver weight, and decreases body weight in chickens (Davidson et al., 1993, see also Wingfield and Romero, 2000).

Other centrally acting peptides such as neuropeptide Y (NPY) and beta-endorphin increases food intake in reproductively active and winter white-crowned sparrows (Richardson et al., 1995; Maney and Wingfield, 1998b). Conversely, central injections of CRF inhibits it (Richardson et al., 2000). How these peptides interact with other hormones of the HPA axis remains unknown.

Promote Escape (Irruptive) Behavior, or Find a Refuge

Facultative movements in response to perturbation factors vary according to species. Some seek shelter, whereas others may leave. In birds (Ketterson and King, 1977; Steube and Ketterson, 1982; Astheimer et al., 1992; Freeman et al., 1980), and rats (File and Day, 1972; Moorecroft, 1981) food restriction and/or corticosterone treatment increases locomotor activity. In fish (Robinson and Pitcher, 1989) and humans (Grande, 1964) the same treatment results in reduced activity levels. Perch-hopping activity, a measure of facultative movements, decreases in response to corticosterone implants if food resources are ad libitum, that is, behavior consistent with the "take it" strategy (Astheimer et al., 1992; Wingfield and Ramenofsky, 1997, 1999). However, if food was removed to simulate a perturbation, perch-hopping activity and "escape-behavior" increased compared with control birds. These behaviors are more consistent with the "leave it" strategy. Corticotropin also may be involved (Wingfield and Romero, 2000).

Increase Gluconeogenesis

Glucocorticosteroids have well established actions to decrease hepatic glucose output, increase fat and enhance protein degradation to produce glucose, and elevate gluconeogenic enzyme activity in mammals and other vertebrates (Chester-Jones et al., 1972), as well as potentiate the action of catecholamines and glucagon on glucose formation (McMahon et al., 1988). More recently it appears that glucocorticoids act on substrate delivery for hepatic gluconeogenesis (Goldstein et al., 1993) and maintaining hepatic glycogen availability in conjunction with catecholamines and glucagon (Fujiwara et al., 1996).

Also in birds corticosterone treatment increases plasma levels of glucose, triglyceride, protein, and uric acid (Davidson et al., 1993). Furthermore, corticosterone treatment may elevate free fatty acids and triglyceride formation because many migratory birds utilize fatty acids and glycerol rather then glucose as fuel for flight (Gray et al., 1990). Corticosterone treatment of birds resulted in atrophy of the flight muscles and a large increase in subcutaneous fat depots, suggesting that protein had been broken down for fat formation. Lipoprotein lipase activity in adipose tissue and plasma levels of glycerol and free fatty acid did not change but lipoprotein lipase activity was greater suggesting that although muscle proteins were mobilized, this was not at the expense of enzymes that regulate lipid uptake (Gray et al.,

1990; Wingfield and Silverin, 1986; Davidson et al., 1983; Le Ninan et al., 1988, see Wingfield and Romero, 2000 for review).

Promote Night Restfulness

Corticosterone treatment of great tits, *Parus major,* and black-headed gulls, *Larus ridibundus,* tends to increase oxygen consumption during the day (Hissa and Palokangas, 1970; Palokangas and Hissa, 1971), decreases nocturnal energy expenditure in chickens (Mitchell et al., 1986), but has no effect on diurnal metabolic rate in the pigeon, *Columba livia* (Hissa et al., 1980). To determine if corticosterone might also increase metabolic rate associated with facultative physiological and behavioral responses to unpredictable events, white-crowned sparrows, and pine siskins, *Spinus pinus,* given subcutaneous implants of corticosterone showed uniformly low oxygen consumption throughout the observation period at night compared with episodic increases in oxygen consumption in controls (Buttemer et al., 1991). The result was an overall energy savings of about 20% overnight an opposite result to effects of corticosterone during the day.

Glucocorticosteroids reduce time spent in rapid eye movement (REM) sleep in mammals with a corresponding increase in slow-wave sleep time (reviewed in McEwen et al., 1993), although it remains to be seen whether "energy savings" result. More research is needed in this area and it would be particularly informative to compare effects in species that seek a refuge rather than emigrate.

Promote Recovery on Return to the Normal Life History Stage

Individuals return to the normal life history stage after a perturbation factor passes (Jacobs, 1996), and a recovery period follows in which energy reserves are replenished. Estrous cycling or renesting will resume (e.g., Wingfield and Moore, 1987). Hormonal involvement in recovery from an ELHS is likely but remains largely unknown, although the role glucocorticosteroids play in recovery from an emergency requiring activation of the immune system is well established (e.g., Munck et al., 1984).

Corticosterone treatment in male song sparrows elevates foraging behavior over control birds (Astheimer et al., 1992), especially after 24 hours food deprivation. Corticosterone-implanted birds dramatically increase rates of refeeding compared with control birds during the recovery phase. These data suggest that glucocorticosteroids may facilitate recovery from acute stress (see Wingfield and Romero, 2000 for review).

APPLICATION OF THE ALLOSTASIS CONCEPT AND ENERGY MODELS AT INDIVIDUAL LEVEL

Having summarized the general concepts of allostasis, allostatic load, and overload in relation to animals in their natural habitat and the neuroendocrine and endocrine mechanisms that underlie adaptation, we can now explore other ways in which the models can provide a framework at population and individuals levels. It is possible to introduce further terms to characterize allostatic load as a function of LHS or other factors such as parasite load. For example, EMi could be modeled for additional energy required to migrate, or EMo for the additional energy required to molt. All could be included into the same equation to describe allostatic load. EE, EI, and EO can include a host of potential events that could increase allostatic load. Examples are social status (subordinates do not have the same access to food and other resources as dominants), parasite load and disease in general, bad weather, predator pressure, human disturbance, endocrine disruption from pollutants, and so on. Some of these will be addressed next.

There have been several field studies of hormonal responses to LPFs that can be modeled by Figures 8.7 through 8.11. A number of LPFs have been identified including obvious ones such as storms, low temperature, and so on. Others may not be so obvious. For example, increased predator pressure can have marked effects on glucocorticosteroid secretion. In tropical stonechats, *Saxicola torquata axillaris,* breeding in East Africa, males in pairs with nests near to fiscal shrikes, *Lanius collaris,* a predator, had higher baseline corticosterone levels in blood and lower body condition that males distant from shrike territories. Proximity of shrikes was also correlated with a delay in production of a second brood (Scheuerlein et al., 2001). Similarly, pied flycatchers exposed to predation attempts had increased corticosterone levels in plasma (Silverin, 1998).

The well known 10-year snowshoe hare, *Lepus americanus,* population cycle appears to be driven by exposure to predators. Hare populations increase followed by a similar rise in numbers of predators, particularly the lynx, *Lynx canadensis,* and then a population crash occurs. This cycle is apparently not because the increasing population outstrips food supplies, nor is it caused by increased social interactions (Krebs et al., 2001). As numbers of predators increase the risk of predation (by far the major cause of mortality) forces hares into habitat that provides shelter from predators but is poor in food. Hares must then forage in open areas where predation increases dramatically. Thus, the 10-year cycle appears to be driven by an interaction of food and predation (Krebs et al., 2001).

During a population decline of snowshoe hares in the Yukon, Canada, in the 1990s, hares had higher levels of free cortisol and lower corticosterone-binding globulin (CBG) capacity in blood. There were no differences in baseline plasma testosterone levels. Furthermore, they showed reduced responsiveness to dexamethasone and corticotropin treatment in the years when the population declined (Boonstra et al., 1998), had reduced leucocyte counts in blood, increased glucose mobilization, and higher overwinter loss

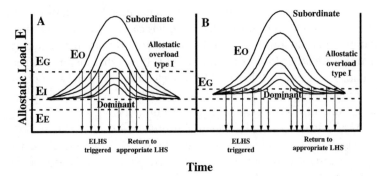

Figure 8.12: The energy models can be used to describe how social status may effect EO. In social groups, dominance status determines access to many resources including food (EG). Therefore, in response to a transitory LPF, EO (and thus allostatic load) increases and then wanes. The degree by which EO increases will be proportional to social status, however. The most dominant individuals will experience less of an increase in EO because they have easier access to EG. Subordinates, on the other hand will have much less access to EG, thus increasing EO and allostatic load. In panel A, EG is high and thus only the most subordinate individuals will accrue allostatic overload that exceeds EG. More dominant animals with greater access to EG will be able to weather the LPF without EE + EI + EO exceeding EG. In panel B, EG is lower and coupled with EO. It can be seen that even dominants may eventually reach allostatic overload that exceeds EG thus triggering an emergency life history stage (ELHS). Note that in both scenarios, the subordinates will trigger an ELHS before more dominant individuals, even when EG is low.

of body weight, compared with hares in years when the population was low. High predation risk, not population density or poor nutrition, accounted for chronic stress and impaired reproductive function. Stress physiology and reproductive function did not improve until predation risk declined (Boonstra et al., 1998). This example is consistent with the concept of allostatic overload type II (McEwen and Wingfield, 2003).

Allostatic Load and Energy Levels in Relation to Social Status and Individual Differences in Body Condition

Social Status
Position in a social hierarchy can have a profound influence on access to resources – particularly food and shelter. If EO increases following onset of a LPF, then the degree of rise in EO will be greater for subordinates than dominants (Goymann and Wingfield, 2004, Fig. 8.12). Thus, allostatic load is likely to increase more dramatically in subordinates, and, if EG is exceeded, overload occurs and individuals must trigger an ELHS and attempt to reduce that allostatic load (Fig. 8.12A). Only the dominants have access to sufficient resources to be able to resist the effects of EO. It is also possible

to model effects of social status by decreasing EG in response to a LPF (Fig. 8.12B). In this case, subordinates once again gain greatest allostatic load faster and are first to trigger an ELHS. In this case, even though the extent of EO is the same as in Figure 8.12A, lower EG results in all individuals eventually reaching allostatic overload type I and triggering an ELHS. The most dominant individuals are still least affected, however (Fig. 8.12B). In other cases, dominants may incur greater EO because of social conflict. If so, then dominants may trigger an ELHS before subordinates (Goymann and Wingfield, 2004).

There are some field studies that have addressed the issue of hormonal responses to EO and social status. Wintering flocks of Harris's sparrows, *Zonotrichia querula,* showed that during severe weather, average circulating levels of corticosterone were higher than during mild weather. Moreover, circulating levels of corticosterone during the storm were negatively correlated with dominance status (Rohwer and Wingfield, 1981). In wintering populations of European blackbirds, *Turdus merula,* sampled in Bavaria, Germany, severe weather, especially snow, results in the younger birds (most subordinate) emigrating to milder southerly habitat in Italy. These subordinate birds had higher plasma levels of corticosterone than dominants just prior to emigrating (Schwabl et al., 1985). Both of these examples are consistent with predictions of Figure 8.12. For further detailed synthesis of allostatic load and social status, see Goymann and Wingfield (2004).

Body Condition
Individuals vary tremendously in body condition, especially stores of fat. EO and/or EG can be adjusted if the individual can use stored fat temporarily to offset allostatic load. Later when the LPF passes, allostatic load is reduced to a level sustainable by existing resources and preferably that allow replenishment of stored fat (Figs. 8.13 and 8.14). Note that the ELHS is triggered only when fat stores are insufficient to fuel the allostatic overload. In Figure 8.13A, the LPF is relatively mild, and EO increases to an extent that only the leanest individuals (fat scores 0–2) reach allostatic overload type I and trigger an ELHS. As the LPF passes and EO drops below EG, these individuals return to the normal LHS. If the LPF is more intense and prolonged, then EO may increase above EG to an extent that all individuals, regardless of fat score (Fig. 8.13B), go into allostatic overload type I and must trigger an ELHS. However, the individuals with the greatest fat score are able to withstand EO for longer than those with low fat score. Another way to depict this is with decreasing EG (Fig. 8.14). If the LPF results in a short decrease in EG (Fig. 8.14A), then those individuals with greatest fat scores can withstand the period when EG drops below EE + EI, but those

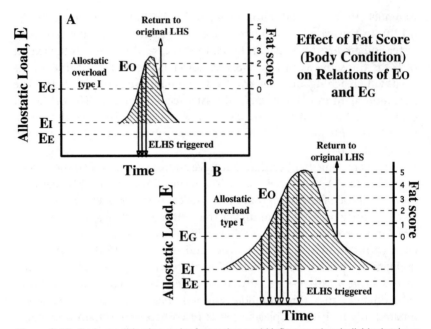

Figure 8.13: Body condition is a major factor that could influence when individuals trigger an emergency life history stage (ELHS) in response to a labile perturbation factor (LPF). In birds, fat score is used as one measure of body condition on a scale of 0 (*no fat*) to 5 (*large bulging fat bodies under the skin of the furculum and abdomen*; see Wingfield and Farner, 1978). If an individual is able to mobilize fat as EO forces it into negative energy balance (allostatic overload), this will effectively buffer EG. The fat score is used to represent the degree by which EG can be buffered. If the individual has no fat (0), then EG will be the same. As fat score increases, then EG can be buffered to a maximum of 5. Thus allostatic load (EE + EI + EO) must be greater for more fat individuals to trigger an ELHS. In panel A, the LPF is brief and EO increases above EG, but only the leanest individuals must trigger an ELHS. Those with the highest fat scores are able to weather the LPF. In panel B, the LPF is more intense and lasts longer. EO increases so that all individuals, the leanest and the fattest, enter allostatic overload and must trigger an ELHS. Note that the individuals with highest fat scores are able to weather the LPF longer. In both scenarios, as the LPF passes and EO declines, allostatic load decreases below EG, and the original life history stage can be resumed.

with low fat scores (0 and 1) will be in allostatic overload type I and trigger an ELHS. If the LPF is severe and prolonged, then EG may decrease to an extent that all except the very fattest individuals (score of 5) are in allostatic overload type I and adopt an ELHS (Fig. 8.14B). Thus, these models can explain great variation in body condition and environmental perturbations using the same framework. Once again the predictions are that glucocorticosteroid secretion increases with allostatic load and when overload type I

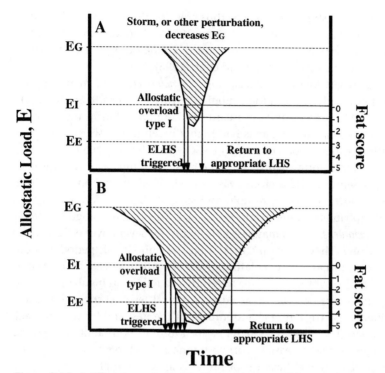

Figure 8.14: A different way to model effects of body condition on when an emergency life history stage (ELHS) is triggered is to decrease EG as a labile perturbation factor (LPF) occurs. In panel A, the LPF is brief but EG still drops below EE + EI, resulting in the equivalent of allostatic overload. If individuals are able to mobilize fat, however, then those with greater fat scores can remain in their life history stage (LHS) and only those with a score of 0 (*no fat*) or a score of 1 trigger an ELHS. In panel B, the LPF is more intense and lasts for longer. In this case, EG drops below even EE, and all individuals except ones with the highest fat score (5) enter allostatic overload and must trigger an ELHS. When EG increases again after the LPF passes, the original, or another appropriate life history stage (LHS), can be resumed. Note again that as EG decreases, the leanest individuals enter allostatic overload and trigger the ELHS first.

occurs, even higher glucocorticosteroid levels trigger an ELHS that then reduces allostatic load to a level sustainable by a combination of resources available (EG) and endogenous reserves of fat.

There are several examples of the role energy stores may play in resisting LPFSs. In some cases, desert and high arctic breeding birds become resistant to potentially stressful effects of increasing allostatic load so that they can breed in severe environments (Wingfield, 1994; Wingfield et al., 1995; Wingfield and Romero, 2000). For short periods, they can also withstand allostatic overload type I, but this can only occur if the individual is able to mobilize stored energy and thus "fuel" additional costs (Astheimer et al., 1995; Wingfield and Ramenofsky, 1999). Therefore, this type of resistance may carry a significant cost, but the trade-off is increased survival and reproductive success (Wingfield et al., 1995).

In Galapagos marine iguanas, *Amblyrhynchus cristatus*, many populations suffer greatly reduced food supplies during the El Niño Southern Oscillation (ENSO) event and abundant food supplies during the opposite phenomenon, La Niña. Iguanas had higher baseline and stressed levels of circulating corticosterone during an ENSO event, but the extent of the increased varied among islands of the Galapagos archipelago. Among individuals, baseline corticosterone was only elevated when body condition dropped below a critical threshold and suggests a direct correlation with survival probability (Romero and Wikelski, 2001). In king penguins, *Aptenodytes patagonicus*, adults may fast voluntarily for several weeks during the austral winter. They draw on stores of fat throughout this period without any increase in baseline circulating levels of corticosterone. If conditions do not improve, or the mate fails to return and relieve an incubating adult, then as fat stores become depleted to a critical level, corticosterone levels suddenly rise triggering abandonment of the nest and movement back to the ocean (i.e., an ELHS; Cherel et al., 1988). In both adult and young king penguins, the increase in corticosterone as fat stores become depleted appears to trigger mobilization of proteins as an alternate energy source (Cherel et al., 1988; Le Ninan et al., 1988).

Recently, several studies in birds have shown that when individuals in a population are exposed to a standardized acute stressor (such as capture, handling, and restraint), the maximum plasma levels of corticosterone attained (as a measure of responsiveness to a LPF) were negatively correlated with fat score or other measure of body condition (Smith et al., 1994; Wingfield et al., 1994, 1995; Hood et al., 1998). Although such relationships are not found in all populations, they do lend a high degree of support to the models in Figures 8.13 and 8.14. The neuroendocrine mechanisms by which rising glucocorticosteroid levels trigger an ELHS depending on environmental input, social status, and body condition remain unknown. Given this common framework, however, it may be possible to design experiments in many contexts and habitats to explore those mechanisms.

CONCLUSIONS

All organisms must adjust to environmental change. They do this by programmed changes in behavior, physiology, and morphology. These comprise a finite state machine of life history stages, each with a unique repertoire of substages. Additionally, there is an emergency life history stage that is triggered in response to unpredictable events in the environment – labile perturbation factors. These can be potential stresses. State of the individual at any point in the life cycle is determined by the combination of LHSs and substages expressed. Neuroendocrine secretions initiate cascades of responses that regulate the temporal sequence of LHSs and also trigger the ELHS. These complex scenarios can present a bewildering array of states and potential control mechanisms. Using FSM theory, based on state levels, secretion levels, and energy models, a common framework to explain acclimation to environmental challenges is proposed.

A key concept is allostasis – stability through change. Homeostasis is maintained throughout a sequence of LHSs in dramatically different environmental conditions. Allostatic load – the sum of energy required to maintain basic homeostasis and to acclimate to changing environmental conditions – could be a common link for physical, social, internal, and pathological perturbations. A simple energy model allows one to include permanent allostatic load accruing from injury, disease, and other challenges, as well as fluctuating environmental challenges including social interactions. The latter can lead to allostatic overload type I that can be alleviated by the ELHS. The former leads allostatic overload type II that may be irreversible (McEwen and Wingfield, 2003).

One possible common control mechanism is activation of the hypothalamic-pituitary-adrenal axis and the resultant increasing secretion of glucocorticosteroids that triggers different events. Indeed, allostatic load and glucocorticosteroid secretion are positively correlated. A series of energy models provide hypotheses and predictions for a common framework of control in diverse environmental, social, and internal conditions. Field and laboratory experiments to date provide extensive support for these ideas, although much more work needs to be done, especially at the central level and at the level of neuroendocrine mechanisms. As allostatic load increases above the level of resources that can be tapped to maintain homeostasis, allostatic overload occurs, and the ELHS is triggered. This dramatic event redirects the individual away from the normal LHS to a survival mode that is designed to reduce allostatic load (i.e., avoid stress).

The models also raise a number of intriguing questions of particular import for neuroendocrine mechanisms. Does changing sensitivity to glucocorticosteroids determine when and how an individual triggers these events? At what point do acclimation and adaptation end and stress begin? How quickly do glucocorticosteroids rise as allostatic load increases, and what defines the boundaries of levels B and C? How quickly can individuals respond to elevated glucocorticosteroids, behaviorally and physiologically? Answers to these questions await further study.

REFERENCES

Arnold, A. P., Breedlove, S. M. (1985). Organizational and activational effects of sex steroid hormones on vertebrate behavior: A re-analysis. *Horm Behav* 19:469–98.

Astheimer, L. B., Buttemer, W. A., Wingfield, J. C. (1992). Interactions of corticosterone with feeding, activity, and metabolism in passerine birds. *Ornis Scand* 23:355–65.

Astheimer, L. B., Buttemer, W. A., Wingfield, J. C. (1995). Seasonal and acute changes in adrenocortical responsiveness in an arctic-breeding bird. *Horm Behav* 29:442–57.

Bauman, D. E. (2000). Regulation of nutrient partitioning during lactation: Homeostasis and homeorhesis revisited. In: *Ruminant Physiology: Digestion, Metabolism and Growth and Reproduction* (ed. P. J. Cronje), pp. 311–27. New York: CAB.

Bohus, B., De Kloet, E. R. (1981). Adrenal steroids and extinction behavior: Antagonism by progesterone, deoxycorticosterone and dexa-methasone of a specific effect of corticosterone. *Life Sci* 28:433–40.

Boonstra, R., Hik, D., Singleton, G. R., Tinnikov, A. (1998). The impact of predator-induced stress on the snowshoe hare cycle. *Ecol Monog* 79:371–94.

Breuner, C. W., Greenberg, A. L., Wingfield, J. C. (1998). Non-invasive corticosterone treatment rapidly increases activity in Gambel's White-crowned Sparrows (*Zonotrichia leucophrys gambelii*). *Gen Comp Endocrinol* 111:386–94.

Bronson, F. H. (1989). *Mammalian Reproductive Biology*. Chicago: University of Chicago Press, 325 pp.

Brown, L. H., Britton, P. L. (1980). The Breeding Seasons of East African Birds. *East Africa Nat Hist Soc*, Nairobi, 164 pp.

Buttemer, W. A., Astheimer, L. B., Wingfield, J. C. (1991). The effect of corticosterone on standard metabolic rates of small passerines. *J Com Physiol B* 161:427–31.

Cherel, Y., Robin, J. P., Walch, O., Karmann, H., Netchotailo, P., Le Maho, Y. (1988). Fasting in the king penguin. 1. Hormonal and metabolic changes during breeding. *Amer J Physiol* 23:R170–7.

Chester-Jones, I., Bellamy, D., Chan, D. K. O., Follett, B. K., Henderson, I. W., Phillips, J. G., Snart, R. S. (1972). Biological actions of steroid hormones in non-mammalian vertebrates. In: *Steroid in Non-mammalian Vertebrates* (ed. D. R. Idler), pp. 414–80. New York: Academic Press.

Dabbs, J. M. Jr. (1990). Age and seasonal variation in serum testosterone concentrations among men. *Chronobiol Int* 7:245–49.

Dallman, M. F., Bhatnagar, S. (2000). Chronic stress: Role of the hypothalamo-pituitary-adrenal axis. In: *Handbook of Physiology* (ed. B. S. McEwen). New York: Oxford University Press.

Davidson, T. F., Rea, J., Rowell, J. G. (1993). Effects of dietary corticosterone on the growth and metabolism of immature *Gallus domesticus*. *Gen Comp Endocrinol* 50:463–8.

Dawkins, R. (1982). *The Extended Phenotype. The Gene as the Unit of Selection*. Oxford: W. H. Freeman.

de Wilde, J. (1978). Seasonal states and endocrine levels in insects. In: *Environmental Endocrinology* (eds. I. Assenmacher, D. S. Farner), pp. 10–19. Springer-Verlag, Berlin.

DeNardo, D. F., Licht, P. (1993). Effects of corticosterone on social behavior of male lizards. *Horm Behav* 27:184–99.

DeNardo, D. F., Sinervo, B. (1994a). Effects of corticosterone on activity and home range size of free-living male lizards. *Horm Behav* 28:53–62.

DeNardo, D. F., Sinervo, B. (1994b). Effects of steroid hormone interaction on activity and home range size of free-living male lizards. *Horm Behav* 28:273–87.

Denver R. J. (1997). Proximate mechanisms of phenotypic plasticity in amphibian metamorphosis. *Am Zool* 37:172–84.

Denver R. J. (1998). Hormonal correlates of environmentally induced metamorphosis in the western spadefoot toad, *Scaphiopus hammondii. Gen Comp Endocrinol* 110:326–36.

Deviche, P., Heyns, W., Balthazart, J., Hendrick, J.-C. (1979). Inhibition of LH plasma levels by corticosterone administration in the male duckling (*Anas platyrhynchos*). *I R C S Med Sci* 7:622.

Deviche, P., Delius, J. (1981). Short-term modulation of domestic pigeon (*Columbia livia L.*) behaviour induced by intraventricular administration of ACTH. *Z Tierpsychol* 55:335–42.

Drent, R. H., Daan, S. (1980). The prudent parent: Energetic adjustments in avian breeding. *Ardea* 68, 224–52.

Dunlap, K. D., Schall, J. J. (1995). Hormonal alterations and reproductive inhibition in male fence lizards (*Sceloporus occidentalis*) infected with the malarial parasite (*Plasmodium mexicanum*). *Physiol Zool* 68:608–21.

Farhi, L. E. (1987). Exposure to stressful environments, strategy of adaptive responses. In: *Comparative Physiology of Environmental Adaptations*, vol. 2 (ed. P. Dejours), pp. 1–14. Basel: Karger.

File, S. E., Day, S. (1972). Effects of time of day and food deprivation on exploratory behavior in the rat. *Anim Behav* 20:758–62.

Freeman, B. M., Manning, A. C., Flack, I. H. (1980). Short-term stressor effects of food withdrawal on the immature fowl. *Comp Biochem Physiol* 67A:569–71.

Fujiwara, T., Cherrington, A. D., Neal, D. N., McGuiness, O. P. (1996). Role of cortisol in the metabolic response to stress hormone infusion in the conscious dog. *Metab* 45:571–8.

Goldstein, R. E., Wasserman, D. H., McGuinness, O. P., Brooks Lacy, D., Cherrington, A. D., Abumrad, N. N. (1993). Effects of chronic elevation in plasma cortisol on hepatic carbohydrate metabolism. *Am J Physiol* 264:E119–27.

Gorbman, A., Dickhoff, W. W., Vigna, S. R., Clark, N. B., Ralph, C. L. (1983). *Comparative Endocrinology*. New York: John Wiley and Sons.

Goymann, W., Wingfield, J. C. (2004). Allostatic load, social status, and stress hormones – the costs of social status matter. *Anim Behav* 67:591–602.

Grande, F. (1964). Man under caloric deficiency. In: *Handbook of Physiology*, Sec. 4 (eds. D. B., Dill, E. F. Adolph, C. G. Wilbur), pp. 911–37. Washington, D.C.: American Physiology Society.

Gray, J. M., Yarian, D., Ramenofsky, M. (1990). Corticosterone, foraging behavior, and metabolism in dark-eyed juncos, *Junco hyemalis. Gen Comp Endocrinol* 79:375–84.

Hamilton, W. D., Zuk, M. (1982). Heritable true fitness and bright birds: A role for parasites? *Science* 218:384–7.

Hayes, T. B. (1997). Steroids as potential modulators of thyroid hormone activity in anuran metamorphosis. *Am Zool* 37:185–94.

Hiebert, S. M., Salvante, K. G., Ramenofsky, M., Wingfield, J. C. (2000a). Corticos-terone and nocturnal torpor in the rufous hummingbird (*Selasphorus rufus*). *Gen Comp Endocrinol* 120:220–34.

Hiebert, S. M., Ramenofsky, M., Salvante, K., Wingfield, J. C., Gass, C. L. (2000b). Noninvasive methods for measuring and manipulating corticosterone in hum-mingbirds. *Gen Comp Endocrinol* 120:235–47.

Hissa, R., Palokangas, R. (1970). Thermoregulation in the titmouse (*Parus major L.*). *Comp Biochem Physiol* 33:941–53.

Hissa, R., George, J. C., Saarela, S. (1980). Dose-related effects of nor-adrenaline and corticosterone on temperature regulation in the pigeon. *Comp Biochem Physiol* 65C:25–32.

Hood, L. C., Boersma, P. D., Wingfield, J. C. (1998). The adrenocortical re-sponse to stress in incubating Magellanic Penguins (*Spheniscus magellanicus*). *Auk* 115:76–84.

Ingram, D. L., Dauncey, M. J. (1993). Physiological responses to variations in day length. In: *Seasonality and Human Ecology* (eds. S. J. Ulyaszek, S. S. Strickland), *Soc Study Hum Biol Symp* 35:54–64, Cambridge University Press, Cambridge.

Iuvone, P. M., Van Hartesveldt, C. (1977). Diurnal locomotor activity in rats: Effects of hippocampal ablation and adrenalectomy. *Behav Biol* 19:228–37.

Jacobs, J. (1996). Regulation of Life History Strategies Within Individuals in Predictable and Unpredictable Environments. Ph.D. Thesis: University of Washington.

Jacobs, J. D., Wingfield, J. C. (2000). Endocrine control of life-cycle stages: A constraint on response to the environment? *Condor* 102:35–51.

Ketterson, E. D., King, J. R. (1977). Metabolic and behavioral responses to fasting in the white-crowned sparrow (*Zonotrichia leucophrys gambelii*). *Physiol Zool* 50:115–29.

Krebs, C. J., Boonstra, R., Boutin, S., Sinclair, A. R. E. (2001). What drives the 10-year cycle of snowshoe hares? *Biosci* 51:25–35.

Kuenzel, W. J., Beck, M. M., Teruyama, R. (1999). Neural sites and pathways regulating food intake in birds: A comparative analysis to mammalian systems. *J Exp Zool* 283:348–64.

Lack, D. (1968). Ecological Adaptations for Breeding in Birds. Chapman and Hall, London.

Lam, D. A., Miron, J. A. (1994). Global patterns of variation in human fertility. *Ann N Y Acad Sci* 709:9–28.

Le Ninan, F., Cherel, Y., Sardet, C., Le Maho, Y. (1988). Plasma hormone levels in relation to lipid and protein metabolism during prolonged fasting in king penguin chicks. *Gen Comp Endocrinol* 71:331–7.

Levine, R. J. (1994). Male factors contributing to the seasonality of human reproduc-tion. *Ann N Y Acad Sci* 709:29–45.

Maney, D. L., Goode, C. T., Wingfield, J. C. (1997a). Intraventricular infusion of arginine vasotocin induces singing in a female songbird. *J Neuroendocrinol* 9:487–91.

Maney, D. L., Richardson, R. D., Wingfield, J. C. (1997b). Central administration of chicken gonadotropin-releasing hormone II enhances courtship behavior in a female sparrow. *Horm Behav* 32:11–18.

Maney, D. L., Wingfield, J. C. (1998a). Neuroendocrine suppression of female courtship in a wild passerine: Corticotropin-releasing factor and endogenous opioids. *J Neuroendocrinol* 10:593–9.

Maney, D. L., Wingfield, J. C. (1998b). Central opioid control of feeding behavior in the white-crowned sparrow, *Zonotrichia leucophrys gambelii*. *Horm Behav* 33:16–22.

McEwen, B. S. (1998). Stress, adaptation, and disease, allostasis and allostatic load. *Ann N Y Acad Sci* 840:33–44.

McEwen, B. S., Stellar, E. (1993). Stress and the individual. *Arch Intern Med* 153:2093–101.

McEwen, B. S., Wingfield, J. C. (2003). The concept of allostasis in biology and biomedicine. *Horm Behav* 43:2–15.

McMahon, M., Gerich, J., Rizza, R. (1988). Effects of glucocorticoids on carbohydrate metabolism. *Diabetes/Metabolism Rev* 4:17–30.

Micco, D. J., Meyer, J. S., McEwen, B. S. (1980). Effects of corticosterone replacement on the temporal patterning of activity and sleep in adrenalectomized rats. *Brain Res* 200:206–12.

Mitchell, M. A., MacLeod, M. G., Raza, A. (1986). The effects of ACTH and dexamethasone upon plasma thyroid hormone levels and heat production in the domestic fowl. *Comp Biochem Physiol* 85A: 207–15.

Moorcroft, W. H. (1981). Heightened arousal in the 2-week-old rat: The importance of starvation. *Dev Psychobiol* 14:187–99.

Moore, M. C. (1991). Application of organization-activation theory to alternative male reproductive strategies: A review. *Horm Behav* 25:154–79.

Moore, F. L., Miller, L. J. (1984). Stress-induced inhibition of sexual behavior: Corticosterone inhibits courtship behaviors of a male amphibian (*Taricha granulosa*). *Horm Behav* 18:400–10.

Moyer, J. A., Leshner, A. I. (1976). Pituitary-adrenal effects on avoidance-of-attack in mice: Separation of the effects of ACTH and Corticosterone. *Physiol Behav* 17:297–301.

Mrosovsky, N. (1990). *Rheostasis: The Physiology of Change*. New York: Oxford University Press.

Munck, A., Guyre P., Holbrook, N. (1984). Physiological functions of glucocorticosteroids in stress and their relation to pharamacological actions. *Endocr Rev* 5:25–44.

Nelson, R. J. (1999). *An Introduction to Behavioral Endocrinology*. Sunderland, MA: Sinauer Assoc. Inc.

Norris, D. O. (1997). Vertebrate Endocrinology. Third, Edition, Academic Press, New York.

Orchinik, M., Gasser, P., Breuner, C. (2002). Rapid corticosteroid actions on behavior: Cellular mechanisms and organismal consequences. In: *Hormones brain and behavior*, vol. 3 (ed. D. Pfaff), pp. 567–600. New York: Academic Press.

Orchinik, M., Murray, T. F., Moore, F. L. (1991). A corticosteroid receptor in neuronal membranes. *Science* 252:1848–51.

Palokangas, R., Hissa, R. (1971). Thermoregulation in young black-headed gulls (*Larus ridibundus L.*). *Comp Biochem Physiol* 38A:743–50.

Richardson, R. D., Boswell, T., Raffety, B. D., Seeley, R., Wingfield, J. C., Woods, S. C. (1995). NPY increases food intake in white-crowned sparrows: Effect in short and long photoperiods. *Am J Physiol* 268:R1418–22.

Richardson, R. D., Boswell, T., Woods, S. C., Wingfield, J. C. (2000). Intracerebroventricular corticotropin releasing factor decreases food intake in white-crowned sparrows. *Physiol Behav* 70:1–4.

Robinson, C. J., Pitcher, T. J. (1989). The influence of hunger and ration level on shoal density, polarization and swimming speed of herring, *Clupea harengus L. J Fish Biol* 34:631–33.

Rogers, C. M., Ramenofsky, M., Ketterson, E. D., Nolan, V. Jr., Wingfield, J. C. (1993). Plasma corticosterone, adrenal mass, winter weather, and season in non-breeding populations of dark-eyed juncos (*Junco hyemalis hyemalis*). *Auk* 110:279–85.

Rohwer, S., Wingfield, J. C. (1981). A field study of social dominance; plasma levels of luteinizing hormone and steroid hormones in wintering Harris' sparrows. *Z Tierpsychol* 47:173–83.

Romero, L. M., Dean, S. C., Wingfield, J. C. (1998). Neurally active peptide inhibits territorial defense in wild birds. *Horm Behav* 34:239–47.

Romero, L. M., Wikelski, M. (2001). Corticosterone levels predict survival probabilities of Galapagos marine iguanas during El Niño events. *Proc Nat Acad Sci U S A* 98:7366–70.

Rosetta, L. (1993). Seasonality and fertility. In: *Seasonality and Human Ecology* (eds. S. J. Ulyaszek, S. S. Strickland), *Soc Study Human Biol Symp* 35:65–75, Cambridge University Press, Cambridge.

Sapolsky, R. M., Romero, L. M., Munck, A. U. (2000). How do glucocorticosteroids influence stress responses? Integrating permissive, suppressive, stimulatory and preparative actions. *Endocr Rev* 21:55–89.

Sapolsky, R. M. (1987). Stress, social status, and reproductive physiology in free-living baboons. In: *Psychobiology of Reproductive Behavior: An Evolutionary Perspective* (ed. D. Crews), pp. 291–322. Englewood Cliffs, New Jersey: Prentice-Hall, Inc.

Sapolsky, R. M., Romero, L. M., Munck, A. U. (2000). How do glucocorticosteroids influence stress responses? Integrating permissive, suppressive, stimulatory and preparative actions. *Endocrine Rev* 21:55–89.

Scheuerlein, A., Van't Hof, T. J., Gwinner, E. (2001). Predators as stressors? Physiological and reproductive consequences of predation risk in tropical stonechats (*Saxicola torquata axillaris*). *Proc Roy Soc Lond Ser B* 268:1575–82.

Schulkin, J. (2003). *Rethinking homeostasis*. Cambridge MA: M.I.T. Press.

Schwabl, H., Wingfield, J. C., Farner, D. S. (1985). Influence of winter on behavior and endocrine state in European blackbirds (*Turdus merula*) *Z Tierpsychol* 68:244–52.

Schwartz, M. W. (2001). Brain pathways controlling food intake and body weight. *Exp Biol Med* 226:978–81.

Short, R. V. (1994). Human reproduction in an evolutionary context. *Ann N Y Acad Sci* 709:416–25.

Silverin, B. (1986). Corticosterone binding proteins and behavioral effects of high plasma levels of corticosterone during the breeding period in the pied flycatcher. *Gen Comp Endocrinol* 64:67–74.

Silverin, B. (1998). Behavioral and hormonal responses of the pied flycatcher to environmental stressors. *Anim Behav* 55:1411–20.

Smith, G. T., Wingfield, J. C., Veit, R. R. (1994). Adrenocortical response to stress in the common diving-petrel, *Pelecanoides urinatrix*. *Physiol Zool* 67:526–37.

Sterling, P., Eyer, J. (1988). Allostasis: A new paradigm to explain arousal pathology. In: *Handbook of Life Stress, Cognition and Health* (ed. S. Fisher, J. Reason), pp. 629–49. New York: John Wiley & Sons.

Stuebe, M. M., Ketterson, E. D. (1982). A study of fasting in tree sparrows (*Spizella arborea*) and dark-eyed juncos (*Junco hyemalis*): Ecological implications. *Auk* 99:299–308.

Tempel, D. L., Leibowitz, S. F. (1989). PVN steroid implants: Effects on feeding patterns and macronutrient selection. *Brain Res Bull* 23:553–60.

Veldhuis, H. D., De Kloet, E. R., Van Zoest, I., Bohus, B. (1982). Adrenalectomy reduces exploratory activity in the rat: A specific role of corticosterone. *Horm Behav* 16:191–8.

Wingfield, J. C. (1994). Modulation of the adrenocortical response to stress in birds. In: *Perspectives in Comparative Endocrinology* (ed. K. G. Davey, R. E. Peter, S. S. Tobe), pp. 520–8. Ottawa: National Research Council Canada.

Wingfield, J. C., Ramenofsky, M. (1997). Corticosterone and facultative dispersal in response to unpredictable events. *Ardea* 85:155–66.

Wingfield, J. C., Breuner,C., J. Jacobs. (1997a). Corticosterone and behavioral responses to unpredictable events. *J Endocrinol* 267–78.

Wingfield, J. C., Jacobs, J., Hillgarth, N. (1997b). Ecological constraints and the evolution of hormone-behavior interrelationships. *Ann N Y Acad Sci* 807:22–41.

Wingfield, J. C., Breuner, C., Jacobs, J., Lynn, S., Maney, D., Ramenofsky, M., Richardson, R. (1998). Ecological bases of hormone-behavior interactions: The "emergency life history stage." *Am Zool* 38:191–206.

Wingfield, J. C., Hegner, R. E., Dufty, A. M. Jr., Ball, G. F. (1990). The "challenge hypothesis": theoretical implications for patterns of testosterone secretion, mating systems, and breeding strategies. *Am Nat* 136:829–46.

Wingfield, J. C., Farner, D. S. (1978). The endocrinology of a naturally breeding population of the white-crowned sparrow (*Zonotrichia leucophrys pugetensis*). *Physiol Zool* 51:188–205.

Wingfield, J. C., Jacobs, J. D. (1999). The interplay of innate and experiential factors regulating the life history cycle of birds. In: *Proceedings of the 22nd. International Ornithological Congress* (eds. N. Adams, R. Slotow), pp. 2417–43. BirdLife South Africa, Johannesburg.

Wingfield, J. C., Kitaysky, A. S. (2002). Endocrine responses to unpredictable environmental events: stress or anti-stress hormones? *Integr Comp Biol* 42:600–10.

Wingfield, J. C., Moore, M. C. (1987). Hormonal, social, and environmental factors in the reproductive biology of free-living male birds. In: *Psychobiology of Reproductive Behavior: An Evolutionary Perspective* (ed. D. Crews), pp. 149–75. Prentice Hall, New Jersey.

Wingfield, J. C., Moore, M. C., Farner, D. S. (1983). Endocrine responses to inclement weather in naturally breeding populations of white-crowned sparrows. *Auk* 100:56–62.

Wingfield, J. C., O'Reilly, K. M., and Astheimer, L. B. (1995). Ecological bases of the modulation of adrenocortical responses to stress in Arctic birds. Am. Zool. 35: 285–294.

Wingfield, J. C., Ramenofsky, M. (1999). Hormones and the behavioral ecology of stress. In: *Stress Physiology in Animals* (ed. P. H. M. Balm), pp. 1–51. Sheffield: Sheffield Academic Press.

Wingfield, J. C., Romero, L. M. (2000). Adrenocortical responses to stress and their modulation in free-living vertebrates. In: *Handbook of Physiology, Section 7: The Endocrine System, Volume 4: Coping With The Environment: Neural and Endocrine Mechanism* (ed. B. S. McEwen), pp. 211–36. Oxford: Oxford University Press.

Wingfield, J. C., Silverin, B. (1986). Effects of corticosterone on territorial behavior of free-living song sparrows, *Melospiza melodia. Horm Behav* 20:405–17.

Wingfield, J. C., Soma, K. K., Wikelski, M., Meddle, S. L., Hau. M. (2001). Life cycles, behavioral traits and endocrine mechanisms. In: *Avian Endocrinology* (eds. A. Dawson, C. M. Chaturvedi), pp. 3–17. Narosa Publishing House, New Delhi, India.

Wingfield, J. C., Suydam, R., Hunt, K. (1994). Adrenocortical responses to stress in snow buntings and Lapland longspurs at Barrow, Alaska. *Comp Biochem Physiol* 108:229–306.

Commentary   Viability as Opposed to Stability: An Evolutionary Perspective on Physiological Regulation

Michael L. Power

This volume contains a tremendous breadth of scientific research related to physiological regulation, from chronobiology to addiction. The authors approach the target concept, allostasis, from a wide range of perspectives. To the original conception of allostasis as "achieving stability through change" (Sterling and Eyer, 1988) have been added McEwen's conception of allostasis as the "process that maintains homeostasis," the idea of an allostatic state, and the concepts of allostatic load and allostatic overload. Although there is generally broad agreement among the authors concerning the words used to define allostasis, allostatic state, and allostatic load, there are subtle, and some not so subtle, differences in conception of these concepts. For some, allostasis is inherently linked to pathology. In contrast, Sterling emphasizes the adaptive nature of allostatic processes. Some authors are primarily concerned with allostatic regulation in broad terms, whereas others are focused on the related concepts of allostatic state and allostatic load and their contributions to pathology and disease. Feed-forward systems, the induction of neuropeptides by steroids, extensive neural involvement with peripheral physiology, and the interactions between behavior and physiology also figure prominently. Some of the chapters are broadly physiological, whereas others are targeted toward understanding human disease and pathology. The differing perspectives and research foci of the authors present what can be, at times, a confusing, but fascinating, mosaic on allostatic regulation of physiology.

The focus of the research of many of the authors in this volume is understanding the pathogenesis of the diseases of modern humans living in developed nations, especially adult-onset diseases. Their focus is on pathology and dysregulation of physiology, and they advance knowledge concerning the physiology of a large, long-lived adult organism that reproduces infrequently and rarely faces difficult life-or-death circumstances. The model organism of their research endeavors is not in danger of being eaten and

343

faces trivial challenges in foraging for food, but does live and compete under circumstances far removed from its evolutionary history. Thus, although
the contention of Sterling and Eyer (1988) that modern human life is more
stressful is open to debate, the challenges of modern life do not necessarily
match the evolved mechanisms for coping with stress.

The goal of this commentary is to examine the concepts of allostasis,
allostatic state, and allostatic load from the evolutionary and comparative
perspectives. The focus is not on human beings nor on disease. Perspectives from other authors are examined in light of evolutionary biology and
comparative physiology. The commentary starts with a brief description of
homeostasis and a discussion of why many authors have proposed alternate
terms to account for physiology that does not comfortably fit within a strict
concept of homeostasis. The concept of allostasis is then discussed, and the
allostatic and homeostatic perspectives are compared. Next there is a brief
consideration of the related terms *allostatic state* and *allostatic load* and their
relation to disease and pathology. There are many ways to categorize physiological responses. One broad, general categorization distinguishes among
those that are reactive, anticipatory, or programmed. Programmed changes
in physiology include seasonal and circadian variation (chronobiology) and
coordinated changes in physiology due to changes in "life history stage"
(Wingfield, this volume). Calcium metabolism during pregnancy and lactation is used as an example of a coordinated change in physiology, and
circadian variation in body temperature is used as an example of chronobiology. Finally, the physiology of fear is offered as an example of feed-forward
physiological mechanisms that are inherent in the allostatic perspective
but considered "unstable" and "rare in nature" from the homeostatic
perspective.

HOMEOSTASIS

Physiology is the biological science of essential life processes. In biology,
evolution and genetics are fundamental organizing principles. In physiology, homeostasis occupies a similar, although lesser, role. Put simply, for
an organism to survive, its internal environment must remain within certain constraints. The internal environment must be buffered from extreme
changes in the external environment. To achieve this necessary condition,
animals have evolved myriad adaptations, from cell membranes to complex central nervous systems, that cope with internal and external challenges. These adaptations serve to keep the "internal milieu" (reasonably)
constant.

A key aspect of physiology for Cannon (1935) was that "the organs
and tissues are set in a fluid matrix.... So long as this personal, individual
sack of salty water, in which each one of us lives and moves and has his

being, is protected from change, we are freed from serious peril." For Cannon, homeostasis was "the stable state of the fluid matrix" and homeostatic mechanisms and processes were those that maintained the constancy of the fluid matrix. A key mechanism for maintaining homeostasis, in Cannon's view, was the "sympathico-adrenal mechanism."

The concept of homeostasis has evolved and matured beyond the 'stability of the fluid matrix,' but it continues to be perceived as a (some would say the) fundamental explanatory principle of regulatory physiology. The concept of stability, of resistance to change, remains fundamental to homeostasis. Homeostatic systems resist change, and when perturbed, function to return the parameters of the system to within a range of appropriate values, often referred to as a "setpoint." Restraint, negative feedback, is the hallmark of homeostatic processes. Sterling, in this volume, remarks that the concept has been used to justify therapeutic interventions that attempt to return a physiological parameter (such as blood pressure) from an "abnormal" value to within the accepted range. It can be cogently argued that this approach (among many other factors) has contributed to the increased longevity now experienced worldwide. Sterling also presents a convincing argument that this approach can misinterpret the significance of "abnormal" values, however, and overemphasizes stability to the detriment of physiological adaptation.

Many scientists (e.g., Bauman and Currie, 1980; Sterling and Eyer, 1988; Mrosovsky, 1990; McEwen, 1998; Schulkin 1999, 2003b; Sapolsky, 2001, among others) have pointed out weaknesses in the homeostatic perspective and instances of physiological regulation that do not comfortably fit within homeostasis. Advances in technology and knowledge have made it clear that much within the "internal milieu" is constantly changing and adapting. Many physiological parameters do not remain constant, but continually adapt to circumstances. Some of the changes are programmed, such as circadian or seasonal rhythms or the physiological changes associated with pregnancy and lactation. Others are acute responses to challenges. This is not an inherent contradiction of homeostasis; Cannon (1935) himself quoted Richet: "We are only stable because we constantly change." But the concept of homeostasis must either be stretched, or other terms and concepts must be added to the lexicon of physiological regulation.

Mrosovsky (1990) proposed the term "rheostasis" to describe circumstances in which physiological setpoints are changed and then maintained or defended at the new level. Bauman and Currie (1980) proposed the term "homeorhesis" to describe alterations of physiology to meet demand states (e.g., reproduction). Moore-Ede (1986) suggested that circadian rhythms of physiological parameters could be incorporated into homeostasis by labeling them "predictive" homeostasis as opposed to the "reactive" homeostasis that meets acute, unpredictable challenges. Sterling and Schulkin, in defense of the allostatic perspective, have emphasized how the central coordination of physiology, anticipatory physiological responses, and the interplay of physiology and behavior appear to be given short shrift by the classic homeostatic paradigm. For Schulkin, the positive induction of neuropeptides by steroids to maintain central motive states is a key element of physiology that appears contradictory to the emphasis on negative restraint inherent in classic homeostasis. McEwen focuses on the related, but not identical, concept of allostatic load and offers it as a more powerful explanatory principle than homeostasis for the understanding of the etiology of pathology and disease.

THE EVOLUTIONARY AND ORGANISMAL PERSPECTIVE

Homeostasis is not synonymous with regulatory physiology, and stability is perhaps a misleading word when considering evolved physiological adaptations. Physiological systems serve the survival and reproductive capabilities of the organism (fitness). Stability is not the goal of all the evolved, physiological systems of an organism. A truly stable organism, in the strictest (and unreasonable) sense, would go extinct. It would be eaten or simply fail to reproduce. To paraphrase Goldstein (this volume), activities of daily life are associated with continual alterations in the apparent steady states of effectors and monitored parameters, directed by the brain and coordinated by multiple effector systems. Perhaps a better term than stability is viability, defined as the capability of success or ongoing effectiveness; in an evolutionary context, this means the ability to pass on genetic material. Regulatory physiology functions to enable an organism to achieve and maintain viability. Maintaining viability requires an organism to change its physiological state over its lifetime, on a daily, a seasonal, or an age-appropriate basis and in cases of acute need or challenge. Thus, physiological regulation to maintain viability requires regulation of setpoints, or even abandonment of setpoints. There must be physiological processes that are not homeostatic and that oppose, at least temporarily, stability. The term "allostasis" has been proposed for such processes (Sterling and Eyer 1988; Schulkin 2003b). As Schulkin (2003b, p. 21) put it, allostasis is "the process by which an organism achieves internal viability through bodily changes of state."

WHAT IS ALLOSTASIS?

The concept of allostasis was proposed to account for regulatory systems that appeared to fall outside of the classic concept of homeostatic processes, for example, regulatory systems in which there are varying setpoints or no obvious setpoints at all (e.g., fear) or in which the behavioral and physiological responses are anticipatory and do not simply reflect feedback from a monitored parameter. McEwen (1998) extended the concept to regulatory systems that were vulnerable to physiological overload, with the resulting development of disease. The term "allostatic load" refers to strain on physiology and regulatory capacity due to sustained activation of regulatory systems. The term "allostatic state" refers to chronic activation of regulatory systems either due to dysregulation or dysfunction of physiology or to competing, opposing demands.

Homeostasis and allostasis can be considered as complementary components of physiological regulation. Homeostatic processes maintain or regulate physiology around a setpoint, and allostatic processes change the state of the animal, including changing or abandoning physiological setpoints. New setpoints may or may not then be defended, and they may or may not be stable, but they represent an adaptive response to challenges to viability. Homeostatic processes are associated with negative restraint, or resistance to perturbations. Allostatic processes are associated with positive induction, perturbing the system, and changing the animal's state. The science of regulatory physiology is the study of the regulation of physiological systems that maintain an internal state (i.e., around setpoints; homeostatic regulation) and the regulation of the internal states themselves, which includes changing or even abandoning setpoints (allostatic regulation). This regulation is achieved through multiple, coordinated effector systems that can be local, systemic (hormonal), or central nervous system derived, using both negative restraint and positive induction mechanisms, to maintain viability (ability to pass on genetic material) of the organism. I would modify Sterling and Eyer's (1988) original definition of allostasis from "achieving stability through change" to "achieving *viability* through change," and define homeostasis as "achieving viability through resistance to change." Thus, neither has primacy in regulatory physiology.

An important concept within allostatic regulation is the concept of centrally coordinated physiology. Much of homeostatic regulation centers on local feedback (usually negative). For many authors (e.g., Sterling, Schulkin) allostatic regulation is intimately linked to regulation and coordination of physiology via the central nervous system. Anticipatory, feed-forward systems are a consistent feature of allostatic mechanisms. Many of the examples of allostatic regulation provided by Schulkin (2003b) involve the induction of neuropeptides by steroid hormones, as well as the concept that the hormones that regulate peripheral physiology in response to a challenge are also involved in changing central motive states of the brain, and thus induce behaviors that aid the animal to meet the challenge. The linkage of peripheral physiology with central motive states and the concept that behavior and physiology act together to preserve viability are central to Schulkin's conception of allostatic regulation.

Sterling also emphasizes anticipatory responses and coordination by the central nervous system as a central tenet of allostasis. He argues that the concept of homeostasis inappropriately leads to considering parameters of peripheral physiology without regard to behavior and neural physiology. Deviations from normal values are then considered "errors," which homeostatic mechanisms should function to "correct." Rather than "correcting errors," Sterling argues that physiology anticipates challenges and, through central coordination of physiology, acts to "prevent errors." Prolonged deviations from "normal" values represent not an error but an attempted adaptation to circumstances. If the prolonged deviation from normal values represents a threat to health, in the allostatic view the better response is to "change the circumstances" instead of pharmacologically attempting to "clamp" the parameter to normal.

To be fair, many scientists have investigated and discussed aspects of physiology that are both anticipatory and homeostatic in nature. Homeostasis can be predictive (Moore-Ede, 1986) as well as reactive. The conception of allostasis outlined above is similar to Mrosovsky's (1990) "rheostasis." It is broader in that Mrosovsky restricted rheostasis to changes in a regulated level that is then defended, and allostasis does not require a new level that is defended or stable. Many circadian phenomena are considered both anticipatory and homeostatic, at least under a broad conception of homeostasis (Moore-Ede, 1986). The central nervous system is involved in regulatory physiology for both homeostatic and allostatic processes. The distinction between homeostasis and allostasis can often be difficult to make.

ALLOSTATIC STATE

Koob and LeMoal (this volume) use the concept of an allostatic state to describe the circumstances of drug addiction. They describe a dysregulation of the reward system due to chronic use of drugs, with a concomitant hyperexcitability state of the brain that, in the addict, becomes regulated by the abuse of drugs. The setpoint for drug use is consistently ramped up, such that more and more of the drug is required for less and less time of "reward." Thus, drug addiction is associated with a chronic activation of neural systems and a continued alteration of setpoints in their regulation–an allostatic state.

Schulkin (2003b) devotes a whole chapter of his book on allostasis to drug addiction. He links the neurophysiology of drug addiction to allostasis in three ways: (1) the increased activation of a number of neural systems (chronic arousal), (2) the dysregulation of the reward system, and (3) the usually destructive physical and physiological consequences of long-term drug abuse, which he calls a vulnerability to allostatic overload. A valid criticism is that although the evidence links drug addiction to the concepts of allostatic state and allostatic load and overload, it has not been demonstrated that these concepts, despite their names, are restricted to allostatic processes, nor that allostatic processes are inherently more vulnerable to dysregulation and "load" than homeostatic processes.

ALLOSTATIC LOAD AND OVERLOAD

The concept of allostasis originally derived from the attempt by Sterling and Eyer (1988) to explain patterns of modern morbidity and mortality with socioeconomic circumstances considered to be stressful. Boulos and Rosenwasser in this volume state that "implicit in Sterling and Eyer's model is the notion that allostatic regulation entails a price." Although Sterling, with his emphasis on the adaptive nature of allostatic regulation, likely would disagree with that statement, many authors in this volume appear to accept it. Allostatic load is the intellectual formulation of that "price."

The chronic activation of regulatory systems can cause wear and tear on tissues and organs. Inherent in the concept of allostatic systems is that they should be activated over a limited duration (Schulkin, 2003b). They act to change an animal's internal state, and perhaps to maintain that state over a short period of time, but they are not meant to be "constantly on." Fear is a good example. There are significant adaptive advantages to being fearful in certain contexts, but a chronic, sustained state of fearfulness is maladaptive. The activation of the hormones of stress enables an animal to respond to challenges, but it also imposes a cost, especially if the stress hormones remain elevated for a significant duration. The cost to the animal of chronic activation of a regulatory system is allostatic load, and if the cost progresses to tissue breakdown and the development of pathology, an organism is in a state of allostatic overload.

Why is allostatic load the proper term? In many cases, such as excessive fear and anxiety, the allostatic nature of the physiology is transparent. In many cases, however, this phenomenon could be more simply termed regulatory or physiological load. It does not appear to be restricted to allostatic processes, unless one weds allostasis to pathology and contends that all chronic activation of regulatory systems outside of "normal" ranges is by default "allostatic." Some authors (e.g., McEwen) would support that formulation of the concept. Indeed, this is the concept of an allostatic state; chronic activation of regulatory systems and alterations of setpoints (Schulkin, 2003b). Why, however, should these costs be exclusive to allostatic processes, and why must all allostatic processes impose a cost?

For example, consider the regulation of serum-ionized calcium under chronic low calcium intake, a condition that might reasonably be said to have a high prevalence in the United States. Under conditions of chronic low dietary calcium intake, 1,25 dihydroxyvitamin D and parathyroid hormone (PTH) will be elevated. The actions of these hormones increase calcium absorption from the intestines, and decreases excretion via the kidney. Calcium may be lost from bone, as the rates of osteoclast and osteoblast activity are changed. A calcium "appetite" may (or may not) develop. These processes and others comprise the homeostatic system that regulates serum-ionized calcium within a relatively narrow range. If calcium intake remains low, these processes will remain (over)activated, and may impose a load that can lead to a lessening of health. The potential harmful effects include low bone mineral, possibly elevated blood pressure, and susceptibility to premenstrual syndrome (PMS), among others (Power et al., 1998). Is this a case of homeostatic or allostatic load? Is the chronically elevated PTH symptomatic of an allostatic state?

Although the terminology can be disputed, the concepts of allostatic state, load, and overload appear to have value. Physiological regulation (whether considered homeostatic, allostatic, or neither) certainly has the potential to exert a "load" on an organism. There are costs to being viable. Many physiological systems are designed as short-term responses; chronic activation of those systems may cause more harm than good. When physiological and behavioral regulation fails to "solve" the challenge facing the animal, the costs can accumulate and intensify.

ALLOSTASIS AND PATHOLOGY

Maintaining a homeostatic state is important for health in an adult, non-reproductive animal, and improving the overall adult population's health rather than viability (under my definition) is the ultimate goal of much of the research described here. Take, for example, Singer and colleague's chapter in this volume. In their opening paragraph, they discuss the rate of comorbidities occuring in men and women in their eight and ninth decades of life. The concept of viability might seem to have little relevance here. Instead, the concept of allostatic load, regulatory systems suffering progressive breakdown due to chronic activation outside of the normal operating range, is an intriguing, and potentially powerful tool for understanding adult onset diseases, especially those of later life, and in understanding, predicting, and treating these patients.

Understanding human disease and pathogenesis is the raison d'être for the research of many of the authors in this volume. Pathological conditions certainly can enhance understanding of physiology, and understanding the physiology is vital to understanding – and, it is hoped treating – pathology. Tieing allostatic regulation too tightly to disease and pathology paradoxically both narrows and broadens the concept, however, reducing its usefulness as an explanatory tool in understanding physiological regulation. The concepts of allostatic state, load, and overload may be inherently linked with pathology, but they are not synonymous with allostatic physiological regulation.

Sterling (this volume) offers an interesting viewpoint on the value of the allostatic perspective in understanding pathology. He notes that the homeostatic perspective can be perceived as modeling the organism as "error-correcting," and pathology as occurring due to failure to "correct" an error of internal physiology. This leads to therapeutic interventions conceived to "correct" the abnormal parameter for the organism – in other words, to force the parameter back into the normal range. Sterling offers the wisdom that organisms regulate their physiology to adapt to circumstances and challenges. An "abnormal" value for a parameter likely indicates an attempted adaptation to perceived challenges to viability. In Sterling's view, the allostatic perspective directs therapy away from "clamping a parameter at a particular value toward restoring adaptive fluctuation." In many ways Sterling's dichotomous conception of the homeostatic and allostatic clinical perspectives are analogous to the discussions of pharmacological treatment versus behavioral or social modification and treating disease versus preventative care to lessen the development of disease. This is a creative tension, and both perspectives are valuable to health care.

PROGRAMMED, SEASONAL, AND CIRCADIAN CHANGES IN PHYSIOLOGY

Physiology is adaptive, but what is adaptive changes depending on external and internal circumstances. Sometimes those changes are predictable, occurring over regular cycles. Sometimes the physiological changes are required because the organism has entered a new life history stage (e.g., become reproductive or is molting in anticipation of migrating). Sometimes physiological change is required due to acute, unpredictable challenges. Wingfield (this volume) lays out a framework for understanding an organism's state changes due to age, seasonal or daily cycles or acute disturbance from the external environment. He emphasizes that change is perhaps the real "constant" in an animal's physiology. The profound physiological shifts many animals employ between life history stages in response to season, age, or severe environmental disturbance would seem excellent examples of allostatic regulation

Wingfield's life history stages and emergency life history stages are appropriate for many animals but perhaps are less useful for humans. As Wingfield says, adult humans have only one life history stage (although there are substages, such as in female reproduction). We don't hibernate or undertake seasonal migrations that require physiological change. Wingfield's attempt at giving allostatic load an ecological interpretation using energy balance (McEwen and Wingfield, 2003; this volume) is ultimately unconvincing (Dallman 2003; Walsberg, 2003). Certainly, if the energetic cost of a physiological adaptation exceeds the animal's ability to extract energy from the environment, over time the adaptation either must be abandoned or the animal must sacrifice body condition. Although an argument can be made that this is an example of allostatic load (a simpler term would be energetic cost), it certainly does not represent the full concept. Many adult-onset diseases in humans that are proposed to derive from allostatic load are diseases of plenty, not of want. They are prevalent in the developed nations and linked with excess consumption (e.g., obesity).

There has been much discussion of how the concept of homeostasis can encompass cyclical and programmed changes in physiology. Some have argued that for circadian rhythms, it is the curve of the whole cycle that is "defended" (Halberg, 1953). Thus deviations from the setpoint are relative to the expected value for the time of day (or season). Of course there must be some mechanism to drive the changes in defended level. Among the many distinctions being made concerning physiological regulation, a convincing argument can be made that the differences among cyclical, programmed, and reactive physiological processes are at least as important as whether they are homeostatic or allostatic.

CALCIUM METABOLISM DURING PREGNANCY AND LACTATION

Pregnancy and lactation represent significantly elevated demand states for female mammals. Over a 9-month gestation, a human fetus will accumulate a mean of 30 g of calcium, roughly equal to 3% of maternal skeletal calcium. After birth, the human infant accumulates calcium at the rate of approximately 200 mg/day. For an infant that is exclusively breast-fed for 6 months, this translates to approximately 60 g of calcium transferred from the mother. The mother's calcium metabolism changes during pregnancy and lactation to meet these demands. This is an example of a programmed change in physiology, what Bauman and Currie (1980) called homeorhesis.

During human pregnancy, maternal calcium metabolism differs from the nonpregnant state. Total serum calcium falls, but that change can be accounted for by the decrease in serum albumin concentration, which translates to a decrease in protein-bound calcium. Serum-ionized calcium concentration remains unchanged (Kovacs and Kronenberg, 1998). Intestinal absorption of calcium is markedly increased, presumably because of increased active transport in response to an increase in 1,25-dihydroxyvitamin D, some of which is placental in origin (Kovacs and Kronenberg, 1998). Urinary excretion of calcium is also increased, due both to the increased intestinal absorption and the increased renal plasma flow that results from expanded plasma volume (Cross et al., 1995; Ritchie et al., 1998). Serum calcitonin is elevated, which may serve to protect the maternal skeleton from excessive resorption of calcium, or even to facilitate accretion of calcium in bone (Kovacs and Kronenberg, 1998). There is a biphasic pattern of bone remodeling, with bone loss in early pregnancy, but bone gain in late pregnancy (Purdie et al., 1988). Serum PTH is apparently unchanged (Ritchie et al., 1998), although some data show an increase from the first to the third trimester (Davis et al., 1988; Rasmussen et al., 1990). Serum PTH-related peptide (PTHrp) does increase over pregnancy.

After parturition, calcium metabolism changes again. Despite the even greater calcium demands of lactation, the calciotropic hormones are not elevated (Ritchie et al., 1998). Intestinal calcium absorption is not significantly different from the nonpregnant, nonlactating state (Specker et al., 1994; Kalkwarf et al., 1996). Urinary excretion of calcium, however, is reduced. Renal calcium conservation does appear to play a role in meeting the calcium requirements of lactation. However, most of the calcium requirement appears to be met by withdrawing calcium from bone (Sowers et al., 1995a). During lactation, women lose 4–7% of their bone mineral (Kalkwarf et al., 1997). In the case of the laboratory rat, a much smaller animal that raises a much larger litter over a short lactation, this bone mineral loss can be 30–50% (Brommage and Deluca, 1985). Calcium supplementation of lactating women does not ameliorate this bone mineral loss; instead, the extra dietary calcium appears to be excreted in the urine (Fairweather-Tait et al., 1995; Prentice et al., 1995). Women with low bone mineral density can develop osteoporotic fractures during lactation. After the return of menses, however, bone mineral is rapidly regained (Kalkwarf et al., 1997), and there is no evidence of any long-term pathology, even if a second pregnancy occurs during that period (Sowers et al., 1995b; Laskey and Prentice, 1997).

Is this homeostatic or allostatic regulation? Or is Bauman and Currie's conception of homeorhesis superior to both in this instance? It is certainly physiological regulation of the internal milieu that directly affects viability. The lactating animal maintains some aspects of its internal milieu constant, while making a short-term sacrifice of other aspects of its internal state. Over the course of pregnancy and lactation there is change, but over a longer period, there is no biologically meaningful change (except in the fundamental unit of fitness, the production of offspring). Does that mean that this aspect of physiological regulation can be termed either allostatic or homeostatic, depending on whether one considers the whole animal (changed to enhance its viability), a physiological parameter (serum ionized calcium, held constant as is necessary for survival), or the time scale one considers (during lactation, changes to promote viability; over a whole cycle, e.g., until the next pregnancy, maintains stability)? Or is this an example of physiological regulation that is outside of both homeostasis and allostasis?

During both gestation and lactation, the unit of physiological regulation could be considered the mother and offspring, instead of either alone. The offspring are dependent on the mother, and physiological costs to the mother result in gains for the offspring. A strict conception of homeostasis does not account for the imperatives of reproductive fitness.

BODY SIZE AND HOMEOSTASIS

Human beings are large animals. The main (nutritional) cost to being large is that the organism requires more nutrients. A main advantage of large body size is that the allometry of most nutrient requirements is less than 1, and the storage capacity is frequently 1 or greater. Thus, for example, energy requirements increase at an allometry of approximately 0.75, and large animals generally are able to carry a higher percentage of their weight in body fat. Therefore, large animals generally have a longer "starvation" time than do small animals (McNab, 2002). Shrews must eat constantly, humans a few times per day, and tigers once per day or less.

Bone mineral content is another example. Large animals typically have a larger proportion of weight accounted for by the skeleton. Thus, their ability to rely on bone calcium to maintain serum ionized calcium homeostasis is greater. Part of the greater percentage loss of bone mineral during lactation by the rat is due to its large litter size and short lactation, but it is also due in part to its smaller size, and hence its relatively smaller store of calcium.

An implication of this relatively greater allometry of nutrient stores to requirements is that larger animals should have an easier time maintaining homeostasis. Small animals have to "work harder" to maintain stability of the internal milieu. They have less time before viability becomes compromised.

CIRCADIAN VARIATION IN BODY TEMPERATURE

Even body temperature, a canonical example of homeostatic regulation, can exhibit what can be broadly considered to be allostatic regulation,

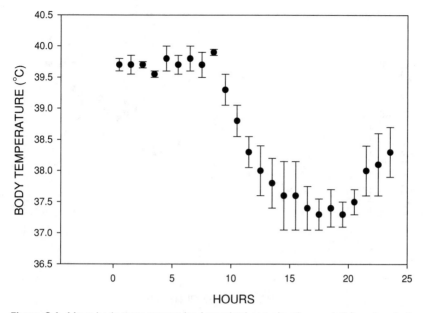

Figure C.1: Mean body temperature (and standard error bars) recorded from two individuals within a social group of four golden lion tamarins (*Leontopithecus rosalia*) over a 24-hour period in their home cage. Room temperature maintained at 27°C.

although many will disagree. For example, golden lion tamarins (*Leontopithecus rosalia*), small neotropical monkeys, undergo a daily rhythm of body temperature. During daytime, their active period, body temperature is regulated within a narrow (39.7 ± 0.2°C) range. When the group retires at dusk (the entire social group of 4–10 animals typically sleep huddled together) their body temperatures rapidly drop several degrees (Fig. C.1), in conjunction with a 30% drop in resting metabolic rate (Thompson et al., 1994). Body temperature is more variable during sleep (Fig. C.1), with a range of 3–4°C, and is correlated with ambient temperature (Thompson et al., 1994; Fig. C.2). When subjected to cold stress, however, the monkeys will elevate their metabolic rate, and body temperature does not drop below 5°C lower than their active phase body temperature. At dawn, body temperature and metabolic rate rapidly return to the active phase norms, and body temperature is again tightly regulated. Thus, golden lion tamarins have a daily cycle of temperature regulation: a tightly controlled period during their active phase (classic homeostatic regulation), a rapid change to lower body temperature at the end of the active phase (allostatic regulation – changing setpoint), an inactive period phase characterized by a lower body temperature that can vary over a much larger range but that has a lower limit that is

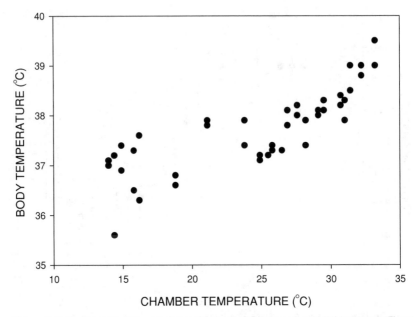

Figure C.2: Body temperature during night sleep of the same two individuals as in Figure C.1, but this time when the group was in a water-jacketed temperature control chamber. Water temperature was varied from approximately 12°C to 34°C. The chamber temperature was recorded via a thermalcouple inserted into the nest box within which the group was sleeping. Data were collected from eight separate nights over a one month period.

defended (relaxed homeostatic regulation), and a rapid return to active period temperature and regulation at the end of the inactive period (allostatic regulation).

Boulos and Rosenwasser in this volume argue against this type of example of allostatic regulation. They cogently note that the circadian pattern of body temperature is constant and defended, and this entire body temperature cycle can either fit under a broad concept of homeostasis or be more properly considered as an example of "chronobiology." In other words, this is neither strictly homeostasis nor allostasis, but a third category that contains clocklike physiological regulation.

There is evidence in favor of this view. Even when placed under completely dark conditions, metabolic rate and body temperature of golden lion tamarins and other marmosets and tamarins is higher during daylight hours than during the night (Power, 1991; Power et al., 2003). When these animals are kept under artificial light conditions, there is an anticipatory decline in body temperature (Hampton, 1973; Hetherington, 1978; Thompson et al., 1994), heart rate (Hampton, 1973; Schnell and Wood, 1995), vocalizations, and activity (personal observation) prior to "lights out" and a parallel increase in those parameters prior to "lights on." However, if the lights do not go off as anticipated, vocalizations and activity resume (personal observation). Unfortunately, body temperature and metabolic rate were not measured under these circumstances, but the presence of feeding, defecating, and other behaviors implies that the animals had not entered their usual semitorpid state for that time period. In the wild, cold temperatures, rain, and early darkness (due to clouds) cause golden lion tamarins to retire to their sleeping sites before the expected time (Ruiz-Miranda, personal communication). Again, it is not known if this behavioral adaptation is accompanied by the physiological response of lowered metabolism and body temperature. So, unsurprisingly, there is evidence both for a "clock" and reaction to external stimuli in regulating this physiology and behavior. Fitting this pattern of body temperature, energy metabolism, and behavior under the concept of homeostasis would emphasize its ultimate, evolutionary adaptiveness (reducing energy expenditure, predation risk, and so forth). Focusing on its allostatic characteristics would emphasize the proximate mechanisms (as yet unknown) that act to modify the setpoints of metabolism, body temperature, and behavior. Regardless, the clocklike aspects of the physiology are essential for understanding both the mechanisms and the functions.

CHRONOBIOLOGY AND ALLOSTATIC LOAD

In their chapter of this volume, Boulos and Rosenwasser present evidence for allostatic load associated with shift work. Being active during the time period of normal inactivity (sleep) for our species is difficult and is associated with greater rates of accidents, pathology, and disease. The circadian rhythm can be considered a "homeostatic state" (i.e., a pattern that the organism attempts to preserve) and attempting to work opposite to it a deviation from "stability" that can have detrimental consequences. Thus, although Boulos and Rosenwasser would argue that circadian rhythms are neither homeostatic nor allostatic, they see a value in the concept of allostatic load.

COMPETING IMPERATIVES

In a thought example, consider the body temperature of a gazelle being chased by a predator, such as a leopard. The gazelle experiences fear, with an increase in cortisol production that mobilizes resources to enable it to engage in intense muscle activity related to flight. Because of this activity, body temperature rises and will continue to rise until the leopard catches and kills the animal, the escape attempt is successful and the animal can cease its flight and homeostatic mechanisms to reduce its body temperature can successfully act, or the animal overheats and collapses (in which case the leopard again likely improves its viability at the expense of the gazelle). The regulation of body temperature is not the imperative – the survival of the organism is the imperative. The physiological and behavioral reactions of the animal were not to serve "homeostasis" but to serve survival, a necessary, but not sufficient, condition for viability. If the animal overheated, then the tactic failed to protect viability, but not attempting it would have ensured failure.

Again, it can be cogently argued that during the chase homeostatic mechanisms were employed in an attempt to regulate body temperature and that they increased the time that the fear-flight processes can operate before organismal failure. Eventually, however, if the threat to viability is not escaped, the imperative for flight will overwhelm them. This is an example of competing homeostatic (temperature control, blood pH, blood oxygen content, and so forth) and allostatic (the fear-flight response) mechanisms in operation. From an organismal biologist's perspective, this situation would be the rule rather than the exception. In most of the contexts within which we study animals (e.g., growth, reproduction, competition, aggression, feeding biology, social behavior, and so forth), there are competing physiological and behavioral systems. Animals are constantly balancing conflicting imperatives. Often stability – viability through resistance to change – is the successful strategy, but this is not always the case.

THE CENTRAL NERVOUS SYSTEM AND PERIPHERAL PHYSIOLOGY

The concept of allostasis has consistently been linked with the role of the central nervous system (CNS) in physiological regulation and with peripheral physiology (Sterling and Eyer, 1988; Schulkin 2003a, 2003b). A central neural response network that regulates stress responses has been proposed (Schulkin et al., 1994) and has found favor even among those who question the utility of the term allostasis (e.g., Dallman, 2003).

From an organismal view, physiological regulation requires CNS involvement. A principal function of the CNS is the coordination of responses to and anticipation of challenges. Behavior and physiology are not separated within the organism, regardless of their frequent separation among academic departments. As in the earlier example, in real-world situations animals are constantly balancing competing imperatives. It can be argued that the resolution and prioritization of those competing imperatives is a principal function of the CNS.

ALLOSTASIS, FEAR, AND FEED-FORWARD SYSTEMS

Fear has been suggested to be a canonical example of allostatic regulation (Schulkin, 2003b) and chronic fear and anxiety a canonical example of allostatic load. Fear has no "setpoint" per se, unless one argues that animals attempt to regulate fear by keeping it to a minimum (e.g., by avoiding or fleeing from fearful stimuli). It is a central state of the brain that combines peripheral physiology with extensive neural involvement. Within the brain there are feed-forward systems that potentiate fear and predispose the organism to experience fear and behave in fearful ways. Fear is intimately linked with behavior and action. Fear is adaptive. It motivates an animal to perform behaviors that may be disruptive to the internal milieu in the short term but that serve viability (survival) in the long term. It serves to heighten alertness and vigilance. It enables animals to expend considerable amounts of energy in a short period of time. It motivates animals to attempt to remove themselves from the stimuli that create the fear. Fear, in small doses, is good.

On the other hand, chronic fear and anxiety have well-documented destructive effects on tissue in the CNS and many of the body's organs. Chronic fear and anxiety are proposed to produce disease and pathology in people. Rosen and Schulkin's chapter lays out the physiology of fear and its relation to pathological anxiety.

Is fear a good example of an allostatic system? Is it impossible to fit fear into a homeostatic framework? Does fear have a setpoint? An argument can be made that the lack of fear is the preferred state of most animals and that animals are motivated to act in ways that reduce fear. Thus, one could make a (strained) homeostatic argument regarding fear. Fear is a result of a perception by an animal that its viability is threatened, and the animal responds by attempting to alleviate the threat. The behaviors and physiological processes invoked could be said to serve the purpose of returning the state of the brain to the nonfearful state. Thus low or no fear is the setpoint the animal attempts to defend. Fear itself is a response to external conditions that motivates homeostatic processes that act to attempt to eliminate the fear.

This is perhaps not very satisfying; however, it is not clear that viewing fear from an allostatic perspective enhances understanding. The mechanisms underlying fear and animals' behavioral and physiological responses to fear can be studied and understood without addressing either homeostasis or allostasis.

This segues to an important point. There are a number of important concepts raised in these chapters that deserve serious consideration regardless of whether the terms allostasis or homeostasis are applied to them or not. For example, in the case of fear, a positive feedback system has been identified in the amygdala, wherein cortisol has a positive feedback effect on corticotropin-releasing hormone (CRH) production. Allostasis has been associated with feed-forward mechanisms by some authors (e.g., Schulkin, 2003b), while homeostasis has been associated with negative restraint, or negative feedback mechanisms. There have been statements in the literature that positive feedback systems are "inherently unstable" and are "rare in nature" (Goldstein, 1995). The allostatic perspective demands that those statements be reexamined.

POSITIVE FEEDBACK SYSTEMS ARE NOT INHERENTLY UNSTABLE

Goldstein (this volume) lays out a scheme to understand much of physiological regulation via multiple circuits of "homeostats" and their associated effector molecules. The circuits are composed of negative and positive feedback loops. Although Goldstein allows positive feedback loops, he specifically states that they must be "restrained within negative feedback loops, or else they will be unstable."

A positive feedback system that is amplified, and not nested within a constraining system, is inherently unstable. In other words, if each response generates a greater associated response, then the system will eventually exceed its real-world capacities. However, this is not the only type of positive feedback system that can be defined. The simplest definition of a positive feedback system is a system in which the release or production of an effector generates a response, and that response induces the release or production of the effector. As long as the induced release-production of the effector is less than the initial release-production, the system is damped and will eventually return to the steady state. Consider the elegant findings of Cook (2002), who measured CRH and cortisol in the amygdala of sheep exposed to a dog. The sight of the dog produced a sharp spike of amygdala CRH that rapidly returned to baseline. Shortly afterward, cortisol levels rose in the amygdala, followed by a parallel rise in CRH. The peak of amygdala CRH that followed the rise in cortisol was lower than the peak following introduction of the dog. Thus, the induction of CRH by cortisol (which was itself at least partly induced by the first CRH spike) is not unstable. It does, however, ensure that CRH levels in the amygdala will be elevated above baseline for a time period beyond that associated with the acute, precipitating event.

Thus in real-world situations, positive feedback physiological systems are not inherently unstable. An analogy would be to contrast the behavior of a bouncing ball under the heuristic conditions of an introductory physics class (no air resistance, an incompressible ball and surface) to that of the real world. In the first instance, conservation of energy would demand that the ball bounces forever, with the sum of kinetic and gravitational potential energy being always equal to the initial conditions. In the real world, however, air resistance and the compression of the ball and the ground on impact translates some of that kinetic energy into heat, with the result that the ball reaches progressively lower heights with each bounce, until at last it stops. In real-world physiology, there will be "waste," cell death, competing receptors and effector molecules, and behavior. In the example of the positive feedback between CRH and cortisol in the amygdala, there is no a priori reason to assume that the cortisol induced by amygdala CRH is sufficient to indefinitely sustain the higher level of amygdala CRH. Chronic elevation of CRH in the amygdala is observed, but it is generally associated with repeated or chronic stress, the frustration of behavioral strategies to cope with stress, or both. For example, in Cook's (2002) work, sheep in pens that had a secondary pen out of sight from the dog to which they could retreat did not show a change in baseline amygdala CRH. Sheep in pens without a way to cope behaviorally with the stress of seeing the dog developed higher baseline amygdala CRH.

In this example, the sheep without a sheltered pen to retreat to were in an allostatic state and suffered from allostatic load, reflected in their elevated amygdala baseline CRH; those sheep that could hide from the dog appeared to avoid any "cost" to the allostatic physiological processes related to being afraid of a dog. The physiological "load" was not due to the allostatic nature of the response, but to the inability of the response to resolve the challenge.

CONCLUSIONS

Whole-body physiological regulation includes homeostatic and allostatic processes, and perhaps mixtures and other systems that defy simple classification. The terminology is only important in so far as it helps in understanding the organism and the physiology being investigated. The terms allostasis, allostatic state, and allostatic load will gain or lose favor based on the insights they bring to individual investigators, but the concepts they encompass are real.

Animals are constantly maintaining certain states while changing others. Viability cannot exist without change. Physiological regulation enables animals to achieve and maintain viability. In many instances, stability, or resistance to change, will be the optimal strategy. In other instances, the animal must change its state and abandon homeostasis, at least in the short term. These state changes can be predictable and programmed, such as seasonal, circadian, or life-cycle changes (Wingfield's life history stages). They can also be responses to unpredictable challenges that require substantial adaptation (Wingfield's emergency life history stages). They can be as simple (and complex) as the coordinated changes that allow us to stand up from a chair and not pass out. In all cases, we are talking about evolved, physiological adaptations that enhanced fitness in the organism's evolutionary past.

Anticipatory, feed-forward systems are vital to regulatory physiology, and physiology is not merely reactive. The CNS is intimately involved in physiological regulation. Behavior and physiology should not be separated. Positive feedback systems exist and are a part of normal physiology. Achieving and maintaining viability costs. The activation of regulatory systems can have short- and long-term costs. If regulatory systems are chronically activated, either due to competing imperatives or the failure of the physiology to resolve the challenge, these costs can accumulate and lead to a lessening of health. All these statements are part of the allostatic perspective. Whether the concept of allostasis is required to encompass these ideas within regulatory physiology is another matter.

REFERENCES

Bauman, D. E., Currie, W. B. (1980). Partitioning of nutrients during pregnancy and lactation: A review of mechanisms involving homeostasis and homeorrhesis. *J Dairy Sci* 1514–29.

Brommage, R., Deluca, H. F. (1985). Regulation of bone mineral loss during lactation. *Am J Physiol* 248(2pt.1):E182–7.

Cannon, W. B. (1935). Stresses and strains of homeostasis. *Am J Med Sci* 189:1–14.

Cook, C. J. (2002). Glucocorticoid feedback increases the sensitivity of the limbic system to stress. *Physiol Behav* 75:455–64.

Cross, N. A., Hillman, L. S., Allen, S. H., Krause, G. F. (1995). Changes in bone mineral density and markers of bone remodeling during lactation and postweaning in women consuming high amounts of calcium. *J Bone Miner Res* 10:1312–20.

Dallman, M. F. Stress by any other name . . . ? (2003). *Horm Behav* 43:18–20.

Davis, O. K., Hawkins, D. S., Rubin, L. P., Posillico, J. T., Brown, E. M., Schiff, I. (1988). Serum parathyroid hormone (PTH) in pregnant women determined by an immunoradiometric assay for intact PTH. *J Clin Endocrinol Metab* 67:850–2.

Fairweather-Tait, S., Prentice, A., Heumann, K. G., Jarjou, L. M. A., Stirling, D. M., Wharf, S. G., Turnland, J. R. (1995). Effect of calcium supplements and stage of lactation on the calcium absorption efficiency of lactating women accustomed to low calcium intakes. *Am J Clin Nutr* 62:1188–92.

Goldstein, D. S. (1995). Stress as a scientific idea: Homeostatic theory of stress and distress. *Homeostasis* 36:117–215.

Halberg, F. (1953). Some physiological and clinical aspects of 24-hour periodicity. *Lancet* 73: 20–32.

Hampton, J. K. Jr. (1973). Diurnal heart rate and body temperature in marmosets. *Am J Phys Anthropol* 38:339–42.

Hetherington, C. M. (1978). Circadian oscillations of body temperature in the marmoset, *Callithrix jacchus*. *Lab Anim* 12:107–8.

Kalkwarf, H. J., Specker, B. L., Heubi, J. E., Vieira, N. E., Yergey, A. L. (1996). Intestinal calcium absorption of women during lactation and after weaning. *Am J Clin Nutr* 63:526–31.

Kalkwarf, H. J., Specker, B. L., Bianchi, D. C., Ranz, J., Ho, M. (1997). The effect of calcium supplementation on bone density during lactation and after weaning. *N Engl J Med* 337:523–8.

Kovacs, C. S., Kronenberg, H. M. (1998). Maternal-fetal calcium and bone metabolism during pregnancy, puerperium, and lactation. *Endocr Rev* 18:832–72.

Laskey, M. A., Prentice A. (1997). Effects of pregnancy on recovery of lactational bone loss. *Lancet* 349:1518–9.

McEwen, B. S. (1998). Stress, adaptation, and disease: Allostasis and allostatic load. *Ann N Y Acad Sci* 840:33–44.

McEwen, B. S., Wingfield, J. C. (2003). The concept of allostasis in biology and biomedicine. *Horm Behav* 43:2–15.

McNab, B. K. (2002). *The Physiological Ecology of Vertebrates: A View from Energetics.* Ithaca, N Y: Comstock Publishing Associates Cornell University Press.

Moore-Ede, M. C. (1986). Physiology of the circadian timing system: Predictive versus reactive homeostasis. *Am J Physiol* 250:R737–52.

Mrosovsky, N. (1990). *Rheostasis: The Physiology of Change.* Oxford: Oxford University Press.

Power, M. L. (1991). *Digestive Function, Energy Intake, and the Response to Dietary Gum in Captive Callitrichids* (Ph.D. diss., University of California at Berkeley).

Power, M. L., Heaney, R. P., Kalkwarf, H. J., Pitkin, R. M., Repke, J. T., Tsang, R. C., Schulkin, J. (1998). The role of calcium in health and disease. *Am J Obstet Gynecol* 181:1560–9.

Power, M. L., Tardif, S. D., Power, R. A., Layne, D. G. (2003). Energy metabilism in Goeldi's monkey (*Callimico goeldii*) is similar to that of other callitrichids. *Am J Primatol* 60:57–67.

Prentice, A., Jarjou, L. M., Cole, T. J., Stirling, D. M., Dibba, B., Fairweather-Tait, S. (1995). Calcium requirements of lactating Gambian mothers: Effects of a calcium supplement on breast-milk calcium concentration, maternal bone mineral content, and urinary calcium excretion. *Am J Clin Nutr* 62:58–67.

Purdie, D. W., Aaron, J. E., Selby, P. L. (1988). Bone histology and mineral homeostasis in human pregnancy. *Br J Obstet Gynaecol* 95:849–54.

Rasmussen, N., Frolich, A., Hornnes, P. J., Hegedus, L. (1990). Serum ionized calcium and intact parathyroid hormone levels during pregnancy and postpartum. *Br J Obstet Gynaecol* 97:857–9.

Ritchie, L. D., Fung, E. B., Halloran, B. P., Turnlund, J. R., Van Loan, M. D., Cann, C. E., King, J. C. (1998). A longitudinal study of calcium homeostasis during human pregnancy and lactation and after resumption of menses. *Am J Clin Nutr* 67:693–701.

Sapolsky, R. M. (2001). Physiological and pathophysiological implications of social stress in mammals. In: *Coping with the Environment: Neural and Endocrine Mechanisms* (ed. B. S. McEwen, H. M. Goodman). New York: Oxford University Press.

Schnell, C. R., Wood, J. M. (1995). Measurement of blood pressure and heart rate by telemetry in conscious unrestrained marmosets. *Lab Anim.* 29:258–61.

Schulkin, J. (1999). Corticotropin-releasing hormone signals adversity in both the placenta and the brain: Regulation by glucocorticoids and allostatic overload. *J Endocrinol* 349–56.

Schulkin, J. (2003a). Allostasis: A neural behavioral perspective. *Horm Behav* 43:21–7.

Schulkin, J. (2003b). *Rethinking Homeostasis: Allostatic Regulation in Physiology and Pathophysiology*. Cambridge: MIT Press.

Sowers, M. F., Eyre, D., Hollis, B. W., Randolph, J. F., Shapiro, B., Jannausch, M. L., Crutchfield, M. (1995a). Biochemical markers of bone turnover in lactating and nonlactating postpartum women. *J Clin Endocrinol Metab* 80:2210–16.

Sowers, M. F., Randolph, J. F., Shapiro, B., Jannaush, M., Crutchfield, M. (1995b). A prospective study of bone density and pregnancy following extended lactation with bone loss. *Obstet Gynecol* 85:285–90.

Specker, B. L., Vieira, N. E., O'Brien, K. O., Ho, M. L., Heubi, J. E., Abrams, S. A., et al. (1994). Calcium kinetics in lactating women with low and high calcium intakes. *Am J Clin Nutr* 59:593–9.

Sterling, P., Eyer, J. (1988). Allostasis: A new paradigm to explain arousal pathology. In: *Handbook of Life Stress, Cognition, and Health* (ed. S. Fisher, and J. Reason), 629–49. New York: John Wiley Sons.

Thompson, S. D. Power, M. L., Rutledge, C. E., Kleiman, D. G. (1994). Energy metabolism and thermoregulation in the golden lion tamarin (*Leontopithecus rosalia*). *Folia Primatol* 63:131–43.

Walsberg, G. E. (2003). How useful is energy balance as a overall index of stress in animals? *Horm Behav* 43:16–17.

Index

acetylcholine esterase production, 180
adaptations
  appropriate/inappropriate, 50–1
  short-term, long-term disruptions and, 3
addiction
  allostatic state and, 348–9
  mechanisms involved in, 40, 47
  tolerance and sensitization in, 159–61
adenosine, 253–4, 255
ADHD (attention deficit, hyperactivity
    disorder) treatment, 56
affect, 37
African-Americans, 43–4
AL. *See* allostatic load
allometry, 354
allostasis
  adaptive nature of, 229
  advantages over homeostasis model, 22
  concept of, 1, 7, 11–12, 68–9, 89, 90, 107,
    165, 346–8
  criteria defining, restricting, 281
  defined, 7, 18, 66, 150, 312, 346
  examples of, 69, 229
  health in, 54, 58
  homeostasis versus, 151, 230–1,
    347–8
  origins of idea, 19–22, 346
  pathology and, 350–1
  phenomena types at question, 277–8
  principles of, 26–34, 228–9
  regulated, 311, 312
  see-saw analogy to, 231
  thermostatic analogy to, 101, 110
  utility of concept, 207
  see also allostatic states; predictive
    regulation
allostasis health model, 58
allostatic load
  allostatic states versus, 79

assessing, 80
categories, Boolean-specified, 126
concept overview, 349–50
contributing factors, 74, 313
defined, 67, 70, 73, 75, 89, 108, 152, 312,
    313, 347
elevated
  criteria for, 142
  hostility and, 116
  social relationships and, 116
endocrine/neuroendocrine secretions and,
    323–5
examples of, 75–8
future study possibilities
  challenge, 131–2
  longitudinal, 132–3, 143–5
  using ^{1}*H-NMR, 138–42
glucocorticoid secretion and, 323–5
home temperature analogy to,
    109–10
indicators of
  possible future, 130–2
  rationale for, 133–7
labile perturbation factors effects on,
    317–21
measurement/scoring of
  approaches summarized, 114
  biomarkers used in, 115, 116
  canonical weight, 120–3, 143
  elevated-risk zone, 115–20, 142–3
  gender-specific, 128–30
  limitations to extant, 137
  recursive partitioning, 123–8, 143
  using ^{1}*H-NMR spectra, 139
regulatory processing costs and, 313
reproductive life history stage effects on,
    321–3
  see also allostatic overload
allostatic load score, 114, 115

365

allostatic overload
  amygdala glucocorticoids role in, 204–6
  defined, 7, 349
  example of, 9–11
  indicators of, 9
  sensitization and, 197–201
  types of, 325
allostatic states
  allostatic load versus, 79
  assessing, 79, 80
  contributing factors, 74, 78
  defined, 7, 67, 75, 152, 153, 347
  examples of, 71, 75–6, 348–9
  maintenance of, 109–10
amygdala
  abnormal responses to aversive stimuli,
      204–5
  basolateral complex, fear and
      glucocorticoid involvement in, 186–9
  blocking NMDA receptors in, 183
  central nucleus' role in fear, 189–90
  corticosterone in, 187–9
  CRH gene expression in, 9
  dendritic changes in, 202–3
  fear circuits involving, 172–3
  fear-related involvement, 172, 173–4, 175–6
  'flashbulb' memory and, 72
  functions of, 37, 202
  glucocorticoid-induced allostatic changes
      in, 201–2
  glucocorticoid-induced morphological
      changes in, 203–4
  glucocorticoids in, role in human allostatic
      overload, 204–6
  glucose metabolism in, depression and, 9
  hyperexcitability in, 199–201
  memory consolidation and, 178–9
  neuronal events from fear conditioning,
      182–6, 187
  projections to cortex, 176
  stria terminalis and, 193–4
  uncertainty states and, 174
anxiety
  amygdala activation and, 174
  amygdala-cortisol interactivity in, 205
  amygdala infusion of corticosterone and,
      181
  brain stress system changes and, 161
  fear versus, 193, 206
  HPA axis in, 205
  neural mechanisms for, 37–8
  pathological
    CREB and, 184
    depression and, 168–9
  sensitization and, 197–9
  stria terminalis' bed nucleus and, 193–4

autoimmune diseases, 86
autonomic nervous system, 243

Bayesian response patterns, 30–1
behavior, neural-endocrine regulation of, 7–8
Bernard, Claude, 17, 99, 100, 150–1
beta adrenergic receptor blockers, 72
beta-endorphin, 326
black-headed gulls, 328
blackbirds, European, 331
blood chemistry, 20–1
blood glucose levels, 29, 33
blood pressure
  anticipation of elevated, 33
  attention demands and, 19–20
  brain's control of, 25
  children's, rise in, 41, 42
  circadian rhythms and, 237
  daily variations in, 23–5, 43, 72
blood supply, 27–8
body fluid depletion, 8
body temperature
  circadian variation in, 354–7
  daily fluctuations, 233
  regulation
    circadian set-point changes and,
        233–4
    daily rhythm ranges, 234–5
    defined, 233
body water regulation, 106–8
bone mineral density, 10
brain
  blood requirements, 28
  circadian rhythm pathways in, 246
  CRH receptor site distribution in, 192
  fear-affecting lesions in, 173
  fear-related circuits in, 172–3
  glucocorticoid-induced morphological
      changes in, 202–4
  reward system of, 159
breast cancer, 271

calcium
  metabolism, 352–4
  regulation, 350
camels, 234–5
Cannon, Walter B., 1–2, 29, 99, 231, 344–5
canonical weight scoring, 120–3, 143
carbohydrates, 20
cardiovascular disease, 268–70
cardiovascular function
  catecholamines and, 84
  glucocorticoids and, 81
catecholamines
  as stress indicators, 136–7
  cardiovascular function and, 84

central nervous system and, 85
dangerous effects of, 70–2
elevated levels, 85
fluid volume and, 84
glucocorticoids and, 83–4, 327
immunity and, 69, 84–5
inflammation and, 84–5
memory formation and, 72
metabolism and, 85
central nervous system
catecholamines and, 85
glucocorticoids and, 81–2
peripheral physiology and, 358
chickens, 326
chronic fatigue syndrome, 87
circadian prediction, 32
circadian rhythms
anticipatory advantages of, 232
as homeostatic states, 357
blood pressure and, 237
brain structure assays for, 246
defined, 232
food-entrainable oscillators, 248
heart rate and, 237
masking effects, 235
molecular genetic bases of, 240–1
non-SCN sources of, 245–8
organ foci of autonomous clock gene
expression, 246–8
pacemaker outputs, 243–5, 255
photic entrainment of pacemaker,
241–2
ubiquity of, 232–3
cognitive functioning, 136
congestive heart failure, 103
coronary artery disease study, 139
corticosteroid-receptor complexes,
186–9
corticosterone
amygdala implants of, 196–7
amygdala infusion of, 181
CRH facilitation by, 195, 196
diet-administered: effects in chickens, 326
elevated
acetylcholine esterase production and,
180
foraging behavior and, 326, 328
low testosterone and, 177
metabolic rate and, 328
oxygen consumption and, 328
parental behavior in birds and, 326
predator pressure and, 329
reproduction behavior suppression and,
326
territorial aggression reduction and,
325–6

fat in chickens and
glucose/fatty acid/triglyceride levels and,
327
implants in sparrows, 326, 328
nongenomic membrane effects of, 194
parent-child separation and, 41
recovery from acute stress and, 328
snow bunting secretion of, 316
social status and, 331
stress, body conditions, and, 334
cortisol
actions of, 9
as biomarker in AL challenge study, 131–2
elevated
depression and, 11, 116, 178, 206–7
evening, 76
fat distribution and, 46
fearfulness in children and, 178
in monkeys, 178
incidence of, 47
long-term effects, 3
osteoporosis and, 116
short-term effects, 3
gluconeogenesis and, 327
inadequate levels, 76
CREB (cyclic AMP-response element binding
protein), 182, 183, 184
CRH (corticotropin-releasing hormone)
anxious depression and, 9
behavioral effects in birds, 326
behavioral inhibition in monkeys and,
178
central activation of, fear and, 192–3
corticosterone facilitation of, fear and, 195,
196
distribution in brain, 191
feed-forward system involving, 152
functional associations, 190–1
increases in cerebrospinal fluid, 204
infusions into stria terminalis, 193–4
response in sheep to threats, 201–2
seizures from, 195–6, 197
startle response facilitation and, 192,
195
CRH family receptors, 192
cytokines
anti-inflammatory, 86
autoimmune diseases and, 86
chronic fatigue syndrome and, 87
defined, 86
elevated levels of, 86
fibromyalgia and, 87
inflammatory, 86, 87–8
measurement problems, 88
oxidative stress and, 87
sleep regulation and, 86–7

dark-eyed juncos, 319, 327
'DASH' study, 55
daylength effects, 32
dendrite length changes, 202–3
depression
    allostatic model and, 161
    amygdala activity in, 204
    anxious, 9–11
    bone mineral density and, 10, 76
    cortisol and, 11, 116, 178, 206–7
    hypercortisolemia in, 204, 205
    osteoporosis and, 116
    pre-frontal cortex in, 171, 172, 204
    shift worker, 273
DHEA, 82–3, 135–6
diabetes (type 2)
    origins, 85
    pathologies associated with, 44
    'thrifty genes' and, 47–8
disease, income and education and,
        228
distress, 107, 109
dopamine release, 38–40
drugs, neural effects of, 38–40

effector adaptation
    circadian, 32–3
    muscles, 32
effectors
    compensatory activation of, 104
    redundancy in, 101, 104
    shared, 104, 105
efficiency of organisms, 26–8
EG, 317
egr-1 (early growth response gene 1), 184–6
elevated-risk zone scoring, 115–20, 142–3
ELHS (emergency life history stages)
    allostatic load reduction by, 318–21, 324
    defined, 309
    hormonal actions in relation to, 325
    responses to
        ACTH and
        corticosterone treatment and, 327
energy requirements model
    basic terms of, 316–7
    endocrine/neuroendocrine bases of, 323–5
    labile perturbation factors in, 317–21
    reproductive life history stage in, 321–3
EO, 317–9
epinephrine
    memory consolidation and, 178–9
    release of, 84
Escheria coli, 134–5
exercise, mental benefits of, 57

fainting reactions, 103

fat
    distribution of, 46
    hunger for, hormones and, 20
fatty acid levels, family stress and, 20–1
fear
    allostatic processes of
        amygdala's role in, 174
        prefrontal cortex in, 174–5
    amygdala involvement in, 172, 173–4,
        175–6
    anxiety versus, 193, 206
    brain circuits involved with, 172–3
    brain lesions affecting, 173
    brain systems and, 179–80
    categorizing, 359
    chronic, 359
    defined, 167–8
    extinction of learned, 181
    general characteristics, 359
    glucocorticoid secretion and, 176–7
    in pathological anxiety, 168, 169
    intracellular events and long-term memory
        of, 183
    learning/memory of
        amygdala's role in, 176, 203
        epinephrine/norepinephrine and, 178–9
        ngfi-b (nerve growth factor induced
            gene-B) involvement in, 188–9
        signal transduction pathways for, 182–6
        speed of, 167
    neuroanatomical schematic of, 170
    normal, 168
    organs engaged in states of, 166
    prefrontal cortex involvement, 171
    response sustaining mechanisms, 176
    sensitizing factors, 198
    see also anxiety
feed-forward mechanisms, 8, 151–2
fibrinogen concentration, 46
fibromyalgia, 87
Fisher rats, 76
fluid volume
    catecholamines and, 84
    glucocorticoids and, 81
flycatchers, pied, 326, 329
food-entrainable oscillators, 248
forskolin treatment, 245–6

gastrointestinal disorders, 266–8
General Adaptation Syndrome, 90
genetic factors, 74, 75
GHT (geniculohypothalamic tract), 242
glucocorticoids
    allostatic load and, 323–5
    as AL biomarkers, 135
    cardiovascular function and, 81

catecholamines and, 83–4, 327
central nervous system and, 81–2
character and action of, 186
chronic activation of, 177
chronic insufficiency of, 82
cognitive functioning and, 136
DHEA and, 82–3, 135–6
elevated levels of, 70, 72–3, 76
fear conditioning and, 180–1
fear feed-forward regulation by, 188, 189
fluid volume and, 81
functions of, 70, 81, 177
glucose utilization and, 177
high-levels before conditioning, 181
HPA activation restraint by, 194
immunity and, 81
inflammation and, 81
measurement problems, 88
memory and, 72, 179
metabolic effects, 81, 327
night restfulness and, 328
reproduction and, 82
secretion of
    allostatic load and, 323–5
    fear and, 176–7
    predator pressure and, 329–30
    see also corticosterone; cortisol
gluconeogenesis, 327
glutamate receptors, 182–3, 187
golden lion tamarins, 355–7
great tits, 328

Hcrt (hypocretin), 253
health, allostatic, 54, 58
health care, 57–8
heart disease, social conditions and, 19
heart rate, 237
heroin, 159–61
heterostasis, 3–4
hippocampus
    adrenal steroid receptors in, 136
    extracellular glutamate levels in, 76–7
    neuronal shrinkage in, 199, 203–4
    shrinkage of, 73, 136, 273
homeorrhesis, 238–9, 310, 345
homeostasis
    allostasis versus, 151, 230–1, 347–8
    body size and, 354
    Cannon's view of, 99, 231, 344–5
    defined, 67, 150, 310–11, 344–5
    examples of, 1, 2
    inadequacy of model, 22–6, 345
    initial model of, 17–18, 230
    'predictive,' 4, 345–6
    'reactive,' 4
    19th century ideas of, 1–2

homeostatic systems principles, 102–4
homeostats
    defined., 101–3
    resetting, 101
hormone secretions
    levels of, 313–5
    roles of, 304
HPA (hypothalamic pituitary adrenal) axis, 9,
        194, 205
hypercortisolemia, 204, 205
hyperglycemia, 51–2, 108
hypertension
    allostatic view of, 41–4
    current treatment recommendations, 55
    fat consumption and, 20
    homeostatic treatment model, 51, 52–4
    incidence of, 40, 150
    salt consumption and, 20
    social conditions and, 19
hypervigilance. See vigilance
hypocretin, 253
hyposatisfaction
hypothalamus
    anterior, 252–3
    lateral, 253
    posterior, 253
hypothermia, 106–7

IGL (intergeniculate leaflet) projection, 242
iguanas, 334
IL-6 (interleukin-6), 131
illness incidence
    genetic abnormalities and, 48
    race and, 19
immediate early genes, 184–6
immune system
    catecholamines and, 84–5
    effects of stress on, 77–8
    glucocorticoids and, 81
    impairments of, 271–2
inflammation
    catecholamines and, 84–5
    glucocorticoids and, 81
insomnia, 276–7
insulin, 133
insulin resistance, 45–6, 133–4
internal milieu, 4–5, 68
intervention
    demand distribution and responses, 52
    higher-level, 57–8
    low-level mechanism, problems with,
        51–4
    most successful, 55–6

kindling, 199–201
king penguins, 334

labile perturbation factors (LPF)
    defined, 303
    effects on allostatic load, 317–21
    responses to
        body condition and, 331–4
        social status and, 330–1
    types of, 309
lactation, calcium metabolism during,
    352–4
leptin deficiency, 45, 85
Lewis rats, 76
life history stages (LHS)
    concept, review of, 351–2
    durations of, 307–8
    examples of, 306, 307
    hormone roles in, 304
    human singularity of, 308–9
    labile perturbations in, 317–21
    levels of, 309–10
    overlap of, 308
    patterns of, 304–6
    phases of, 306
    reproductive, 321–3
    substages of, 306
    temporal sequence boundaries of, 307
    see also ELHS (emergency life history
        stages)
lobotomy, 35, 36–7
LPF. See labile perturbation factors (LPF)

MacArthur Study of Successful Aging, 115
masking effects, 235
mediators
    immediate positive effects of, 70, 72
    inadequate responses of, 76
    pathophysiology from, 75
    patterns of release, 71, 73–6
    protection versus damage from, 76
    summarized, 74
    systemic versus local, 73
melanopsin, 241
melatonin secretion, 243, 245
Melville, Herman, 50
memory
    consolidation mechanisms, 179
    CREB and, 184
    'flash-bulb,' 72
    mediators in formation of, 69
mental disorders
    in shift workers, 272–4
    pharmacotherapy for, 54, 56
    see also depression
metabolic syndrome
    allostatic view of, 46–8
    defined, 44
    low-level treatments for, 54

metabolism
    catecholamines and, 85
    glucocorticoids and, 81, 327
metabonomics, 137–8
methylphenidate, 56
metyrapone
milieu intérieur, 99, 100
mortality
    employment status and, 45
    marital status and, 45
    occupational, per time of day, 274, 275
    social organization and, 39
    social relationships disruption and, 228
motivation, 159, 324–5

natural killer cell activity, 272
negative feedback, 101
nervous system, evolution of, 5
nest-building, 4
ngfi-b (nerve growth factor induced gene-B)
NMR spectroscopy
    basics of, 138
    recursive partitioning of spectra, 140
    terminal node spectra, 141
    use in AL assessments, 138–42
norepinephrine release, 84
NPY, 47
nutrient needs and replenishment, 33–4, 354

obesity
    cultural disruption and, 48
    homeostatic treatment model, 51, 54
    incidence of, 44, 48
    pathological sequelae, 44
    'thrifty genes' and, 47–8
opiates, 159–61
orexin, 253

pain placebos, 57
Parkinson's disease, 57
Per genes, 240–1, 245
petrels, 319, 321
pharmacotherapy, 54, 56
phosphate regulation, 6–7
physiological systems
    capacities per loads on, 26–7
    levels of functioning of, 310–3
physiology
    brain control of, 19–21
    defined, 344
    kinds of change in, review of, 351–2
    long-term needs versus short term
        demands, 357–8
    'stability' in, 346
Pine Siskins, 328
PKA (protein kinase A), 183

poikilostasis, 310, 311–12
positive feedback systems, 103–4, 360–1
post-traumatic stress disorder, 161, 205–6
predictive homeostasis, 4, 345–6
predictive regulation
  behavioral neural adaptation and, 37–8
  examples of, 24–5, 28–9
  levels of, 32
  time course of, 31–2
prefrontal cortex
  anxiety/depressive disorders and, 204
  'demands' of, 38
  integrative function of, 35–7
  roles of, 171–2, 174–5
pregnancy
  calcium metabolism during, 352–4
  nutrient partitioning regulation during,
    238–9
  risks from shift work during, 270–1
preindustrial communities, 48–9
public speaking stress, 75

raphe projection, 242
rats
  adrenalectomized, 4–5, 180
  anxiety in, corticosterone and, 181
  parathyroidectomized, 4–5
  phosphate-deprived, 6
  pre-frontal cortex in, 171
  responses to extreme temperatures, 4
  SCN efferent pathways in, 244
  urine ^{1}*H-NMR spectrum, 138
'reactive' homeostasis, 4
receptor downregulation, 33
recursive partitioning scoring, 123–8, 143
regulatory systems
  chronic activation of, 7
  circadian versus homeostatic, 279–81
  input/output curves per load, 30
  optimal functioning mode, 9
reproduction
  energy needs and allostatic load during,
    321–3
  glucocorticoids and, 82
  seasonal variations in, 308–9
  women's problems with, 270–1
restlessness, 38, 56
restraint stress, 76–7
rheostasis
  defined, 4, 25–6, 237, 310, 311, 345
  programmed, 237–8, 239
  reactive, 238, 239
  second order, 238
RHT (retinohypothalamic) projection, 241,
    242
Richter, Curt, 4

Ritalin, 56
rough-skinned newt, 326
rufous hummingbird, 320

salt
  hunger for, 20, 41
  hypertension and, 20
satisfaction
  best, 50
  fleetingness of, 50
  in industrial societies, 50
  in preindustrial societies, 49
  neural mechanisms for, 38–40
  socioeconomic status and, 47
  see also hyposatisfaction
SCN (suprachiasmatic nucleus)
  afferent systems to, 241–2
  anatomy of, 240
  efferents from, 243–5, 255
  functions of, 240, 244–5
  neuronal components, 240
Selye, Hans, 3, 90, 99–101, 107
sensitization
  allostatic overload and, 197–201
  as neural process, 199–201
  early-life-induced, 198–9
sensor adaptation
  rate of, per input change rate, 31–2
  sensitivity, 31
  transduction range, 29–31
serotonin, 37–8, 255
set point, 23, 345
sheep, CRH response to threats in, 201–2
shift work
  ability to tolerate, 256
  accidents risk during, 274
  age of workers performing, 256
  circadian rhythm adaptation during, 257–60
  circadian rhythm adjustment rates during,
    260
  commuting accidents risk, 274
  domestic/social effects of, 264–6
  health effects
    breast cancer, 271
    cardiovascular, 268–70
    gastrointestinal, 266–8
    immune system, 271–2
    mental, 272–4
    reproductive, 270–1
  incidence of, 256
  models of, 277
  research in, challenges of, 257
  stress from, 256, 274–5, 276
  see also shift workers
shift workers
  adrenaline excretion by, 258, 259

shift workers (*cont.*)
  body temperature studies, 257–8
  circadian reentrainment incompleteness,
    260
  eating habits, 267–8
  EEG recordings of, 262–3
  light exposure effects on, 260
  primary complaints of, 261
  salivary melatonin and cortisol level
    rhythms of, 259–60
  sleep disturbances in, 261–4, 275–6
  sleep length per sleep onset time, 261, 262
  sleep timing, metabolic effects of, 268
shrikes, fiscal, 329
side-blotched lizards, 326
skeletal muscle, 27–8
skin immunity, 77–8
sleep
  circadian-homeostatic influences on
    interactions between, 254–6
    separating, 250
  cytokines and regulation of, 86–7
  deprivation of, 76, 86, 249, 272
  naps, 249, 263
  NREM (non-rapid-eye-movement), 249–50
  parameters of, circadian variations in,
    249–50
  pathways for circadian control of, 244
  physiological/hormonal parameter profiles
    during, 235–7
  processes regulating, 248–9
  REM, glucocorticosteroids and, 328
  two-process regulation model, 250–1
  see also insomnia; sleep-arousal
sleep-arousal
  adenosine's role in, 253–4, 255
  basal forebrain circuits involved in, 252–3
  lateral hypothalamus circuits involved in,
    253
  posterior hypothalamus circuits involved
    in, 253
snow bunting, 316
snowshoe hares, 329–30
social phobics, 206
social relationships, 116–18
social status, 330–1
societies
  industrial, 49–50
  preindustrial, 48–9

sparrows
  Harris', 331
  song, 325, 326, 328
  white-crowned, 323, 326, 328
startle response
  bases for, 189, 193
  CRH facilitation of, 192, 195
  unconditioned, interference with,
    193
steroid functions, 7–8
stonechats, tropical, 329
stress
  behavioral responses to, 68
  defined, 67, 89–90
  homeostatic theory of, 104–7
  medical/psychological consequences,
    108–9
  repeated, effects of, 75
  Selyes theory of, 99–101, 107
  stressor-intensity/effector-system models,
    100
stress responses, failure to turn off,
    75–6
stressors
  effects on ACTH/epinephrine/
    norepinephrine, 102
  neuroendocrine responses and, 106
stria terminalis, 193–4
stroke, social conditions and, 19
symmorphosis, 26–7
Syndrome X. *See* metabolic syndrome
system challenges, adaptive responses to, 2

testosterone, corticosterone and,
    177
therapeutic communities, 55–6
thermoregulation. *See* body temperature
TNF alpha levels, 87

ulcers, peptic, 267, 268
uncertainty states, 174
urocortins, 191

vigilance
  African-American, 43–4
  long-term effects of, 41–4
VTA (ventral tegmental area), 38

waist-to-hip ratio, 134